FUNDAMENTALS OF

Automotive Maintenance and Light Repair

SECOND EDITION **TASKSHEET MANUAL**

Kirk VanGelder

ASE Certified Master Technician & LI
ASE Evaluation Team Leader
Certified Automotive Service Instructor
Vancouver, Washington, USA

JONES & BARTLETT
LEARNING

World Headquarters
Jones & Bartlett Learning
5 Wall Street
Burlington, MA 01803
978-443-5000
info@jblearning.com
www.jblearning.com

Production Credits
General Manager: Kimberly Brophy
VP, Product Development: Christine Emerton
Product Owner: Kevin Murphy
Product Development Manager: Amanda Brandt
Project Specialist: Brooke Haley
Marketing Manager: Amanda Banner
VP, Manufacturing and Inventory Control: Therese Connell
Composition: Integra Software Services Pvt. Ltd.
Project Management: Integra Software Services Pvt. Ltd.
Cover Design: Scott Moden
Text Design: Integra Software Services Pvt. Ltd.
Rights & Media Specialist: Maria Leon Maimone
Cover Image: © E+/Getty Images
Printing and Binding: McNaughton & Gunn
Cover Printing: McNaughton & Gunn

6048

Printed in the United States of America

23 22 21 20 19   10 9 8 7 6 5 4 3 2 1

Contents

General Safety

Student/Intern information:

Name ____________________ Date __________ Class ________________

Vehicle used for this activity:

Year ____________ Make ________________________ Model________________

Odometer________________ VIN________________________________

<table>
<tr><td rowspan="2">Learning Objective/Task</td><td>CDX Tasksheet Number</td><td>ASE Education Foundation Reference Number; Priority Level</td></tr>
<tr><td></td><td></td></tr>
<tr><td>• Identify general shop safety rules and procedures.</td><td>A0001</td><td>A0A1; RS</td></tr>
<tr><td>• Identify marked safety areas.</td><td>A0006</td><td>A0A6; RS</td></tr>
<tr><td>• Identify the location and the types of fire extinguishers and other fire safety equipment; demonstrate knowledge of the procedures for using fire extinguishers and other fire safety equipment.</td><td>A0007</td><td>A0A7; RS</td></tr>
<tr><td>• Identify the location and use of eye wash stations.</td><td>A0008</td><td>A0A8; RS</td></tr>
<tr><td>• Identify the location of the posted evacuation routes.</td><td>A0009</td><td>A0A9; RS</td></tr>
<tr><td>• Locate and demonstrate knowledge of safety data sheets (SDS).</td><td>A0015</td><td>A0A15; RS</td></tr>
<tr><td>• Draw a diagram of the shop and label all parts.</td><td>N/A</td><td>N/A</td></tr>
</table>

Time off____________

Time on____________

Total time____________

Materials Required

- Program's shop policy and other safety information
- SDS book
- Fire extinguisher(s) from the shop
- Safety Data Sheets for the products listed

Some Safety Issues to Consider

- Comply with personal and environmental safety practices associated with clothing; eye protection; hand tools; power equipment; proper ventilation; and the handling, storage, and disposal of chemicals/materials in accordance with local, state, and federal safety and environmental regulations.
- Shop rules and procedures are critical to your safety. Please give these your utmost attention.
- Marked safety areas play an important role in maintaining a safe work environment. Always understand and heed marked safety areas.
- Fire blankets are an important component of fire safety in the shop. Always know their location, purpose, and use.

- Fire extinguishers come in a variety of types and sizes. It pays to understand the differences before they are needed.
- Eye wash stations are an important component of shop safety. Always know their location, purpose, and use.
- Pre-planned evacuation routes are an important component of shop safety. Always know the location of all evacuation routes for your shop.
- Being able to locate information in an SDS is critical to your safety and health. Make sure you are familiar with their location and use.

Performance Standard

0−No exposure: No information or practice provided during the program; complete training required

1−Exposure only: General information provided with no practice time; close supervision needed; additional training required

2−Limited practice: Has practiced job during training program; additional training required to develop skill

3−Moderately skilled: Has performed job independently during training program; limited additional training may be required

4−Skilled: Can perform job independently with no additional training

Student/Intern information:

Name _________________________________ Date ____________ Class _________________________

Vehicle used for this activity:

Year _______________ Make _____________________________ Model_____________________________

Odometer____________________________ VIN____

▶ TASK Identify general shop safety rules and procedures.

MLR
AOA1

Time off________________

Time on________________

Total time________________

CDX Tasksheet Number: A0001

1. **List the location(s) of the following items.**

    a. **Program's general shop rules and policies:**

    b. **Safety data sheet (SDS) book:**

    c. **Procedure for operation of a fire extinguisher:**

    d. **Procedure for operation of a vehicle hoist:**

2. **List the shop's policy for the wearing of safety glasses while in the shop:**

3. **List the shop's policy for driving vehicles:**

4. **List the shop's policy for clothing in the shop:**

5. **List the shop's policy for jewelry in the shop:**

6. **Pass the shop's safety test and record your score here:** _________________

7. **Have your supervisor/instructor verify satisfactory completion of this procedure, any observations found, and any necessary action(s) recommended.**

Performance Rating

CDX Tasksheet Number: A0001

0	1	2	3	4

Supervisor/instructor signature ____ Date _______________

Student/Intern information:

Name _________________________ Date __________ Class _________________________

Vehicle used for this activity:

Year _____________ Make _________________________ Model _________________________

Odometer _________________________ VIN _________________________

▶ TASK Identify marked safety areas.

MLR
A0A6

CDX Tasksheet Number: A0006

1. **Research the uniform color code system used to designate safety areas in a shop. List each color and its designation:**

   a. **Color:** _________________ **designates:** _________________
   b. **Color:** _________________ **designates:** _________________
   c. **Color:** _________________ **designates:** _________________
   d. **Color:** _________________ **designates:** _________________

2. **Label all of the marked safety areas on the diagram of the shop at the end of this section.**

3. **Have your supervisor/instructor verify satisfactory completion of this procedure, any observations found, and any necessary action(s) recommended.**

Time off _________________

Time on _________________

Total time _________________

Performance Rating

CDX Tasksheet Number: A0006

0	1	2	3	4

Supervisor/instructor signature _________________________ Date __________

Student/Intern information:

Name _________________________________ Date ____________ Class _________________________

Vehicle used for this activity:

Year ______________ Make __________________________ Model_______________________

Odometer________________________ VIN____

▶ TASK Identify the location and the types of fire extinguishers and other fire safety equipment; demonstrate knowledge of the procedures for using fire extinguishers and other fire safety equipment.

MLR
AOA7

Time off______________

Time on______________

Total time______________

CDX Tasksheet Number: AOOO7

1. **Research the types, location, and use of fire extinguishers.**
   a. **List the different types of fire extinguishers (dry chemical, CO_2, etc.) available:**

      i. **Type: _______________ ; for use on what class of fire: _______________**

      ii. **Type: _______________ ; for use on what class of fire: _______________**

      iii. **Type: _______________ ; for use on what class of fire: _______________**

      iv. **Type: _______________ ; for use on what class of fire: _______________**

   In the above list, put a star next to the type(s) of fire extinguishers in this shop.

2. **List the steps for proper use of a fire extinguisher:**

3. **Research the location, purpose, and use of fire blankets in your shop.**
   a. **Describe the purpose of a fire blanket (under what circumstances should a fire blanket be used?):**

b. Describe how a fire blanket puts out a fire:

4. **Label the location of all fire extinguishers on the diagram of the shop at the end of this section.**

5. **Label the location of the fire blanket(s) on the diagram of the shop at the end of this section.**

6. **Have your supervisor/instructor verify satisfactory completion of this procedure, any observations found, and any necessary action(s) recommended.**

Performance Rating

CDX Tasksheet Number: A0007

0	1	2	3	4

Supervisor/instructor signature ____ Date _______________

Student/Intern information:

Name _________________________________ Date ___________ Class _________________________

Vehicle used for this activity:

Year ______________ Make _____________________________ Model_____________________________

Odometer_____________________________ VIN____

▶ TASK Identify the location and use of eye wash stations.

MLR
AOA8

Time off_________________

Time on_________________

Total time_________________

CDX Tasksheet Number: AOOO8

1. **Research the location, purpose, and use of eye wash stations in your shop.**

   a. **Describe the purpose of an eye wash station:**

   b. **Describe the proper use of an eye wash station (including time):**

   c. **What type of eye injury would use of an eye wash station NOT be appropriate for?**

2. **Label the location of the eye wash station(s) on the diagram of the shop at the end of this section.**

3. **Have your supervisor/instructor verify satisfactory completion of this procedure, any observations found, and any necessary action(s) recommended.**

Performance Rating

CDX Tasksheet Number: A0008

0	1	2	3	4

Supervisor/instructor signature ____ Date ______________

Student/Intern information:

Name _________________________________ Date ____________ Class _____________________

Vehicle used for this activity:

Year _______________ Make ___________________________ Model_____________________

Odometer_____________________ VIN____

▶ **TASK** Identify the location of the posted evacuation routes.

Time off_______________

Time on_______________

Total time_______________

CDX Tasksheet Number: A0009

1. **Research the location and purpose of evacuation routes.**

   a. **Describe the purpose of evacuation routes:**

2. **Label the evacuation route(s) on the diagram of the shop at the end of this section. Also label the location of the posted evacuation route diagram(s).**

3. **Have your supervisor/instructor verify satisfactory completion of this procedure, any observations found, and any necessary action(s) recommended.**

Performance Rating

CDX Tasksheet Number: A0009

0	1	2	3	4

Supervisor/instructor signature ___________________________________ Date _______________

Student/Intern information:

Name _________________________________ Date ____________ Class _______________________

Vehicle used for this activity:

Year _______________ Make ____________________________ Model__________________________

Odometer__________________________ VIN____

▶ TASK Locate and demonstrate knowledge of
safety data sheets (SDS).

MLR
A0A15

Time off_____________

Time on_____________

Total time_____________

CDX Tasksheet Number: A0O15

1. **List the safety precautions when handling motor oil:**

2. **List the flash point of gasoline/petroleum:** ___________________ **°F/°C**

3. **List the firefighting equipment needed to put out a gasoline/petroleum fire:**

4. **List the first aid treatment for battery acid in the eyes:**

5. **List the first aid treatment for ingestion of antifreeze (ethylene glycol):**

6. **Label the location of the SDS books on the diagram of the shop at the end of this section.**

7. **Have your supervisor/instructor verify satisfactory completion of this procedure, any observations found, and any necessary action(s) recommended.**

Student/Intern information:

Name _________________________________ Date _____________ Class _____________________

Vehicle used for this activity:

Year _______________ Make _____________________________ Model___________________________

Odometer___________________________ VIN____

▶ TASK Draw a diagram of the shop and label all parts. **Additional Task**

Time off_______________

Time on_______________

CDX Tasksheet Number: N/A

Total time_______________

1. **On a full-sized piece of paper, draw a diagram of the shop and label each of the following:**

   a. **Marked safety areas (floor lines, etc.)** _____________ **(X when finished)**
   b. **Location of the fire blanket(s)** _____________ **(X when finished)**
   c. **Location of all fire extinguishers** _____________ **(X when finished)**
   d. **Location of the exhaust fan switch(es)** _____________ **(X when finished)**
   e. **Location of the electrical disconnect switch(es)** ____ **(X when finished)**
   f. **Location of the eye wash station(s)** _____________ **(X when finished)**
   g. **Location of the SDS books** _________________ **(X when finished)**
   h. **Evacuation route(s)** _________________ **(X when finished)**
   i. **Location of the master air shut-off valve** _________ **(X when finished)**
   j. **Location of the overhead door switch(es)** _________ **(X when finished)**

2. **Have your supervisor/instructor verify satisfactory completion of this task.**

Performance Rating

CDX Tasksheet Number: N/A

0	1	2	3	4

Supervisor/instructor signature ____ Date _____________

Personal Safety

<hr>

Student/Intern information:

Name _______________________________ Date ____________ Class _____________________

Vehicle used for this activity:

Year ______________ Make _______________________ Model_____________________

Odometer__________________ VIN_______________________________________

Learning Objective/Task	CDX Tasksheet Number	ASE Education Foundation Reference Number; Priority Level
• Comply with the required use of safety glasses, ear protection, gloves, and shoes during lab/shop activities.	A0010	AOA10; RS
• Identify and wear appropriate clothing for lab/shop activities.	A0011	AOA11; RS
• Secure hair and jewelry for lab/shop activities.	A0012	AOA12; RS

Time off_____________

Time on_____________

Total time_____________

Materials Required
- Regular safety glasses
- Gloves
- Shoes

Some Safety Issues to Consider
- Comply with personal and environmental safety practices associated with clothing; eye protection; hand tools; power equipment; proper ventilation; and the handling, storage, and disposal of chemicals/materials in accordance with local, state, and federal safety and environmental regulations.
- Shop rules and procedures are critical to your safety. Please give these your utmost attention.

Performance Standard
0–No exposure: No information or practice provided during the program; complete training required

1–Exposure only: General information provided with no practice time; close supervision needed; additional training required

2–Limited practice: Has practiced job during training program; additional training required to develop skill

3–Moderately skilled: Has performed job independently during training program; limited additional training may be required

4–Skilled: Can perform job independently with no additional training

Student/Intern information:

Name ___________________________________ Date ____________ Class ___________________________

Vehicle used for this activity:

Year ______________ Make _________________________________ Model_______________________________

Odometer_______________________________ VIN____

▶ **TASK** Comply with the required use of safety glasses, ear protection, gloves, and shoes during lab/shop activities.

MLR
AOA10

Time off________________

Time on________________

Total time________________

CDX Tasksheet Number: AOO10

1. **Describe the safety glasses policy for your shop (be specific):**

2. **Describe the policy related to ear protection in the shop (be specific):**

3. **Describe the policy related to gloves for your shop (be specific):**

4. **Describe the policy related to work shoes for your shop (be specific):**

5. **These tasks require observation of the student over a prolonged period. Ask your instructor to give you a date for your evaluation.**

   a. **Write that date here:** ___________________________________

6. **Continue with your projects, complying with the safe use of safety glasses, gloves, shoes, clothing, and hair containment during all lab/shop activities.**

7. On or after that date, have your instructor verify satisfactory completion of these tasks.

Student/Intern information:

Name _______________________________ Date ___________ Class ___________________

Vehicle used for this activity:

Year _______________ Make ______________________________ Model_______________________

Odometer_________________________ VIN____

▶ **TASK** Identify and wear appropriate clothing for
lab/shop activities.

MLR
AOA11

Time off________________

Time on________________

Total time________________

CDX Tasksheet Number: AOO11

1. **Describe the clothing requirements for your shop (be specific):**

2. **These tasks require observation of the student over a prolonged period. Ask
your instructor to give you a date for your evaluation.**

   a. **Write that date here:** _______________________________

3. **Continue with your projects, complying with the safe use of safety glasses,
gloves, shoes, clothing, and hair containment during all lab/shop activities.**

4. **On or after that date, have your instructor verify satisfactory completion of
these tasks.**

Performance Rating

CDX Tasksheet Number: AOO11

☐	☐	☐	☐	☐
0	**1**	**2**	**3**	**4**

Supervisor/instructor signature ____ Date ___________

Student/Intern information:

Name _________________________________ Date _____________ Class _______________________

Vehicle used for this activity:

Year _______________ Make _________________________ Model___________________________

Odometer_______________________ VIN____

▶ TASK Secure hair and jewelry for lab/shop activities.

CDX Tasksheet Number: A0012

MLR
A0A12

Time off____________

Time on____________

Total time____________

1. **List your shop's policy concerning securing hair in the shop (be specific):**

2. **List your shop's policy concerning jewelry in the shop (be specific):**

3. **These tasks require observation of the student over a prolonged period. Ask your instructor to give you a date for your evaluation.**

   a. **Write that date here:** _________________________________

4. **Continue with your projects, complying with the safe use of safety glasses, gloves, shoes, clothing, and hair containment during all lab/shop activities.**

5. **On or after that date, have your instructor verify satisfactory completion of these tasks.**

Performance Rating

CDX Tasksheet Number: A0012

☐	☐	☐	☐	☐
0	**1**	**2**	**3**	**4**

Supervisor/instructor signature ___________________________________ Date ___________

Tool Safety

Student/Intern information:

Name _________________________________ Date ___________ Class _____________________

Vehicle used for this activity:

Year _______________ Make _______________________________ Model___________________________

Odometer___________________________ VIN____

Learning Objective/Task	CDX Tasksheet Number	ASE Education Foundation Reference Number; Priority Level
• Identify tools and their usage in automotive applications.	A0016	A0B1; RS
• Identify standard and metric designation.	A0017	A0B2; RS
• Utilize safe procedures for handling of tools and equipment.	A0002	A0A2; RS
• Demonstrate safe handling and use of appropriate tools.	A0018	A0B3; RS
• Demonstrate proper cleaning, storage, and maintenance of tools and equipment.	A0019	A0B4; RS
• Demonstrate proper use of precision measuring tools (e.g., micrometer, dial-indicator, dial-caliper).	A0020	A0B5; RS

Time off______________

Time on______________

Total time______________

Materials Required
- Standard toolkit and other tools as required
- Precision measuring tools: micrometer, dial-indicator, dial-caliper, etc.

Some Safety Issues to Consider
- Comply with personal and environmental safety practices associated with clothing; eye protection; hand tools; power equipment; proper ventilation; and the handling, storage, and disposal of chemicals/materials in accordance with local, state, and federal safety and environmental regulations.
- Tools allow us to increase our productivity and effectiveness. However, they must be used according to the manufacturer's procedures. Failure to follow those procedures can result in serious injury or death.

Performance Standard

0–No exposure: No information or practice provided during the program; complete training required

1–Exposure only: General information provided with no practice time; close supervision needed; additional training required

2–Limited practice: Has practiced job during training program; additional training required to develop skill

3–Moderately skilled: Has performed job independently during training program; limited additional training may be required

4–Skilled: Can perform job independently with no additional training

---

Student/Intern information:

Name ________________________________ Date ____________ Class ________________________

Vehicle used for this activity:

Year ______________ Make ____________________________ Model__________________________

Odometer__________________________ VIN____

▶ **TASK** Identify tools and their usage in automotive applications.

MLR
AOB1

Time off________________

Time on________________

Total time________________

CDX Tasksheet Number: A0016

1. **Using the following list, describe the specific function/purpose of each of the following tools and any disadvantages/problems with using the tool:**

    a. **Open-end wrench:**

    b. **Box-end wrench:**

    c. **Socket:**

    d. **Ratchet:**

e. Torque wrench:

f. Slotted screwdriver:

g. Phillips screwdriver:

h. Tap:

i. Die:

j. Feeler blade:

k. Line wrench/flare-nut wrench:

l. Allen wrench:

m. Torx screwdriver or socket:

n. Hacksaw:

o. Oil filter wrench:

p. Compression gauge:

q. DVOM/DMM:

r. Test light:

s. Diagonal side cutters:

t. Locking pliers:

u. Needle-nose pliers:

v. Brake spoon:

w. Micrometer:

x. Dial indicator:

y. Antifreeze hydrometer:

z. Snap-ring pliers:

3. **Have your supervisor/instructor verify satisfactory completion of this procedure, any observations found, and any necessary action(s) recommended.**

Performance Rating

CDX Tasksheet Number: A0016

0	1	2	3	4

Supervisor/instructor signature ____ Date _______________

Student/Intern information:

Name ___________________________ Date ___________ Class ___________________________

Vehicle used for this activity:

Year ___________ Make ___________________________ Model ___________________________

Odometer ___________________________ VIN ___________________________

▶ TASK Identify standard and metric designation.

CDX Tasksheet Number: AOO17

Time off___________

Time on___________

Total time___________

1. **Complete the following conversions from metric to standard and vice versa using the conversion charts below.**

Volume

Volume is the amount of space occupied by a three-dimensional object. The metric system uses liters (l) or cubic centimeters (cc or cm^3). The Imperial system uses gallons (gal) and quarts (qt) for "wet" volume and cubic feet (ft^3) for "dry" volume. You'll need to determine volume any time you fill a vehicle's reservoir with liquid. This includes petrol/gasoline, coolant, oil, transmission fluid, or lubricant.

Volume Conversions	
Imperial-Imperial	4 US qt = 1 US gal 1 ft^3 = 7.48 US gal 1 ft^3 = 6.22 UK gal
Metric-Metric	1 L = 1000 cc 1 cc = 0.001 L
Imperial-Metric	1 $in.^3$ = 16.387 cc 1 US gal = 3.78 L 1 UK gal = 4.54 L 1 US qt = 0.95 L
Metric-Imperial	1 L = 61.0237 $in.^3$ 1 L = 0.035 ft^3 1 L = 0.26 US gal 1 L = 0.21 UK gal 1 L = 1.05 qt

2. **Knowledge Check: Convert the following:**
   a. **3.0 L = ___________________ $in.^3$**
   b. **350 $in.^3$ = ___________________ L**
   c. **3.0 gal = ___________________ L**
   d. **9 L = ___________________ gal**

Mass

Mass is a unit or system of units by which a degree of heaviness is measured. The metric system uses grams (g), kilograms (kg), and tonnes (t). The Imperial system uses ounces (oz), pounds (lb), and tons (T). In the workshop, you will use these measurements to determine the lifting capacity of equipment like hydraulic and engine hoists and floor jacks.

Mass Conversions	
Imperial–Imperial	16 oz = 1 lb 2000 lb = 1 T
Metric–Metric	1000 g = 1 kg 1000 kg = 1 t
Imperial–Metric	1 oz = 28.3 g 1 lb = 453 g 2.2 lb = 1 kg 1 T = 0.907 t
Metric–Imperial	1 t = 1.10 T

3. **Knowledge Check: Convert the following:**
    a. **8 oz =** _________________________ **g**
    b. **475 g =** _________________________ **oz**
    c. **6.6 lb =** _________________________ **kg**
    d. **4500 kg =** _________________________ **lb**

Torque

Torque is the twisting force applied to a shaft. The metric system uses the Newton meter (Nm). The Imperial system uses the inch-pound (in-lb) and the foot-pound (ft-lb). Vehicle manufacturers specify torque settings for key fasteners on the engine and wheels. You will need to follow the specifications or you could strip threads or break bolts. Torque is also an important concept when discussing engine performance. A foot-pound (ft-lb) is the twisting force applied to a shaft by a lever 1 foot long with a 1-pound mass on the end. A Newton meter (Nm) is the twisting force applied to a shaft by a level 1 meter long with a force of 1 Newton applied to the end of the lever. (1N is equivalent to the force applied by a mass of 100.)

Torque Conversions	
Imperial–Imperial	12 in-lb = 1 ft-lb 1 in-lb = 0.08 ft-lb
Imperial–Metric	1 ft-lb = 1.34 N·m
Metric–Imperial	1 N·m = 0.74 ft-lb 1 N·m = 8.8 in-lb

4. **Knowledge Check: Convert the following:**
    a. **48 in-lb =** _________________________ **ft-lb**
    b. **15 ft-lb =** _________________________ **in-lb**
    c. **65 ft-lb =** _________________________ **N·m**
    d. **142 N·m =** _________________________ **ft-lb**

Pressure

Pressure is a measurement of force per unit area. The metric system uses kilopascals (kPa) and bar. The Imperial system uses pounds per square inch (psi) and atmospheres. *Vacuum* is a term given to a pressure that is less than atmospheric pressure. The Imperial system measures vacuum in inches of mercury (" Hg) or inches of water. The metric system measures vacuum in millimeters of mercury (mm Hg). You'll need to understand pressure conversions when filling tires and replacing air-conditioning refrigerants or using a vacuum gauge.

Pressure Conversions	
Imperial-Imperial	14.7 psi = 1 atmosphere 1" Hg = 14" H_2O 0" Hg = 1 atmosphere
Metric-Metric	100 kPa = 1 bar
Imperial-Metric	1 psi = 6.89 kPa 1 atmosphere = 101.3 kPa 1" Hg = 25.4 mm Hg 1 atmosphere = 1.013 bar

5. **Knowledge Check: Convert the following:**
    a. 14.7 psi = ____________________ kPa
    b. 650 kPa = ____________________ psi
    c. 22 psi = ____________________ bar
    d. 5.5 bar = ____________________ psi

6. **Have your supervisor/instructor verify satisfactory completion of this procedure, any observations found, and any necessary action(s) recommended.**

Performance Rating

CDX Tasksheet Number: AOO17

0	1	2	3	4

Supervisor/instructor signature ____ Date ____________

Student/Intern information:

Name _________________________________ Date _____________ Class _________________________

Vehicle used for this activity:

Year _______________ Make ___________________________ Model_______________________________

Odometer_______________________ VIN____

▶ TASK Utilize safe procedures for handling of tools
and equipment.

MLR
AOA2

Time off________________

Time on________________

Total time________________

CDX Tasksheet Number: A0002

1. **These tasks will require observation of the student over a prolonged period after the initial check. Ask your instructor to give you a date for your evaluation.**

   a. **Write that date here:** _____________________

2. **Continue with your other projects, utilizing safe handling of tools and equipment, including proper cleaning, maintenance, and storage of the tools and equipment, until the date of your evaluation.**

3. **On or after that date, have your instructor verify satisfactory completion of each task.**

Performance Rating

CDX Tasksheet Number: A0002

0	1	2	3	4

Supervisor/instructor signature ____ Date _______________

▶ TASK Demonstrate safe handling and use of appropriate tools.

MLR
AOB3

Time off________________

Time on________________

Total time________________

CDX Tasksheet Number: AOO18

1. **These tasks will require observation of the student over a prolonged period after the initial check. Ask your instructor to give you a date for your evaluation.**

   a. **Write that date here:** ____________________

2. **Continue with your other projects, demonstrating safe handling and proper cleaning, maintenance, and storage of the tools until the date of your evaluation.**

3. **On or after that date, have your instructor verify satisfactory completion of each task.**

Performance Rating

CDX Tasksheet Number: AOO18

0	1	2	3	4

Supervisor/instructor signature ____ Date ____________

Student/Intern information:

Name _________________________________ Date ____________ Class _________________________

Vehicle used for this activity:

Year _______________ Make ___________________________ Model_____________________________

Odometer_______________________ VIN____

▶ TASK Demonstrate proper cleaning, storage, and maintenance of tools and equipment.

MLR
AOB4

Time off_______________

Time on_______________

Total time_______________

CDX Tasksheet Number: AOO19

1. **These tasks will require observation of the student over a prolonged period after the initial check. Ask your instructor to give you a date for your evaluation.**

   a. **Write that date here:** ___________________________

2. **Continue with your other projects, demonstrating safe handling and proper cleaning, maintenance, and storage of the tools until the date of your evaluation.**

3. **On or after that date, have your instructor verify satisfactory completion of each task.**

Performance Rating

CDX Tasksheet Number: AOO19

0	1	2	3	4

Supervisor/instructor signature ____ Date ______________

Student/Intern information:

Name ___________________________ Date ___________ Class ___________________

Vehicle used for this activity:

Year _____________ Make _____________________ Model_____________________

Odometer_____________________ VIN_____________________________________

▶ **TASK** Demonstrate proper use of precision measuring tools
(e.g., micrometer, dial-indicator, dial-caliper). **MLR AOB5**

Time off_____________

Time on_____________

CDX Tasksheet Number: A0020

Total time_____________

1. **These tasks will require observation of the student over a prolonged period after the initial check. Ask your instructor to give you a date for your evaluation.**

   a. **Write that date here:** _____________________

2. **Continue with your other projects, demonstrating safe handling and proper cleaning, maintenance, and storage of the tools until the date of your evaluation.**

3. **On or after that date, have your instructor verify satisfactory completion of each task.**

Performance Rating

CDX Tasksheet Number: A0020

0	1	2	3	4

Supervisor/instructor signature _____________________________________ Date ___________

Vehicle Protection and Jack and Lift Safety

Student/Intern information:

Name ___________________________ Date __________ Class ________________

Vehicle used for this activity:

Year ____________ Make ___________________ Model________________

Odometer________________ VIN_______________________________

Learning Objective/Task	CDX Tasksheet Number	ASE Education Foundation Reference Number; Priority Level
• Identify purpose and demonstrate proper use of fender covers, mats.	A0022	A0C2; RS
• Utilize proper ventilation procedures for working within the lab/ shop area.	A0005	A0A5; RS
• Identify and use proper placement of floor jacks and jack stands.	A0003	A0A3; RS
• Identify and use proper procedures for safe lift operation.	A0004	A0A4; RS
• Ensure vehicle is prepared to return to customer per school/ company policy (floor mats, steering wheel cover, etc.).	A0026	A0D1; RS

Time off____________

Time on____________

Total time____________

Materials Required
- Vehicle
- Fender, seat, steering wheel, and carpet covers
- Floor jack
- Jack stand(s)
- Wheel chocks
- Vehicle hoist
- Shop rag
- Possible cleaning supplies

Some Safety Issues to Consider
- Comply with personal and environmental safety practices associated with clothing; eye protection; hand tools; power equipment; proper ventilation; and the handling, storage, and disposal of chemicals/materials in accordance with local, state, and federal safety and environmental regulations.
- Floor jacks and jack stands must have a capacity rating higher than the load lifted.
- Floor jacks and jack stands must be used on a hard, level surface. Do not attempt to lift or support a vehicle under any other conditions.
- Jacks are designed to lift a vehicle, not to support it. Always use jack stands once the vehicle has been lifted.

- Have the jack stands ready, and at hand, prior to lifting the vehicle.
- Vehicle hoists are important tools that increase productivity and make the job easier. But they also can cause severe injury or death if used improperly. Make sure you follow the hoist's and vehicle manufacturers' operation procedures. Also, make sure you have your supervisor's/instructor's permission to use a vehicle hoist.
- It is critical that the vehicle be returned to the customer in proper working order. Double-check your work before releasing the vehicle to the customer.

Performance Standard

0–No exposure: No information or practice provided during the program; complete training required

1–Exposure only: General information provided with no practice time; close supervision needed; additional training required

2–Limited practice: Has practiced job during training program; additional training required to develop skill

3–Moderately skilled: Has performed job independently during training program; limited additional training may be required

4–Skilled: Can perform job independently with no additional training

▶ TASK Identify purpose and demonstrate proper use of fender covers, mats.

MLR
AOC2

Time off________________

Time on_________________

Total time_______________

CDX Tasksheet Number: A0022

1. **Identify the purpose of the following items:**

   a. **Fender cover:**

   b. **Seat cover:**

   c. **Steering wheel cover:**

   d. **Carpet cover/floor mat:**

2. **Properly prepare a vehicle for service or repair, using the above covers.**

3. Have your supervisor/instructor verify satisfactory completion of this
    procedure, any observations found, and any necessary action(s)
    recommended.

Performance Rating

CDX Tasksheet Number: A0022

0	1	2	3	4

Supervisor/instructor signature _______________________________________ Date _______________

Student/Intern information:

Name _________________________________ Date _____________ Class _________________________

Vehicle used for this activity:

Year _______________ Make _____________________________ Model_____________________________

Odometer_________________________ VIN____

▶ **TASK** Utilize proper ventilation procedures for working within the lab/shop area.

MLR
AOA5

Time off_______________

Time on_______________

Total time_______________

CDX Tasksheet Number: A0005

1. **List the OSHA personal exposure limit for carbon monoxide over an 8-hour work period:** _______________ **ppm**

2. **Properly position a vehicle in a work stall.**

3. **Properly connect the exhaust extraction system to the vehicle exhaust.**

4. **Turn on, or verify that the extraction system is on.**

5. **With your instructor's permission, start the vehicle and verify the extraction equipment is secure and operating properly.**

6. **Have your instructor verify your previously given answer and your proper exhaust extraction usage. Supervisor/instructor's initials:** _______________

7. **Turn off the vehicle, return the exhaust hoses to their proper storage places, and shut off the extraction system if it isn't being used anymore.**

8. **Have your supervisor/instructor verify satisfactory completion of this procedure, any observations found, and any necessary action(s) recommended.**

Performance Rating

CDX Tasksheet Number: A0005

0	1	2	3	4

Supervisor/instructor signature ____ Date _____________

Student/Intern information:

Name _________________________________ Date ____________ Class _________________________

Vehicle used for this activity:

Year _______________ Make _________________________ Model_____________________________

Odometer_____________________________ VIN____

▶ TASK Identify and use proper placement of floor jacks and jack stands.

MLR
AOA3

CDX Tasksheet Number: A0003

1. **Research the jacking and lifting procedures for this vehicle in the appropriate service manual.**

    a. **Draw a diagram of the vehicle's lift points:**

2. **Check to make sure the vehicle is on a hard, level surface. If not, move it to a safe location.**

3. **Install wheel chocks. Prepare the floor jack and stands for use.**

4. **Lift and support one end of the vehicle on jack stands according to the manufacturer's procedure.**

    a. **Have your instructor initial to verify proper jack stand placement:**

    ____

5. **Lift the vehicle and remove the jack stands. Return the jack and stands to their proper storage places.**

6. **Return the vehicle to its beginning condition and return any tools you used to their proper locations.**

Time off____________________

Time on____________________

Total time____________________

7. Have your supervisor/instructor verify satisfactory completion of this
    procedure, any observations found, and any necessary action(s)
    recommended.

Performance Rating

CDX Tasksheet Number: A0003

| 0 | 1 | 2 | 3 | 4 |

Supervisor/instructor signature ____ Date _______________

Student/Intern information:

Name _________________________________ Date ___________ Class _________________________

Vehicle used for this activity:

Year _______________ Make ___________________________ Model___________________________

Odometer_________________________ VIN____

▶ TASK Identify and use proper procedures for safe lift operation.

MLR
AOA4

Time off________________

Time on________________

Total time________________

CDX Tasksheet Number: A0004

1. **Position the vehicle in proper relation to the lift, taking into consideration the center of gravity of the vehicle.**

   > **NOTE** Check the vehicle for unusual loading, such as heavy loads in the trunk or truck bed. If you find this situation, notify your instructor immediately.

2. **Position the lift arms in the proper location as specified by the manufacturer.**

3. **Raise the lift until one of the arms lightly contacts the lift point. Check the position of the lift arms to make sure they are in contact with (or just about to contact) the proper points.**

   > **NOTE** Make sure the lift arms are not touching or pinching anything they shouldn't be in contact with, including the rocker panel, running boards, and fuel or brake lines.

4. **If the arms are in the proper position, raise the vehicle a few inches off the ground. Using a strong part of the vehicle, moderately shake the vehicle to make sure it is stable.**

   > **NOTE** If the vehicle shifts position at all or is out of balance, lower the vehicle and reset the lift arms or reposition the vehicle.

5. **If the vehicle is stable, lift the vehicle to the height indicated by your instructor and engage the locks or lower the lift onto the locks. Instructor/ supervisor initials: _____________**

6. **Verify that there are no obstacles under the vehicle, and that all doors are closed. Lower the vehicle and move the lift arms out of the way of the vehicle.**

7. **Return the vehicle to its beginning condition and return any tools you used to their proper locations.**

8. **Have your supervisor/instructor verify satisfactory completion of this procedure, any observations found, and any necessary action(s) recommended.**

Performance Rating

CDX Tasksheet Number: A0004

0	1	2	3	4

Supervisor/instructor signature ____ Date ______________

▶ TASK Ensure vehicle is prepared to return to customer per school/company policy (floor mats, steering wheel cover, etc.).

MLR
AOD1

Time off________

Time on________

Total time________

CDX Tasksheet Number: A0026

> **NOTE** A properly protected vehicle and good work habits will make this task much easier.

1. **Double check that all work has been completed. Nothing can be missing, loose, or leaking.**

   a. **Student initial when completed:** _________________

2. **If your instructor deems it necessary, test-drive the vehicle to be sure of proper repair and operation of the vehicle.**

   a. **Have your instructor initial here:** _________________

3. **Double check that all tools are put away and stored properly.**

   a. **Student initial when completed:** _________________

4. **Remove all fender covers, seat covers, floor covers, and steering wheel covers. Return them to their storage place or dispose of them properly, depending on the type of cover.**

   a. **Student initial when completed:** _________________

5. **Check the exterior of the vehicle for greasy fingerprints or grime. Clean with an appropriate cleaner. Follow your shop's policies on this procedure.**

   a. **Student initial when completed:** _________________

6. **Check the following interior locations for dirt or greasy spots. Clean with an appropriate cleaner. Follow your shop's policies on this procedure:**

   a. **Carpet and floor mats. Student initial when completed:** _____________
   b. **Seats. Student initial when completed:** _________________
   c. **Steering wheel and parking brake handle. Student initial when completed:** _________________
   d. **Door panel and handles. Student initial when completed:** _____________

7. **If the vehicle is ready to return to the customer, the vehicle may need to be moved out of the shop. Get your instructor's permission to move the vehicle to the customer pick-up area.**

   a. **Have your instructor initial here:** _________________

8. **Return to your work stall and clean up the floor, benches, and related area.**

   a. **Student initial when completed:** _________________

9. **Have your supervisor/instructor verify satisfactory completion of this procedure, any observations found, and any necessary action(s) recommended.**

Performance Rating

CDX Tasksheet Number: A0026

0	1	2	3	4

Supervisor/instructor signature ____ Date _______________

Vehicle, Customer, and Service Information

Student/Intern information:

Name _________________________________ Date _____________ Class _________________________

Vehicle used for this activity:

Year _______________ Make _____________________________ Model_____________________________

Odometer_____________________________ VIN ____

Learning Objective/Task	CDX Tasksheet Number	ASE Education Foundation Reference Number; Priority Level
• Proper vehicle identification information: Define the use and purpose of the VIN, engine numbers, and data code; locate the VIN, and apply knowledge of VIN information.	N/A	N/A
• Identify information needed and the service requested on a repair order.	A0021	AOC1; RS
• Demonstrate use of the three Cs (concern, cause, and correction).	A0023	AOC3; RS
• Review vehicle service history.	A0024	AOC4; RS
• Complete work order to include customer information, vehicle identifying information, customer concern, related service history, cause, and correction.	A0025	AOC5; RS

Time off________________

Time on________________

Total time________________

Materials Required
- Vehicle
- Service information
- Completed repair order assigned by your instructor
- Several repair orders for the same vehicle from repairs/services performed over an extended period of time
- Scheduled maintenance chart for this vehicle
- Blank repair order

Some Safety Issues to Consider
- Comply with personal and environmental safety practices associated with clothing; eye protection; hand tools; power equipment; proper ventilation; and the handling, storage, and disposal of chemicals/materials in accordance with local, state, and federal safety and environmental regulations.

Performance Standard

0—No exposure: No information or practice provided during the program; complete training required

1—Exposure only: General information provided with no practice time; close supervision needed; additional training required

2—Limited practice: Has practiced job during training program; additional training required to develop skill

3—Moderately skilled: Has performed job independently during training program; limited additional training may be required

4—Skilled: Can perform job independently with no additional training

Student/Intern information:

Name _________________________________ Date _____________ Class _________________________________

Vehicle used for this activity:

Year _______________ Make _____________________________ Model_________________________________

Odometer_______________________________ VIN____

▶ TASK Proper vehicle identification information: Define the use and purpose of the VIN, engine numbers, and data code; locate the VIN, and apply knowledge of VIN information.

Additional Task

Time off_____________

Time on_____________

Total time_____________

CDX Tasksheet Number: N/A

1. **Research the location and description of the VIN in an appropriate service manual.**

   a. **Location of VIN:** ____
   b. **VIN:** ____

2. **Using the VIN on the vehicle assigned to you, identify the following information. (Write both corresponding VIN code designation and what it represents.)**

   a. **Country of origin:** ____
   b. **Location of manufacturing plant:** _________________________________
   c. **Passenger restraint system:** ______________________________________
   d. **Engine designation:** ____
   e. **Model year:** ____

3. **Which position in the VIN is the model year?** _________________________________

4. **Which position in the VIN is the engine designation?** _________________________

5. **Have your supervisor/instructor verify satisfactory completion of this procedure, any observations found, and any necessary action(s) recommended.**

Performance Rating

CDX Tasksheet Number: N/A

0	1	2	3	4

Supervisor/instructor signature ____ Date _____________

Student/Intern information:

Name ________________________________ Date ____________ Class ________________________

Vehicle used for this activity:

Year ______________ Make ______________________________ Model________________________

Odometer__________________________ VIN____

▶ TASK Identify information needed and the service requested on a repair order.

MLR
AOC1

Time off______________

Time on______________

Total time____________

CDX Tasksheet Number: A0021

1. **Familiarize yourself with the assigned repair order. Locate and list the following information below.**

    a. **Date:** ______________________________
    b. **Customer:** ______________________________
    c. **Address:** ____
    d. **Daytime phone number:** ______________________________
    e. **Year:** ______________________________
    f. **Make:** ______________________________
    g. **Model:** ______________________________
    h. **Color:** ______________________________
    i. **License and state:** ______________________________
    j. **Odometer reading:** ______________________________
    k. **VIN:** ______________________________
    l. **Customer concern(s)/service requested:**

2. **Did the customer sign the repair order authorizing the repairs?**
    Yes: __________ **No:** __________

3. **Return the sample repair order to its proper storage place.**

4. **Have your supervisor/instructor verify satisfactory completion of this procedure, any observations found, and any necessary action(s) recommended.**

Student/Intern information:

Name _________________________________ Date ___________ Class _______________________

Vehicle used for this activity:

Year _______________ Make ___________________________ Model_______________________

Odometer_______________________ VIN____

▶ **TASK** Demonstrate use of the three Cs (concern, cause, and correction).

MLR
AOC3

Time off_______________

Time on_______________

Total time_______________

CDX Tasksheet Number: A0023

1. **Using the following scenario, write up the three Cs as listed on most repair orders. Assume that the customer authorized the recommended repairs.**

   A customer complains that his vehicle is leaving what looks like oil spots on the landlord's driveway after he ran over something in the road a few days ago. You check the engine oil find that it is about 1/2 a quart low, but looks pretty clean, like it was changed recently. The engine oil life monitor indicates 92% oil life remaining. You safely raise and secure the vehicle on the hoist. While visually inspecting the underside of the vehicle, you notice oil dripping off the engine oil drain plug. Checking the torque of the drain plug shows that the drain plug isn't loose. Closer inspection reveals a shiny spot on the aluminum oil pan near the drain plug. There is a small crack in the oil pan that is seeping oil and is dripping slowly off the drain plug.

2. **Concern/complaint:**

3. **Cause:**

4. **Correction:**

5. Have your supervisor/instructor verify satisfactory completion of the
    previous answers, and any necessary action(s) recommended.

Performance Rating

CDX Tasksheet Number: A0023

0	1	2	3	4

Supervisor/instructor signature ____ Date _______________

Student/Intern information:

Name _________________________________ Date ___________ Class _________________________

Vehicle used for this activity:

Year _______________ Make _______________________ Model_____________________________

Odometer_____________________ VIN____

▶ TASK Review vehicle service history.

MLR
AOC4

CDX Tasksheet Number: A0024

Time off________________

Time on________________

Total time________________

1. **Familiarize yourself with the repair history as listed on the repair orders and answer the following questions.**

   a. **What was the first date this vehicle was serviced?** ___________________

   b. **What was the last date this vehicle was serviced?** ___________________

   c. **What was the most major repair performed?**

   d. **Was this vehicle ever returned for the same problem more than once? Yes: __________ No: __________**

      i. **If so, for what and how many times?** ___________________________

   e. **Compare this list to the scheduled maintenance chart and list any missed maintenance tasks between the first service and the last service:**

2. Have your supervisor/instructor verify satisfactory completion of this procedure, any observations found, and any necessary action(s) recommended.

Performance Rating

CDX Tasksheet Number: A0024

| 0 | 1 | 2 | 3 | 4 |

Supervisor/instructor signature _______________________________________ Date _______________

▶ TASK Complete work order to include customer information, vehicle identifying information, customer concern, related service history, cause, and correction.

MLR
AOC5

Time off________________

Time on________________

Total time________________

CDX Tasksheet Number: A0025

1. **Use your company's repair order to complete this task.**

   Fred Smith brings in a 2008 Hyundai Santa Fe AWD, with a 3.3 L engine, automatic transmission, 72,426 miles on the odometer, silver paint, VIN 5NMSH73E28H192794. It needs some work before going on a 3000-mile trip. He would like an estimate of repairs needed and has agreed to let your technician inspect the vehicle while you write up the repair order. He gives you the following information:

   a. **Home address: 1234 NE Main Street, Anytown, CA 13579**
   b. **Cell phone: (111) 222-1234**
   c. **Work phone: (111) 333-4567**
   d. **Vehicle license number: CDX – 111**

2. **The customer listed the following concerns/complaints:**

   a. **Small oil leak from under the engine**
   b. **Small coolant leak from under the engine**
   c. **Squealing noise coming from the front brakes**

3. **The technician found the following conditions:**

   a. **Both valve covers have leaking gaskets.**
   b. **The water pump is leaking from the shaft and the bearing is worn.**
   c. **The front brake pads are worn down to the wear indicators, the rotors are a bit under specifications, and the calipers are starting to seep brake fluid past the caliper piston seal and need to be replaced.**

4. **Complete the repair order as if all tasks were completed, including parts, their cost, and labor.**

5. Have your supervisor/instructor verify satisfactory completion of this
    procedure, any observations found, and any necessary action(s)
    recommended.

Performance Rating

CDX Tasksheet Number: A0025

0	1	2	3	4

Supervisor/instructor signature ____ Date _______________

SRS and ABS Safety

Student/Intern information:

Name _________________________________ Date _____________ Class _________________________

Vehicle used for this activity:

Year _______________ Make _____________________________ Model_______________________

Odometer_______________________ VIN____

Learning Objective/Task	CDX Tasksheet Number	ASE Education Foundation Reference Number; Priority Level
• Demonstrate awareness of the safety aspects of supplemental restraint systems (SRS), electronic brake control systems, and hybrid vehicle high voltage circuits.	A0013	AOA13; RS
• Demonstrate awareness of the safety aspects of high voltage circuits (such as high-intensity discharge [HID] lamps, ignition systems, injection systems, etc.).	A0014	AOA14; RS

Time off_________________

Time on_________________

Total time_________________

Materials Required

- Vehicle with SRS and anti-lock brake system (ABS)
- Special tools or equipment, if specified

Some Safety Issues to Consider

- Comply with personal and environmental safety practices associated with clothing; eye protection; hand tools; power equipment; proper ventilation; and the handling, storage, and disposal of chemicals/materials in accordance with local, state, and federal safety and environmental regulations.
- Deployment of the SRS could cause serious injury or death. Follow all the manufacturer's procedures before working on this system.
- The ABS may hold brake fluid under extremely high pressure. Follow the manufacturer's procedures to relieve this pressure before working on this system.
- Hybrid and electric vehicles have batteries with VERY high voltage, which can kill you if handled improperly. Always get permission from your supervisor/instructor before working on, or near, the system. Also, ALWAYS follow the manufacturer's procedure when servicing a hybrid or electric vehicle.

Performance Standard

0—No exposure: No information or practice provided during the program; complete training required

1—Exposure only: General information provided with no practice time; close supervision needed; additional training required

2—Limited practice: Has practiced job during training program; additional training required to develop skill

3—Moderately skilled: Has performed job independently during training program; limited additional training may be required

4—Skilled: Can perform job independently with no additional training

Student/Intern information:

Name ________________________________ Date ____________ Class ________________________

Vehicle used for this activity:

Year ______________ Make ________________________________ Model________________________

Odometer________________________ VIN ____

▶ TASK Demonstrate awareness of the safety aspects of supplemental restraint systems (SRS), electronic brake control systems, and hybrid vehicle high voltage circuits.

MLR
AOA13

Time off________________

Time on________________

Total time________________

CDX Tasksheet Number: AOO13

1. **Research the following procedures for a hybrid vehicle in the appropriate service information.**

    a. **List the precautions when working around or on the SRS on this vehicle:**

    b. **List the steps to disable the SRS system on this vehicle:**

    c. **List the steps to enable the SRS on this vehicle:**

d. List the precautions when working on or around the electronic brake
    control system on this vehicle:

e. Identify the high-voltage circuit wiring on this vehicle. What color is
    the wire conduit?

f. List or print out the high-voltage disable procedure for this vehicle.

2. Have your supervisor/instructor verify satisfactory completion of this task.

Performance Rating

CDX Tasksheet Number: A0013

0	1	2	3	4

Supervisor/instructor signature ____ Date ______________

Student/Intern information:

Name _________________________________ Date ____________ Class _________________________

Vehicle used for this activity:

Year ________________ Make _________________________ Model_________________________

Odometer________________________ VIN_______________________________________

▶ TASK Demonstrate awareness of the safety aspects of
high voltage circuits (such as high-intensity
discharge [HID] lamps, ignition systems,
injection systems, etc.)

MLR
AOA14

CDX Tasksheet Number: AOO14

1. **Using appropriate service information, identify system voltage and safety precautions associated with high-intensity discharge headlights, ignition systems, and injection systems.**

    a. **HID lamp voltage:** _________________________ **volts**
    b. **List the safety precautions required when working on HID systems:**

    c. **Maximum secondary ignition system voltage:** _________________________ **volts**
    d. **List the safety precautions required when working around ignition systems:**

    e. **Injection system voltage (on a vehicle with a high-voltage injection system):** _________________________ **volts**
    f. **List the safety precautions required when working around high-voltage injection systems:**

2. **Have your supervisor/instructor verify satisfactory completion of this task.**

Performance Rating

CDX Tasksheet Number: A0014

0	1	2	3	4

Supervisor/instructor signature _________________________________ Date _____________

1A. Engine Repair General

Learning Objective/Task	CDX Tasksheet Number	ASE Education Foundation Reference Number; Priority Level
• Research vehicle service information, including fluid type, vehicle service history, service precautions, and technical service bulletins.	A1001	A1A1; P-1
• Verify operation of the instrument panel engine warning indicators.	A1002	A1A2; P-1
• Inspect engine assembly for fuel, oil, coolant, and other leaks; determine necessary action.	A1003	A1A3; P-1
• Install engine covers using gaskets, seals, and sealers as required.	A1004	A1A4; P-1
• Verify engine mechanical timing.	A1005	A1A5; P-2
• Perform common fastener and thread repair, to include: remove broken bolt, restore internal and external threads, and repair internal threads with thread insert.	A1006	A1A6; P-1
• Identify service precautions related to service of the internal combustion engine of a hybrid vehicle.	A1007	A1A7; P-2

Materials Required

- Blank work order
- Vehicle with available service history records
- Depending on the type of concern, special hand tool/diagnostic tools may be required. See your supervisor/instructor for instructions to identify what tools may be required.
- Vehicle service information database

Safety Considerations

- When running any vehicles in the shop, make sure you use the shop's exhaust ventilation system to discharge all exhaust gas safely outside.
- Extreme caution must be exercised when working around rotating components.
- Lifting equipment such as vehicle jacks and stands, vehicle hoists, and engine hoists are important tools that increase productivity and make the job easier. However, they can also cause severe injury or death if used improperly. Make sure you follow the manufacturer's operation procedures. Also, make sure you have your supervisor's/instructor's permission to use any particular type of lifting equipment.
- Comply with personal and environmental safety practices associated with clothing; eye protection; hand tools; power equipment; proper ventilation; and the handling, storage, and disposal of chemicals/materials in accordance with local, state, and federal safety and environmental regulations.

CDX Tasksheet Number: A1001

Student/Intern Information

Name _________________________________ Date ____________ Class _________________________

Vehicle, Customer, and Service Information

Vehicle used for this activity:

Year ________________ Make ___________________________ Model ___________________________

Odometer _________________________________ VIN _________________________________

Materials Required

- Blank work order
- Vehicle with available service history records
- Depending on the type of concern, special hand tool/diagnostic tools may be required. See your supervisor/instructor for instructions to identify what tools may be required.
- Vehicle service information database

Task-Specific Safety Considerations

- Shop rules and procedures are critical to your safety. Please give these your utmost attention.

▶ TASK Research vehicle service information, including fluid type, vehicle service history, service precautions, and technical service bulletins.

MLR
A1A1

Time off_____________

Time on_____________

Total time_____________

Student Instructions: Read through the entire procedure prior to starting. Prepare your workspace and any tools or parts that may be needed to complete the task. When directed by your instructor, begin the procedure to complete the task and check the box as each step is finished. Track your time on this procedure for later comparison to the standard completion time (i.e., 'flat rate' or customer pay time).

Procedure:	Step Completed
1. Research vehicle fluid types and list the following criteria:	
a. Determine engine oil type and capacity with oil filter change and list below.	☐

b. Determine transmission fluid type and full capacity and list below.	☐
c. Determine coolant/anti-freeze type and full capacity and list below.	☐
2. Locate and list vehicle service history including:	
a. Determine (if possible) whether any routine maintenance has been completed and list below.	☐
b. Determine outstanding/completed recall work and any service precautions and list below.	☐
c. Determine (if possible) any major repair in service history and list below.	☐
3. Determine TSBs issued for specific vehicle:	
a. Identify one TSB, and list TSB number, vehicle concern, and repair procedure.	☐

Non-Task-Specific Evaluations:	Step Completed
1. Tools and equipment were used as directed and returned in good working order.	☐
2. Complied with all general and task-specific safety standards, including proper use of any personal protective equipment.	☐
3. Completed the task within an appropriate time frame (Recommendation: 1.5 or 2 times flat rate).	☐
4. Left the work space clean and orderly.	☐
5. Cared for customer property and returned it undamaged.	☐

Student signature _________________________ Date _________________________

Comments:

Have your supervisor/instructor verify satisfactory completion of this procedure, note any observations found, and recommend any necessary action(s).

Evaluation Instructions: The scoring box below is intended to act as a guide for both student and instructor. Each criterion listed will help students understand what is expected from them and evaluators articulate the success at a particular task. The scoring is set up to allow a second attempt at each task (see the "Test" and "Retest" columns). Scoring is designed to reward students for correct completion of the task. Points are lost for failure to complete the employability requirements (see "Non-Task-Specific" criteria). When all the criteria are evaluated, tally the points for a total at the bottom of each column.

Tasksheet Scoring

Evaluation Items	Test		Retest	
	Pass	Fail	Pass	Fail
Task-Specific Evaluation	**(1 pt)**	**(O pts)**	**(1 pt)**	**(O pts)**
Determine vehicle fluid types and capacities.				
Determine vehicle service history.				
Determine vehicle service precautions.				
Determine vehicle TSBs.				
Non-Task-Specific Evaluation	**(O pts)**	**(−1 pt)**	**(O pts)**	**(−1 pt)**
Student successfully completed at least three of the non-task-specific steps.				
Student successfully completed all five of the non-task-specific steps.				
Total Score: <total # of points / 4 = %>				

CDX Tasksheet Number: A1002

Student/Intern Information

Name _________________________________ Date ____________ Class _________________________

Vehicle, Customer, and Service Information

Vehicle used for this activity:

Year _______________ Make ___________________________ Model ____________________________

Odometer _______________________________ VIN _______________________________________

> **Materials Required**
> - Blank work order
> - Vehicle with available service history records
> - Depending on the type of concern, special hand tool/diagnostic tools may be required. See your supervisor/instructor for instructions to identify what tools may be required.
> - Vehicle service information database

Task-Specific Safety Considerations
- Shop rules and procedures are critical to your safety. Please give these your utmost attention.
- When running any vehicles in the shop, make sure you use the shop's exhaust ventilation system to discharge all exhaust gas safely outside.

Time off______________

Time on______________

Total time____________

▶ **TASK** Verify operation of the instrument panel engine warning indicators. **MLR A1A2**

Student Instructions: Read through the entire procedure prior to starting. Prepare your work space and any tools or parts that may be needed to complete the task. When directed by your instructor, begin the procedure to complete the task and check the box as each step is finished. Track your time on this procedure for later comparison to the standard completion time (i.e., "flat rate" or customer pay time).

Procedure:	Step Completed
1. Cycle the ignition to the **ON** position but do not start the vehicle.	
a. List all the warning indicators that illuminate when the key is cycled on. Verify that the Check Engine warning indicator is illuminated.	☐

b. **Start** the vehicle. Verify that the warning indicators turn off after the vehicle is started. List any indicators that remain on once the vehicle has been started.	☐
2. List the instrument panel gauges that are equipped with your specific vehicle.	
a. **Start** the vehicle. Verify that the gauges are in proper working condition. List any gauges that are not working properly.	☐
b. Operate the turn signals in the vehicle. Verify the turn indicators are working correctly. If not working properly, list below.	☐
c. Operate the high beam headlights. Verify the high-beam indicator is working correctly. If not working properly, list below.	☐
3. Research the procedure for resetting maintenance indicators.	
a. Identify and list (or print off) the procedure for resetting the maintenance indicator.	☐
b. Perform the maintenance indicator reset procedure. Verify that the reset was performed properly.	☐

Non-Task-Specific Evaluations:	Step Completed
1. Tools and equipment were used as directed and returned in good working order.	☐
2. Complied with all general and task-specific safety standards, including proper use of any personal protective equipment.	☐
3. Completed the task in an appropriate time frame (Recommendation: 1.5 or 2 times flat rate).	☐
4. Left the work space clean and orderly.	☐
5. Cared for customer property and returned it undamaged.	☐

Student signature _______________________________ Date _______________________________

Comments:

Have your supervisor/instructor verify satisfactory completion of this procedure, any observations found, and any necessary action(s) recommended.

Evaluation Instructions: The scoring box below is intended to act as a guide for both student and instructor. Each criterion listed will help students understand what is expected from them and evaluators articulate the success at a particular task. The scoring is set up to allow a second attempt at each task (see the "Test" and "Retest" columns). Scoring is designed to reward students for correct completion of the task. Points are lost for failure to complete the employability requirements (see "Non-Task-Specific" criteria). When all the criteria are evaluated, tally the points for a total at the bottom of each column.

Tasksheet Scoring

	Test		Retest	
Evaluation Items	**Pass**	**Fail**	**Pass**	**Fail**
Task-Specific Evaluation	**(1 pt)**	**(0 pts)**	**(1 pt)**	**(0 pts)**
Student successfully started the vehicle and listed any warning indicators.				
Student verified that all gauges were working appropriately and listed any that were not.				
Student operated the vehicle's turn signals and high beam headlights and noted any that were not working correctly.				
Student reviewed the procedure for resetting the maintenance indicator and performed the maintenance indicator reset procedure.				
Non-Task-Specific Evaluation	**(0 pts)**	**(−1 pt)**	**(0 pts)**	**(−1 pt)**
Student successfully completed at least three of the non-task-specific steps.				
Student successfully completed all five of the non-task-specific steps.				
Total Score: <total # of points / 4 = %>				

Supervisor:

Supervisor/instructor signature ___________________________________ Date _____________________

Comments:

Retest supervisor/instructor signature ________________________________ Date _____________________

Comments:

CDX Tasksheet Number: A1003

Student/Intern Information

Name _________________________________ Date _____________ Class _________________________________

Vehicle, Customer, and Service Information

Vehicle used for this activity:

Year _______________ Make _________________________________ Model _________________________________

Odometer _________________________________ VIN _________________________________

> **Materials Required**
>
> - Blank work order
> - Vehicle with available service history records
> - Depending on the type of concern, special hand tool/diagnostic tools may be required. See your supervisor/instructor for instructions to identify what tools may be required.
> - Vehicle service information database
> - Specialized leak detector equipment may be required. See your supervisor/instructor for instructions to identify what tools may be required.

Task-Specific Safety Considerations

- Comply with personal and environmental safety practices associated with clothing; eye protection; hand tools; power equipment; proper ventilation; and the handling, storage, and disposal of chemicals/materials in accordance with local, state, and federal safety and environmental regulations.
- Shop rules and procedures are critical to your safety. Please give these your utmost attention.
- Eye wash stations are an important component of shop safety. Always know their location, purpose, and use.
- When running any vehicles in the shop, make sure you use the shop's exhaust ventilation system to discharge all exhaust gas safely outside.

Time off_______________

Time on_______________

▶ TASK Inspect engine assembly for fuel, oil, coolant, and other leaks; determine necessary action.

MLR
A1A3

Total time_______________

Student Instructions: Read through the entire procedure prior to starting. Prepare your work space and any tools or parts that may be needed to complete the task. When directed by your instructor, begin the procedure to complete the task and check the box as each step is finished. Track your time on this procedure for later comparison to the standard completion time (i.e., "flat rate" or customer pay time).

Procedure:	Step Completed
1. Research the inspection procedure for leak diagnosis in the appropriate service information.	
a. List or print off and attach to this sheet the proper method of performing a leak inspection.	☐
2. Following the instructions and all the safety precautions outlined in the service information, inspect your designated vehicle's engine for leaking fluid and note any actions needed.	
a. Engine oil leak	☐
b. Fuel leak	☐
c. Coolant leak	☐
d. Transmission/transaxle leak	☐
e. Brake fluid leak	☐

Non-Task-Specific Evaluations:	Step Completed
1. Tools and equipment were used as directed and returned in good working order.	☐
2. Complied with all general and task-specific safety standards, including proper use of any personal protective equipment.	☐
3. Completed the task in an appropriate time frame (Recommendation: 1.5 or 2 times flat rate).	☐
4. Left the work space clean and orderly.	☐
5. Cared for customer property and returned it undamaged.	☐

Student signature ________________________________ Date ________________________________

Comments:

Have your supervisor/instructor verify satisfactory completion of this procedure, any observations found, and any necessary action(s) recommended.

Evaluation Instructions: The scoring box below is intended to act as a guide for both student and instructor. Each criterion listed will help students understand what is expected from them and evaluators articulate the success at a particular task. The scoring is set up to allow a second attempt at each task (see the "Test" and "Retest" columns). Scoring is designed to reward students for correct completion of the task. Points are lost for failure to complete the employability requirements (see "Non-Task-Specific" criteria). When all the criteria are evaluated, tally the points for a total at the bottom of each column.

Tasksheet Scoring

	Test		Retest	
Evaluation Items	**Pass**	**Fail**	**Pass**	**Fail**
Task-Specific Evaluation	**(1 pt)**	**(0 pts)**	**(1 pt)**	**(0 pts)**
Research leak inspection procedure.				
Inspect engine assembly for fluid leaks.				
Correctly identify type of fluid leak.				
Correctly identify source of fluid leak.				
Non-Task-Specific Evaluation	**(0 pts)**	**(−1 pt)**	**(0 pts)**	**(−1 pt)**
Student successfully completed at least three of the non-task-specific steps.				
Student successfully completed all five of the non-task-specific steps.				
Total Score: <total # of points / 4 = %>				

Supervisor:

Supervisor/instructor signature _______________________________ Date _________________

Comments:

Retest supervisor/instructor signature _______________________ Date _________________

Comments:

CDX Tasksheet Number: A1004

Student/Intern Information

Name _______________________________ Date ___________ Class _____________________

Vehicle, Customer, and Service Information

Vehicle used for this activity:

Year _______________ Make _______________________ Model _____________________

Odometer _______________________ VIN _________________________________

Materials Required

- Blank work order
- Vehicle with available service history records
- Depending on the type of concern, special hand tool/diagnostic tools may be required. See your supervisor/instructor for instructions to identify what tools may be required.
- Vehicle service information database.
- Appropriate gaskets and/or RTV sealer

Task-Specific Safety Considerations

- Comply with personal and environmental safety practices associated with clothing; eye protection; hand tools; power equipment; proper ventilation; and the handling, storage, and disposal of chemicals/materials in accordance with local, state, and federal safety and environmental regulations.

Time off______________

Time on______________

Total time______________

▶ **TASK** Install engine covers using gaskets, seals, and sealers as required.

MLR
A1A4

Student Instructions: Read through the entire procedure prior to starting. Prepare your work space and any tools or parts that may be needed to complete the task. When directed by your instructor, begin the procedure to complete the task and check the box as each step is finished. Track your time on this procedure for later comparison to the standard completion time (i.e., "flat rate" or customer pay time).

Procedure:	Step Completed
1. Remove and install engine valve cover. Research the procedure for removing, cleaning, inspection, and installation of the engine valve cover.	
a. List (or print off and attach to this sheet) the proper procedure for removing and installing the engine valve cover.	☐

b. Clean and inspect the engine valve cover using the approved solvents and methods.	☐
c. Identify the sealant or gasket used.	☐
d. Carefully install the valve cover using the recommended procedure. Note: It is best to start all fasteners by hand before tightening any fastener completely.	☐
e. Using the correct sequence, tighten all fasteners to specification.	☐
f. Start engine and verify there are no oil leaks.	☐

Non-Task-Specific Evaluations:	Step Completed
1. Tools and equipment were used as directed and returned in good working order.	☐
2. Complied with all general and task-specific safety standards, including proper use of any personal protective equipment.	☐
3. Completed the task in an appropriate time frame (Recommendation: 1.5 or 2 times flat rate).	☐
4. Left the work space clean and orderly.	☐
5. Cared for customer property and returned it undamaged.	☐

Student signature _______________________ Date _______________________

Comments:

Have your supervisor/instructor verify satisfactory completion of this procedure, any observations found, and any necessary action(s) recommended.

Tasksheet Scoring

		Test		Retest	
Evaluation Items		**Pass**	**Fail**	**Pass**	**Fail**
Task-Specific Evaluation		**(1 pt)**	**(0 pts)**	**(1 pt)**	**(0 pts)**
Research the procedure to remove the engine valve cover.					
Remove valve cover, clean, and inspect.					
Install the proper sealant and gasket and properly install the valve cover.					
Torque all fasteners to specification and verify repairs.					
Non-Task-Specific Evaluation		**(0 pts)**	**(−1 pt)**	**(0 pts)**	**(−1 pt)**
Student successfully completed at least three of the non-task-specific steps.					
Student successfully completed all five of the non-task-specific steps.					
Total Score: <total # of points / 4 = %>					

Supervisor:

Supervisor/instructor signature ________________________________ Date ________________________

Comments:

Retest supervisor/instructor signature ________________________________ Date ________________________

Comments:

CDX Tasksheet Number: A1005

Student/Intern Information

Name _________________________________ Date ___________ Class _________________________

Vehicle, Customer, and Service Information

Vehicle used for this activity:

Year _______________ Make _____________________________ Model _____________________________

Odometer _________________________________ VIN _________________________________

Materials Required

- Blank work order
- Vehicle with available service history records
- Depending on the type of concern, special diagnostic tools may be required. See your supervisor/instructor for instructions to identify what tools may be required.
- Vehicle service information
- Timing light or scan tool
- Spark plug socket
- Compression gauge
- Flashlight/droplight
- Personal protective equipment

Task-Specific Safety Considerations

- When running any vehicles in the shop, make sure you use the shop's exhaust ventilation system to discharge all exhaust gas safely outside.
- Because the vehicle will be running for an extended amount of time, make sure the vehicle cannot move or roll by applying the parking brake and using wheel chocks.
- You will be working under the hood of a running vehicle. Keep your hands and fingers away from moving belts, fans, and other parts.
- Lifting equipment, such as vehicle jacks and stands, vehicle hoists, and engine hoists, are important tools that increase productivity and make the job easier. However, they can also cause severe injury or death if used improperly. Make sure you follow the manufacturer's operation procedures. Also make sure you have your supervisor's/instructor's permission to use any particular type of lifting equipment.

▶ TASK Verify engine mechanical timing.

MLR
A1A5

Time off____________

Time on____________

Student Instructions: Read through the entire procedure prior to starting. Prepare your work space and any tools or parts that may be needed to complete the task. When directed by your instructor, begin the procedure to complete the task and check the box as each step is finished. Track your time on this procedure for later comparison to the standard completion time (i.e., "flat rate" or customer pay time).

Total time____________

Procedure:	Step Completed
1. Research the procedure for verifying correct engine mechanical timing on your vehicle.	
a. List (or print off and attach to this sheet) the procedure for checking the engine mechanical timing. Caution: On most engines, it is crucial that mechanical timing is exact to eliminate the chance of components contacting each other and causing damage while the engine is running.	☐
b. Check for any technical service bulletins (TSBs) that may relate to the task of checking engine mechanical timing. List any TSBs that may apply.	☐
2. Following the procedure, verify that the engine mechanical timing is correct. Note any action that is needed.	
a. Does your vehicle use a timing belt or a timing chain?	☐
b. Are there timing inspection covers used on your vehicle to make this task easier?	☐
c. Explain how you ensured that cylinder 1 was at top dead center.	☐
d. Describe the procedure you used to verify mechanical timing. Did you have any problems with this task? If so, explain.	☐

Non-Task-Specific Evaluations:	Step Completed
1. Tools and equipment were used as directed and returned in good working order.	☐
2. Complied with all general and task-specific safety standards, including proper use of any personal protective equipment.	☐
3. Completed the task in an appropriate time frame (Recommendation: 1.5 or 2 times flat rate).	☐
4. Left the work space clean and orderly.	☐
5. Cared for customer property and returned it undamaged.	☐

Student signature _______________________________ Date _______________________________

Comments:

Have your supervisor/instructor verify satisfactory completion of this procedure, any observations found, and any necessary action(s) recommended.

Evaluation Instructions: The scoring box below is intended to act as a guide for both student and instructor. Each criterion listed will help students understand what is expected from them and evaluators articulate the success at a particular task. The scoring is set up to allow a second attempt at each task (see the "Test" and "Retest" columns). Scoring is designed to reward students for correct completion of the task. Points are lost for failure to complete the employability requirements (see "Non-Task-Specific" criteria). When all the criteria are evaluated, tally the points for a total at the bottom of each column.

Tasksheet Scoring

	Test		Retest	
Evaluation Items	**Pass**	**Fail**	**Pass**	**Fail**
Task-Specific Evaluation	**(1 pt)**	**(0 pts)**	**(1 pt)**	**(0 pts)**
Research the procedure for verifying correct engine mechanical timing on your engine.				
Verify top dead center on cylinder #1.				
Verify type of timing (belt, chain, etc.).				
Verify mechanical timing.				
Non-Task-Specific Evaluation	**(0 pts)**	**(−1 pt)**	**(0 pts)**	**(−1 pt)**
Student successfully completed at least three of the non-task-specific steps.				
Student successfully completed all five of the non-task-specific steps.				
Total Score: <total # of points / 4 = %>				

Supervisor:

Supervisor/instructor signature _______________________________ Date _______________

Comments:

Retest supervisor/instructor signature _______________________________ Date _______________

Comments:

CDX Tasksheet Number: A1006

Student/Intern Information

Name _________________________________ Date ____________ Class _________________________

Vehicle, Customer, and Service Information

Vehicle used for this activity:

Year ______________ Make _____________________________ Model _____________________________

Odometer _________________________________ VIN _____________________________________

Materials Required

- Blank work order
- Vehicle with available service history records
- Depending on the type of concern, special hand tool/diagnostic tools may be required. See your supervisor/instructor for instructions to identify what tools may be required.
- Personal protective equipment
- Drill and drill bits
- Cutting oil
- Tap set, die set, thread chaser
- Center punch
- Bolt/stud extractor tools
- Thread insert repair kit
- Vise grips
- Rust penetrant

Task-Specific Safety Considerations

- Comply with personal and environmental safety practices associated with clothing; eye protection; hand tools; power equipment; proper ventilation; and the handling, storage, and disposal of chemicals/materials in accordance with local, state, and federal safety and environmental regulations.
- Shop rules and procedures are critical to your safety. Please give these your utmost attention.

▶ TASK Perform common fastener and thread repair, to include: remove broken bolt, restore internal and external threads, and repair internal threads with thread insert.

MLR
A1A6

Time off____________

Time on____________

Total time____________

Student Instructions: Read through the entire procedure prior to starting. Prepare your work space and any tools or parts that may be needed to complete the task. When directed by your instructor, begin the procedure to complete the task and check the box as each step is finished. Track your time on this procedure for later comparison to the standard completion time (i.e., "flat rate" or customer pay time).

Procedure:	Step Completed
1. Inspect/remove a broken bolt or stud. **Tip:** If enough stud is exposed, it may be possible to use vise grips to remove the broken piece.	
a. Using the proper procedure, mark the center of the stud with a center punch to allow for the centering of a drill bit.	☐
b. Using the proper procedure, carefully drill through the center of the bolt with the correct size drill bit. Drill until you are able to insert the correct stud extractor.	☐
c. Insert the stud extractor and remove the broken bolt/stud.	☐
2. If applicable, use the proper size thread chaser and attempt to restore the threads. If threads are damaged beyond repair, a threaded insert may be required.	
a. What size and thread pitch did you use as a thread chase?	☐
b. Describe the difference between a thread chaser and a thread tap.	☐
3. Repair internal threads using a threaded insert kit.	
a. Using the correct size drill bit, drill the damaged hole to the correct size. **Tip:** Use a drill stop to prevent drilling too deep into blind holes. Blind holes often lead to areas such as water jackets. Drilling too far could damage other components.	☐
b. What size drill bit did you use?	☐
c. Using the tap provided in the thread repair kit, carefully tap the hole to the correct size. **Tip:** When using a tap, cut new threads 1/2 turn at a time. After cutting 1/2 turn, back the tap up 1 turn and "chip" away the metal you are cutting. This allows the thread tap to cut through the metal without metal chips binding the tap.	☐
d. Using the proper procedure, install the threaded insert using the tools provided. Once installed, a new fastener should thread into the repaired section with minimal effort or binding of the threads.	☐

Non-Task-Specific Evaluations:	Step Completed
1. Tools and equipment were used as directed and returned in good working order.	☐
2. Complied with all general and task-specific safety standards, including proper use of any personal protective equipment.	☐
3. Completed the task in an appropriate time frame (Recommendation: 1.5 or 2 times flat rate).	☐
4. Left the work space clean and orderly.	☐
5. Cared for customer property and returned it undamaged.	☐

Student signature ________________________________ Date ________________________________

Comments:

Have your supervisor/instructor verify satisfactory completion of this procedure, any observations found, and any necessary action(s) recommended.

Evaluation Instructions: The scoring box below is intended to act as a guide for both student and instructor. Each criterion listed will help students understand what is expected from them and evaluators articulate the success at a particular task. The scoring is set up to allow a second attempt at each task (see the "Test" and "Retest" columns). Scoring is designed to reward students for correct completion of the task. Points are lost for failure to complete the employability requirements (see "Non-Task-Specific" criteria). When all the criteria are evaluated, tally the points for a total at the bottom of each column.

Tasksheet Scoring

	Test		Retest	
Evaluation Items	Pass	Fail	Pass	Fail
Task-Specific Evaluation	**(1 pt)**	**(0 pts)**	**(1 pt)**	**(0 pts)**
Use the correct procedure to remove a broken bolt or stud.				
Select the correct size drill bit, tap, and thread chaser.				
Use the correct procedure for tapping new threads.				
The threaded insert accepts the new fastener with minimal binding.				
Non-Task-Specific Evaluation	**(0 pts)**	**(−1 pt)**	**(0 pts)**	**(−1 pt)**
Student successfully completed at least three of the non-task-specific steps.				
Student successfully completed all five of the non-task-specific steps.				
Total Score: <total # of points / 4 = %>				

Supervisor:

Supervisor/instructor signature ________________________________ Date ________________

Comments:

Retest supervisor/instructor signature ________________________________ Date ________________

Comments:

CDX Tasksheet Number: A1007

Student/Intern Information

Name _________________________________ Date ____________ Class _____________________

Vehicle, Customer, and Service Information

Vehicle used for this activity:

Year _______________ Make _________________________ Model ____________________

Odometer _________________________ VIN _________________________________

Materials Required

- Blank work order
- Vehicle with available service history records
- Depending on the type of concern, special hand tool/diagnostic tools may be required. See your supervisor/instructor for instructions to identify what tools may be required.
- Vehicle service information database
- High-voltage personal protective equipment

Task-Specific Safety Considerations

- Comply with personal and environmental safety practices associated with clothing; eye protection; hand tools; power equipment; proper ventilation; and the handling, storage, and disposal of chemicals/materials in accordance with local, state, and federal safety and environmental regulations.
- Shop rules and procedures are critical to your safety. Please give these your utmost attention.
- Marked safety areas play an important role in maintaining a safe work environment. Always understand and heed marked safety areas.
- **Caution:** Working with high voltage can cause severe injury or death. All safety precaution measures must be taken.

▶ TASK Identify service precautions related to service of the internal combustion engine of a hybrid vehicle.

MLR
A1A7

Time off_______________

Time on_______________

Total time_______________

Student Instructions: Read through the entire procedure prior to starting. Prepare your work space and any tools or parts that may be needed to complete the task. When directed by your instructor, begin the procedure to complete the task and check the box as each step is finished. Track your time on this procedure for later comparison to the standard completion time (i.e., "flat rate" or customer pay time).

Procedure:	Step Completed
1. Research service precautions related to working on your specific hybrid vehicle.	
a. List (or print and attach to this sheet) the precautionary measures taken when working on a hybrid vehicle.	☐
b. List the steps necessary to safely disable the high-voltage system of a hybrid vehicle.	☐
c. Determine where the high-voltage battery is located on your specific vehicle.	☐
d. Determine the location of the high-voltage disconnect junction in your vehicle.	☐
e. List the necessary personal protective equipment that is recommended for use when working on your hybrid vehicle.	☐
f. What color are high-voltage wires on hybrid vehicles?	☐

Non-Task-Specific Evaluations:	Step Completed
1. Tools and equipment were used as directed and returned in good working order.	☐
2. Complied with all general and task-specific safety standards, including proper use of any personal protective equipment.	☐
3. Completed the task in an appropriate time frame (Recommendation: 1.5 or 2 times flat rate).	☐
4. Left the work space clean and orderly.	☐
5. Cared for customer property and returned it undamaged.	☐

Student signature _________________________ Date _____________________

Comments:

Have your supervisor/instructor verify satisfactory completion of this procedure, any observations found, and any necessary action(s) recommended.

Evaluation Instructions: The scoring box below is intended to act as a guide for both student and instructor. Each criterion listed will help students understand what is expected from them and evaluators articulate the success at a particular task. The scoring is set up to allow a second attempt at each task (see the "Test" and "Retest" columns). Scoring is designed to reward students for correct completion of the task. Points are lost for failure to complete the employability requirements (see "Non-Task-Specific" criteria). When all the criteria are evaluated, tally the points for a total at the bottom of each column.

Tasksheet Scoring

Evaluation Items	Test		Retest	
	Pass	**Fail**	**Pass**	**Fail**
Task-Specific Evaluation	**(1 pt)**	**(0 pts)**	**(1 pt)**	**(0 pts)**
List precautionary measures needed when servicing a hybrid vehicle, as well as the personal protective equipment required.				
List steps to disable high-voltage systems.				
Locate a high-voltage disconnect junction.				
Locate a high-voltage battery.				
Determine the color of high-voltage wiring.				
Non-Task-Specific Evaluation	**(0 pts)**	**(−1 pt)**	**(0 pts)**	**(−1 pt)**
Student successfully completed at least three of the non-task-specific steps.				
Student successfully completed all five of the non-task-specific steps.				
Total Score: <total # of points / 4 = %>				

Supervisor:

Supervisor/instructor signature _________________________ Date _________________

Comments:

Retest supervisor/instructor signature _________________________ Date _________________

Comments:

Learning Objective/Task	CDX Tasksheet Number	ASE Education Foundation Reference Number; Priority Level
• Adjust valves (mechanical or hydraulic lifters).	A1008	A1B1; P-3
• Identify components of the cylinder head and valve train.	A1009	A1B2; P-1

Materials Required

- Blank work order
- Vehicle with available service history records
- Depending on the type of concern, special hand tool/diagnostic tools may be required. See your supervisor/instructor for instructions to identify what tools may be required.
- Vehicle service information database

Safety Considerations

- Comply with personal and environmental safety practices associated with clothing; eye protection; hand tools; power equipment; proper ventilation; and the handling, storage, and disposal of chemicals/materials in accordance with local, state, and federal safety and environmental regulations.
- Shop rules and procedures are critical to your safety. Please give these your utmost attention.
- When running any vehicles in the shop, make sure you use the shop's exhaust ventilation system to discharge all exhaust gas safely outside.
- Lifting equipment, such as vehicle jacks and stands, vehicle hoists, and engine hoists, are important tools that increase productivity and make the job easier. However, they can also cause severe injury or death if used improperly. Make sure you follow the manufacturer's operation procedures. Also, make sure you have your supervisor's/instructor's permission to use any particular type of lifting equipment.

CDX Tasksheet Number: A1008

Student/Intern Information

Name _________________________________ Date ____________ Class _____________________

Vehicle, Customer, and Service Information

Vehicle used for this activity:

Year _______________ Make _____________________________ Model _____________________

Odometer _________________________________ VIN _______________________________________

Materials Required

- Blank work order
- Vehicle with available service history records
- Depending on the type of concern, special hand tool/diagnostic tools may be required. See your supervisor/instructor for instructions to identify what tools may be required.
- Vehicle service information database
- Personal protective equipment
- General hand tools
- Feeler gauge set

Task-Specific Safety Considerations

- Extreme caution must be exercised when working around rotating components.
- Comply with personal and environmental safety practices associated with clothing; eye protection; hand tools; power equipment; proper ventilation; and the handling, storage, and disposal of chemicals/materials in accordance with local, state, and federal safety and environmental regulations.

▶ **TASK** Adjust valves (mechanical or hydraulic lifters).

MLR
A1B1

<table>
<tr><td>Time off________</td></tr>
<tr><td>Time on________</td></tr>
<tr><td>Total time________</td></tr>
</table>

Student Instructions: Read through the entire procedure prior to starting. Prepare your work space and any tools or parts that may be needed to complete the task. When directed by your instructor, begin the procedure to complete the task and check the box as each step is finished. Track your time on this procedure for later comparison to the standard completion time (i.e., "flat rate" or customer pay time).

Procedure:	Step Completed
1. Research the procedure for performing a valve adjustment for your specific vehicle.	
a. List (or print and attach to this sheet) the procedure for checking and adjusting valve lash.	☐

b. List the specification for valve lash of the intake and exhaust valve. Is the adjustment made when the engine is cold or at operating temperature?	☐
3. Following the procedure, remove the valve cover and begin valve lash measurement of your first cylinder.	
a. What stroke must the specific cylinder be on to make an adjustment to the intake valve? What is the location of the piston in that cylinder?	☐
b. List your measurement for the intake valve lash. Is adjustment required? Make necessary adjustments and check lash once the task is complete.	☐
c. Following the procedure, list your measurement for the exhaust valve lash. Is a valve adjustment required? Make necessary adjustments and check lash once the task is complete.	☐
4. Continue the procedure to complete valve adjustments for your engine.	
a. List all measurements taken before adjustments are made. Continue the procedure and adjust all valves as necessary.	☐

Non-Task-Specific Evaluations:	Step Completed
1. Tools and equipment were used as directed and returned in good working order.	☐
2. Complied with all general and task-specific safety standards, including proper use of any personal protective equipment.	☐
3. Completed the task in an appropriate time frame (Recommendation: 1.5 or 2 times flat rate).	☐
4. Left the work space clean and orderly.	☐
5. Cared for customer property and returned it undamaged.	☐

Student signature ________________________________ Date ________________________________

Comments:

Have your supervisor/instructor verify satisfactory completion of this procedure, any observations found, and any necessary action(s) recommended.

Tasksheet Scoring

	Test		Retest	
Evaluation Items	**Pass**	**Fail**	**Pass**	**Fail**
Task-Specific Evaluation	**(1 pt)**	**(O pts)**	**(1 pt)**	**(O pts)**
Research the procedure for performing a valve adjustment for your specific vehicle.				
List valve lash specifications for the intake and exhaust valve.				
List all valve lash readings prior to making adjustments.				
Perform the valve lash procedure on designated cylinders.				
Non-Task-Specific Evaluation	**(O pts)**	**(−1 pt)**	**(O pts)**	**(−1 pt)**
Student successfully completed at least three of the non-task-specific steps.				
Student successfully completed all five of the non-task-specific steps.				
Total Score: <total # of points / 4 = %>				

CDX Tasksheet Number: A1009

Student/Intern Information

Name _________________________________ Date ___________ Class _______________________

Vehicle, Customer, and Service Information

Vehicle used for this activity:

Year _______________ Make _________________________ Model _____________________

Odometer _______________________ VIN _________________________________

Materials Required

- Blank work order
- Vehicle with available service history records
- Depending on the type of concern, special hand tool/diagnostic tools may be required. See your supervisor/instructor for instructions to identify what tools may be required.
- Vehicle service information database

Task-Specific Safety Considerations

- Comply with personal and environmental safety practices associated with clothing; eye protection; hand tools; power equipment; proper ventilation; and the handling, storage, and disposal of chemicals/materials in accordance with local, state, and federal safety and environmental regulations.

▶ **TASK** Identify components of the cylinder head and valve train. **MLR A1B2**

Time off___________

Time on___________

Total time___________

Student Instructions: Read through the entire procedure prior to starting. Prepare your work space and any tools or parts that may be needed to complete the task. When directed by your instructor, begin the procedure to complete the task and check the box as each step is finished. Track your time on this procedure for later comparison to the standard completion time (i.e., "flat rate" or customer pay time).

Procedure:	Step Completed
1. Identify the following individual components of a cylinder head. Explain the purpose of each component and how it works.	
a. Intake valve	☐

b. Exhaust valve	☐
c. Valve spring	☐
d. Valve spring retainer	☐
e. Valve seal	☐
f. Combustion chamber. List which components are responsible for sealing the combustion chamber, allowing compression to build in a cylinder.	☐
g. Intake port	☐
h. Exhaust port	☐
i. Coolant passage	☐
2. Identify the following individual components of the valve train. Explain the purpose of each component and how it works (excluding valves).	
a. Camshaft	☐
b. Lifter (if applicable)	☐
c. Bucket-type lifter (if applicable). In what type of engine/valve train design would this type of lifter be used?	☐

d. Pushrod (if applicable)	☐
e. Timing chain	☐
f. Rocker arm (if applicable)	☐
3. Explain the difference between an engine that features a camshaft in the engine block and an engine that uses a camshaft in the cylinder head.	☐

Non-Task-Specific Evaluations:	Step Completed
1. Tools and equipment were used as directed and returned in good working order.	☐
2. Complied with all general and task-specific safety standards, including proper use of any personal protective equipment.	☐
3. Completed the task in an appropriate time frame (Recommendation: 1.5 or 2 times flat rate).	☐
4. Left the work space clean and orderly.	☐
5. Cared for customer property and returned it undamaged.	☐

Student signature ________________________ Date ________________________

Comments:

Have your supervisor/instructor verify satisfactory completion of this procedure, any observations found, and any necessary action(s) recommended.

Evaluation Instructions: The scoring box below is intended to act as a guide for both student and instructor. Each criterion listed will help students understand what is expected from them and evaluators articulate the success at a particular task. The scoring is set up to allow a second attempt at each task (see the "Test" and "Retest" columns). Scoring is designed to reward students for correct completion of the task. Points are lost for failure to complete the employability requirements (see "Non-Task-Specific" criteria). When all the criteria are evaluated, tally the points for a total at the bottom of each column.

Tasksheet Scoring

Evaluation Items	Test		Retest	
	Pass	Fail	Pass	Fail
Task-Specific Evaluation	**(1 pt)**	**(0 pts)**	**(1 pt)**	**(0 pts)**
Identify components of the cylinder head; explain its purpose and how it works.				
Identify components of the valve train; explain the purpose and how it works.				
Identify differences between overhead cams and cams in block engines.				
Identify components responsible for cylinder sealing.				
Non-Task-Specific Evaluation	**(0 pts)**	**(−1 pt)**	**(0 pts)**	**(−1 pt)**
Student successfully completed at least three of the non-task-specific steps.				
Student successfully completed all five of the non-task-specific steps.				
Total Score: <total # of points / 4 = %>				

Supervisor:

Supervisor/instructor signature ___________________________ Date ___________________

Comments:

Retest supervisor/instructor signature ___________________________ Date ___________________

Comments:

Learning Objective/Task	CDX Tasksheet Number	ASE Education Foundation Reference Number; Priority Level
• Perform cooling system pressure and dye tests to identify leaks; check coolant condition and level; inspect and test radiator, pressure cap, coolant recovery tank, heater core, and galley plugs; determine necessary action.	A1010	A1C1, P-1
• Inspect, replace, and/or adjust drive belts, tensioners, and pulleys; check pulley and belt alignment.	A1011	A1C2, P-1
• Remove, inspect, and replace thermostat and gasket/seal.	A1012	A1C3, P-1
• Inspect and test coolant; drain and recover coolant; flush and refill cooling system; use proper fluid type per manufacturer specification; bleed air as required.	A1013	A1C4, P-1
• Perform engine oil and filter change; use proper fluid type per manufacturer specification; reset maintenance reminder as required.	A1014	A1C5, P-1
• Identify components of the lubrication and cooling systems.	A1015	A1C6, P-1

Materials Required

- Blank work order
- Vehicle with available service history records
- Depending on the type of concern, special diagnostic tools may be required. See your supervisor/instructor for instructions to identify what tools may be required.

Safety Considerations

- When running any vehicles in the shop, make sure you use the shop's exhaust ventilation system to discharge all exhaust gas safely outside.
- Lifting equipment, such as vehicle jacks and stands, vehicle hoists, and engine hoists, are important tools that increase productivity and make the job easier. However, they can also cause severe injury or death if used improperly. Make sure you follow the manufacturer's operation procedures. Also, make sure you have your supervisor's/instructor's permission to use any particular type of lifting equipment.
- Comply with personal and environmental safety practices associated with clothing; eye protection; hand tools; power equipment; proper ventilation; and the handling, storage, and disposal of chemicals/materials in accordance with local, state, and federal safety and environmental regulations.
- If you need to start the vehicle, you should ensure that the parking brake is firmly applied; if necessary, use wheel chocks to prevent the vehicle from moving when the vehicle is started to verify the completion of these tasks.
- Extreme caution must be exercised when working around rotating components.

CDX Tasksheet Number: A1010

Student/Intern Information

Name _______________________________ Date ____________ Class _______________________

Vehicle, Customer, and Service Information

Vehicle used for this activity:

Year ______________ Make _____________________________ Model _______________________

Odometer _________________________________ VIN _________________________________

Materials Required

- Blank work order
- Vehicle with available service history records
- Depending on the type of concern, special diagnostic tools may be required. See your supervisor/instructor for instructions to identify what tools may be required.
- Cooling system pressure tester
- Coolant dye kit
- Hydrometer, refractometer, litmus-style test strips
- General hand tools
- Personal protective equipment

Task-Specific Safety Considerations

- Comply with personal and environmental safety practices associated with clothing; eye protection; hand tools; power equipment; proper ventilation; and the handling, storage, and disposal of chemicals/materials in accordance with local, state, and federal safety and environmental regulations.
- Shop rules and procedures are critical to your safety. Please give these your utmost attention.
- **Caution:** Serious injury can occur if any part of the cooling system is opened while the vehicle is hot!

▶ TASK Perform cooling system pressure and dye tests to identify leaks; check coolant condition and level; inspect and test radiator, pressure cap, coolant recovery tank, heater core, and galley plugs; determine necessary action.

MLR
A1C1

Time off____________

Time on____________

Total time____________

Student Instructions: Read through the entire procedure prior to starting. Prepare your work space and any tools or parts that may be needed to complete the task. When directed by your instructor, begin the procedure to complete the task and check the box as each step is finished. Track your time on this procedure for later comparison to the standard completion time (i.e., "flat rate" or customer pay time).

Procedure:	Step Completed
1. Begin by checking coolant level and condition. Leak test results may be incorrect if the coolant level is not as specified.	
a. Check the coolant level. Top off coolant if necessary. List your results and any action required.	☐
b. Check the coolant condition using the hydrometer. List the level of freeze protection.	☐
c. Test the pH level of the coolant using the litmus-style test strips. List your results and any action required.	☐
2. Perform a cooling system pressure test.	
a. Install the correct adapter (if needed) to the radiator.	☐
b. Pressurize the cooling system to the recommended pressure. **Tip:** Usually the pressure rating can be found on the radiator pressure cap. Once pressurized, observe the pressure reading and inspect the following areas for leaks. List any action required. Pressure should be stable for at least 15 minutes.	☐
3. Inspect the following components:	
a. Radiator hoses and clamps	☐
b. Engine core plugs	☐
c. Heater core	☐
d. Water pump	☐
e. Thermostat housing	☐
f. Head gasket area	☐
g. Intake manifold area	☐
h. Coolant recovery tank	☐

4. Perform a radiator cap pressure test.	
a. Install the correct adaptor from your pressure test kit to the radiator cap.	☐
b. Pressurize the radiator cap to the recommended pressure. Observe the pressure reading. Evaluate your results and list any action required.	☐

Non-Task-Specific Evaluations:	Step Completed
1. Tools and equipment were used as directed and returned in good working order.	☐
2. Complied with all general and task-specific safety standards, including proper use of any personal protective equipment.	☐
3. Completed the task in an appropriate time frame (Recommendation: 1.5 or 2 times flat rate).	☐
4. Left the work space clean and orderly.	☐
5. Cared for customer property and returned it undamaged.	☐

Student signature ____________________________ Date __________________________

Comments:

Have your supervisor/instructor verify satisfactory completion of this procedure, any observations found, and any necessary action(s) recommended.

Tasksheet Scoring

Evaluation Items	Test		Retest	
	Pass	Fail	Pass	Fail
Task-Specific Evaluation	**(1 pt)**	**(0 pts)**	**(1 pt)**	**(0 pts)**
Use a hydrometer to test coolant freeze protection.				
Identify the coolant level and condition.				
Use a pressure tester and the correct pressure rating to inspect the entire cooling system.				
Use a pressure tester to test the radiator cap.				
Non-Task-Specific Evaluation	**(0 pts)**	**(−1 pt)**	**(0 pts)**	**(−1 pt)**
Student successfully completed at least three of the non-task-specific steps.				
Student successfully completed all five of the non-task-specific steps.				
Total Score: <total # of points / 4 = %>				

Supervisor:

Supervisor/instructor signature _______________________________ Date _________________

Comments:

Retest supervisor/instructor signature _______________________________ Date _________________

Comments:

CDX Tasksheet Number: A1O11

Student/Intern Information

Name _________________________________ Date ___________ Class _________________________

Vehicle, Customer, and Service Information

Vehicle used for this activity:

Year _______________ Make _____________________________ Model _____________________________

Odometer _________________________________ VIN ____

Materials Required

- Blank work order
- Vehicle with available service history records
- Depending on the type of concern, special diagnostic tools may be required. See your supervisor/instructor for instructions to identify what tools may be required.
- Service information
- Serpentine belt removal tool set
- Belt tension tester
- Pulley alignment laser (if available) or straight edge

Task-Specific Safety Considerations

- When running any vehicles in the shop, make sure you use the shop's exhaust ventilation system to discharge all exhaust gas safely outside.
- Lifting equipment, such as vehicle jacks and stands, vehicle hoists, and engine hoists, are important tools that increase productivity and make the job easier. However, they can also cause severe injury or death if used improperly. Make sure you follow the manufacturer's operation procedures. Also, make sure you have your supervisor's/instructor's permission to use any particular type of lifting equipment.
- Comply with personal and environmental safety practices associated with clothing; eye protection; hand tools; power equipment; proper ventilation; and the handling, storage, and disposal of chemicals/materials in accordance with local, state, and federal safety and environmental regulations.
- If you need to start the vehicle, you should ensure that the parking brake is firmly applied; if necessary, use wheel chocks to prevent the vehicle from moving when the vehicle is started to verify the completion of these tasks.
- Extreme caution must be exercised when working around rotating components.

▶ TASK Inspect, replace, and/or adjust drive belts, tensioners, and pulleys; check pulley and belt alignment.

MLR
A1C2

Time off_____________

Time on_____________

Student Instructions: Read through the entire procedure prior to starting. Prepare your work space and any tools or parts that may be needed to complete the task. When directed by your instructor, begin the procedure to complete the task and check the box as each step is finished. Track your time on this procedure for later comparison to the standard completion time (i.e., "flat rate" or customer pay time).

Total time_____________

Procedure:	Step Completed
1. Locate the belt routing diagram.	
a. Draw (or print and attach to this sheet) the belt routing diagram for your specific vehicle.	☐
2. Using the proper procedure, use the serpentine belt removal tool and remove the belt.	☐
a. Inspect the belt for cracks or fraying. Based on your observation, what is your recommendation?	☐
b. Explain what type of tensioner is being used with your specific vehicle.	☐
c. Inspect all engine accessory drive pulleys for binding or looseness. List any further necessary action.	☐
3. Check for pulley alignment.	
a. Mount the pulley alignment laser tool to the crankshaft. A straight-edge may be used in place of a laser tool. Check pulley alignment. List any further necessary action.	☐

<table>
<tr><td colspan="2">4. Install the belt.</td></tr>
<tr><td>a. If an automatic tensioner is being used, install the belt. Check that the belt is installed completely.</td><td>☐</td></tr>
<tr><td>b. If a manual tensioner is being used, install the belt to the recommended tension. What is the specified tension for either a new or used belt?</td><td>☐</td></tr>
<tr><td>c. Use a tension gauge to verify correct installation. List the tension result after installation.</td><td>☐</td></tr>
</table>

Non-Task-Specific Evaluations:	Step Completed
1. Tools and equipment were used as directed and returned in good working order.	☐
2. Complied with all general and task-specific safety standards, including proper use of any personal protective equipment.	☐
3. Completed the task in an appropriate time frame (Recommendation: 1.5 or 2 times flat rate).	☐
4. Left the work space clean and orderly.	☐
5. Cared for customer property and returned it undamaged.	☐

Student signature _______________________________ Date _______________________________

Comments:

Have your supervisor/instructor verify satisfactory completion of this procedure, any observations found, and any necessary action(s) recommended.

Evaluation Instructions: The scoring box below is intended to act as a guide for both student and instructor. Each criterion listed will help students understand what is expected from them and evaluators articulate the success at a particular task. The scoring is set up to allow a second attempt at each task (see the "Test" and "Retest" columns). Scoring is designed to reward students for correct completion of the task. Points are lost for failure to complete the employability requirements (see "Non-Task-Specific" criteria). When all the criteria are evaluated, tally the points for a total at the bottom of each column.

Tasksheet Scoring

	Test		Retest	
Evaluation Items	**Pass**	**Fail**	**Pass**	**Fail**
Task-Specific Evaluation	**(1 pt)**	**(0 pts)**	**(1 pt)**	**(0 pts)**
Locate, print, or draw a belt routing diagram.				
Remove the belt and inspect the tensioner for proper operation.				
Inspect the pulleys for correct alignment.				
Install the belt.				
Non-Task-Specific Evaluation	**(0 pts)**	**(−1 pt)**	**(0 pts)**	**(−1 pt)**
Student successfully completed at least three of the non-task-specific steps.				
Student successfully completed all five of the non-task-specific steps.				
Total Score: <total # of points / 4 = %>				

Supervisor:

Supervisor/instructor signature ________________________________ Date ___________________

Comments:

Retest supervisor/instructor signature ________________________________ Date ___________________

Comments:

CDX Tasksheet Number: A1O12

Student/Intern Information

Name ___________________________ Date __________ Class ___________________

Vehicle, Customer, and Service Information

Vehicle used for this activity:

Year ____________ Make __________________ Model __________________

Odometer __________________ VIN __________________________

Materials Required

- Blank work order
- Vehicle with available service history records
- Depending on the type of concern, special diagnostic tools may be required. See your supervisor/instructor for instructions to identify what tools may be required.
- Personal protective equipment
- General hand tools
- Coolant
- Thermostat and gasket
- Scan tool
- Infrared temperature tool

Task-Specific Safety Considerations

- Comply with personal and environmental safety practices associated with clothing; eye protection; hand tools; power equipment; proper ventilation; and the handling, storage, and disposal of chemicals/materials in accordance with local, state, and federal safety and environmental regulations.
- Lifting equipment, such as vehicle jacks and stands, vehicle hoists, and engine hoists, are important tools that increase productivity and make the job easier. However, they can also cause severe injury or death if used improperly. Make sure you follow the manufacturer's operation procedures. Also, make sure you have your supervisor's/instructor's permission to use any particular type of lifting equipment.
- When running any vehicles in the shop, make sure you use the shop's exhaust ventilation system to discharge all exhaust gas safely outside.
- **Caution:** Serious injury can occur if any part of the cooling system is opened while the vehicle is hot!

▶ **TASK** Remove, inspect, and replace thermostat and gasket/seal. **MLR** *A1C3*

Time off_____________

NOTE: The following steps are in order for diagnosing a vehicle with a coolant temperature-related issue. Allow plenty of time for the vehicle to cool after Step 2 is complete. Refer to your instructor for the preferred method of completing this lab.

Time on_____________

Total time_____________

Procedure:	Step Completed
1. Research the procedure for removal, inspection, and replacement of the thermostat for your specific vehicle.	
a. List (or print and attach to this sheet) the procedure for thermostat removal and inspection.	☐
b. List the temperature rating of your thermostat. Explain what this temperature rating means.	☐
2. Monitor the temperature of the coolant as you bring the vehicle up to operating temperature.	
a. Use a scan tool to monitor the temperature of the coolant as the vehicle warms up. List the temperature reading once the vehicle has reached operating temperature. How does this compare to your specification?	☐
b. Monitor the temperature gauge on the gauge cluster of your vehicle. How does this reading compare to the reading taken from your scan tool?	☐
c. Use the infrared temperature tool to monitor temperature at the upper radiator hose. Record the lowest temperature seen as well as the highest temperature seen. How does this compare to your specification? Explain why there is a temperature fluctuation.	☐
3. Remove the thermostat and replace.	
a. Following proper procedure, remove and replace the thermostat.	☐

b. After installation, select the proper coolant and fill the cooling system.	☐
c. Following proper procedure, bleed air from the cooling system.	☐
d. Verify correct operation of the thermostat as the vehicle warms to operating temperature. Does the vehicle have good heat from the dash vents after thermostat replacement?	☐
Optional: Test the thermostat using boiling water and a thermometer. Using the proper procedure, heat the water and monitor the temperature as the thermostat begins to open. What is the temperature when the thermostat opens? How does this compare to your specification?	☐
e. Clean the thermostat housing sealing surface using the approved method and solvents.	☐
f. Install a new thermostat and gasket/RTV. Tighten housing bolts to specification. List the torque specification for your thermostat housing.	☐

Non-Task-Specific Evaluations:	**Step Completed**
1. Tools and equipment were used as directed and returned in good working order.	☐
2. Complied with all general and task-specific safety standards, including proper use of any personal protective equipment.	☐
3. Completed the task in an appropriate time frame (Recommendation: 1.5 or 2 times flat rate).	☐
4. Left the work space clean and orderly.	☐
5. Cared for customer property and returned it undamaged.	☐

Student signature _________________________________ Date _________________________________

Comments:

Have your supervisor/instructor verify satisfactory completion of this procedure, any observations found,

and any necessary action(s) recommended.

Evaluation Instructions: The scoring box below is intended to act as a guide for both student and instructor. Each criterion listed will help students understand what is expected from them and evaluators articulate the success at a particular task. The scoring is set up to allow a second attempt at each task (see the "Test" and "Retest" columns). Scoring is designed to reward students for correct completion of the task. Points are lost for failure to complete the employability requirements (see "Non-Task-Specific" criteria). When all the criteria are evaluated, tally the points for a total at the bottom of each column.

Tasksheet Scoring

		Test		Retest	
Evaluation Items	**Pass**	**Fail**	**Pass**	**Fail**
Task-Specific Evaluation	**(1 pt)**	**(0 pts)**	**(1 pt)**	**(0 pts)**
Research the thermostat temperature rating and removal procedure.				
Remove the thermostat following the proper procedure.				
Install the thermostat.				
Select the correct coolant and bleed the system of air.				
Non-Task-Specific Evaluation	**(0 pts)**	**(−1 pt)**	**(0 pts)**	**(−1 pt)**
Student successfully completed at least three of the non-task-specific steps.				
Student successfully completed all five of the non-task-specific steps.				
Total Score: <total # of points / 4 = %>				

Supervisor:

Supervisor/instructor signature _______________________________ Date _______________________

Comments:

Retest supervisor/instructor signature _______________________________ Date _______________________

Comments:

CDX Tasksheet Number: A1013

Student/Intern Information

Name _________________________________ Date ____________ Class _________________________

Vehicle, Customer, and Service Information

Vehicle used for this activity:

Year _______________ Make _____________________________ Model _______________________

Odometer _________________________________ VIN _______________________________________

Materials Required

- Blank work order
- Vehicle with available service history records
- Depending on the type of concern, special diagnostic tools may be required. See your supervisor/instructor for instructions to identify what tools may be required.
- Service information database
- Personal protective equipment
- Coolant
- Fluid catch pan
- Hydrometer, refractometer, litmus-style test strips
- General hand tools
- Coolant flush machine
- Air bleed device

Task-Specific Safety Considerations

- Comply with personal and environmental safety practices associated with clothing; eye protection; hand tools; power equipment; proper ventilation; and the handling, storage, and disposal of chemicals/materials in accordance with local, state, and federal safety and environmental regulations.
- Shop rules and procedures are critical to your safety. Please give these your utmost attention.
- **Caution:** Serious injury can occur if any part of the cooling system is opened while the vehicle is hot!

▶ TASK Inspect and test coolant; drain and recover coolant; flush and refill cooling system; use proper fluid type per manufacturer specification; bleed air as required.

MLR
A1C4

Time off____________

Time on____________

Total time____________

Student Instructions: Read through the entire procedure prior to starting. Prepare your work space and any tools or parts that may be needed to complete the task. When directed by your instructor, begin the procedure to complete the task and check the box as each step is finished. Track your time on this procedure for later comparison to the standard completion time (i.e., "flat rate" or customer pay time).

Procedure:	Step Completed
1. Inspect coolant condition:	
a. Using the hydrometer and refractometer, measure the level of freeze protection of the coolant in your specific vehicle. What is the degree of freeze protection? Compare your results and list any necessary action.	☐
b. Using the litmus-style test strips, measure the pH level of the coolant. List your results and any further action needed.	☐
2. Drain and refill the coolant.	
a. Lift the vehicle if needed. Using a catch pan, open the radiator petcock and drain the coolant.	☐
b. Remove the engine block drain and drain the coolant (if applicable).	☐
c. Install the radiator petcock and engine block drain and lower the vehicle.	☐
d. Fill the cooling system with the proper type and amount of coolant. Describe the coolant type and mixture used.	☐
e. If only a drain and refill procedure is being performed, attach the air bleeding tool and remove air from the cooling system following the proper procedure.	☐
3. Flush the cooling system.	
a. Fill the coolant flush machine with the proper type and amount of coolant for your vehicle. Describe the coolant type and mixture used.	☐
b. Using the proper procedure, install the flush machine components and flush the cooling system.	☐
c. Fill the cooling system to the correct level.	☐
d. Using the machine or air bleeding tool, remove all air from the cooling system.	☐
e. Remove the flush machine and seal cooling system. Start the vehicle and allow it to run until operating temperature is achieved. Be sure that the cooling fans begin to run as the vehicle warms up. Watch the temperature gauge on the dash and shut down the vehicle if it begins to overheat.	☐

<table>
<tr><td>f. As the vehicle runs, select the highest heat setting and blower fan speed. Verify that heat is available from vents. If air is trapped in the system, cool air may be felt at the vents. Are there any issues found?</td><td>☐</td></tr>
</table>

Non-Task-Specific Evaluations:	Step Completed
1. Tools and equipment were used as directed and returned in good working order.	☐
2. Complied with all general and task-specific safety standards, including proper use of any personal protective equipment.	☐
3. Completed the task in an appropriate time frame (Recommendation: 1.5 or 2 times flat rate).	☐
4. Left the work space clean and orderly.	☐
5. Cared for customer property and returned it undamaged.	☐

Student signature _______________________ Date _______________________

Comments:

Have your supervisor/instructor verify satisfactory completion of this procedure, any observations found, and any necessary action(s) recommended.

Evaluation Instructions: The scoring box below is intended to act as a guide for both student and instructor. Each criterion listed will help students understand what is expected from them and evaluators articulate the success at a particular task. The scoring is set up to allow a second attempt at each task (see the "Test" and "Retest" columns). Scoring is designed to reward students for correct completion of the task. Points are lost for failure to complete the employability requirements (see "Non-Task-Specific" criteria). When all the criteria are evaluated, tally the points for a total at the bottom of each column.

Tasksheet Scoring

	Test		Retest	
Evaluation Items	**Pass**	**Fail**	**Pass**	**Fail**
Task-Specific Evaluation	**(1 pt)**	**(0 pts)**	**(1 pt)**	**(0 pts)**
Inspect coolant freeze protection and pH level.				
Perform the coolant drain and refill procedure.				
Perform a coolant system flush.				
Select the correct type of coolant, fill the system with coolant, bleed air from the cooling system, and verify correct operation after service.				
Non-Task-Specific Evaluation	**(0 pts)**	**(−1 pt)**	**(0 pts)**	**(−1 pt)**
Student successfully completed at least three of the non-task-specific steps.				
Student successfully completed all five of the non-task-specific steps.				
Total Score: <total # of points / 4 = %>				

Supervisor:

Supervisor/instructor signature ________________________________ Date ________________________

Comments:

Retest supervisor/instructor signature ________________________________ Date ________________________

Comments:

CDX Tasksheet Number: A1014

Student/Intern Information

Name _________________________________ Date ____________ Class _____________________

Vehicle, Customer, and Service Information

Vehicle used for this activity:

Year _______________ Make _________________________ Model _____________________

Odometer _________________________ VIN _________________________________

Materials Required

- Blank work order
- Vehicle with available service history records
- Depending on the type of concern, special diagnostic tools may be required. See your supervisor/instructor for instructions to identify what tools may be required.
- Personal protective equipment
- General hand tools
- Oil filter wrench
- Torque wrench
- Fluid catch pan
- Engine oil and filter
- Service information database

Task-Specific Safety Considerations

- When running any vehicles in the shop, make sure you use the shop's exhaust ventilation system to discharge all exhaust gas safely outside.
- Lifting equipment such as vehicle jacks and stands, vehicle hoists, and engine hoists are important tools that increase productivity and make the job easier. However, they can also cause severe injury or death if used improperly. Make sure you follow the manufacturer's operation procedures. Also, make sure you have your supervisor's/instructor's permission to use any particular type of lifting equipment.
- Comply with personal and environmental safety practices associated with clothing; eye protection; hand tools; power equipment; proper ventilation; and the handling, storage, and disposal of chemicals/materials in accordance with local, state, and federal safety and environmental regulations.
- If you need to start the vehicle, you should ensure that the parking brake is firmly applied; if necessary, use wheel chocks to prevent the vehicle from moving when the vehicle is started to verify the completion of these tasks.
- Extreme caution must be exercised when working around rotating components.

▶ TASK Perform engine oil and filter change; use proper fluid type per manufacturer specification; reset maintenance reminder as required.

MLR
A1C5

Student Instructions: Read through the entire procedure prior to starting. Prepare your workspace and any tools or parts that may be needed to complete the task. When directed by your instructor, begin the procedure to complete the task and check the box as each step is finished. Track your time on this procedure for later comparison to the standard completion time (i.e., "flat rate" or customer pay time).

Procedure:	Step Completed
1. Research the oil change procedure, oil type, and capacity.	
a. List the type of oil used on your specific vehicle. Include API rating and viscosity.	☐
b. List the proper amount of oil to be used in your specific vehicle.	☐
c. List the torque specifications for the oil filter (or oil filter housing) and oil filter drain plug.	☐
2. Prepare the vehicle for an oil change.	
a. Bring the vehicle up to or near operating temperature and then shut the vehicle off. Explain why this is beneficial. **Tip:** Check the oil level before starting the vehicle to ensure safe operation.	☐
b. Open the hood and place fender covers over the fenders of your vehicle.	☐
c. Inspect the oil condition prior to changing the oil. Describe the oil level and note the condition of the oil before it is changed.	☐
d. Using the proper procedure, lift the vehicle to allow for the oil change to be performed.	☐
3. Change the engine oil.	
a. Locate the oil drain plug(s). Find the correct size wrench to remove the plug. What size wrench was used for removal?	☐

b. Locate an oil drain pan. Loosen the drain plug and allow oil to drain into the fluid pan. Be sure to remove the oil drain plug gasket, if applicable.	☐
c. Allow the oil to drain completely. Once drained, clean the oil drain plug and install a new oil drain plug gasket. Tighten the drain plug to specification. What is the torque specification for your drain plug?	☐
4. Change the engine oil filter.	
a. Locate the engine oil filter. What type of filter is used on your specific vehicle?	☐
b. Using a fluid catch pan, remove the oil filter and allow all oil to drain. How long should an oil filter drain for before being crushed and disposed of?	☐
c. Ensure that the oil filter gasket has been removed with the filter. What could happen if the gasket remained on the engine block and a new filter were installed over top of the old gasket?	☐
d. Apply a light coating of fresh oil to the gasket of the new filter. Clean the engine block of all old oil before installing the new filter. Why is a coating of oil necessary?	☐
e. Install the new filter and torque the filter or filter housing to specification. What is the torque specification for your filter?	☐
5. Fill the engine with fresh oil.	
a. Lower the vehicle. Locate the oil fill cap. Using a funnel, fill the engine with the correct type and amount of oil.	☐
b. Remove any oil change materials, tools, or equipment from the under-hood area. Start the vehicle. Observe that the oil pressure warning lamp on the gauge cluster goes out once running for 2 or 3 seconds. Inspect under the vehicle for any signs of oil leaking. Once satisfied, shut off the vehicle.	☐

c. It is good practice to double check your work until you become comfortable with any task that you are doing. Even then, it is still good practice to double check all work. Lift the vehicle once more and inspect the drain plug and filter area to ensure that there are no leaks present. Clean all areas thoroughly before giving the vehicle back to the customer. Lower the vehicle.	☐
d. Inspect the oil level on the dipstick. If the level is correct, close the hood. Wash your hands/arms before entering the vehicle.	☐
e. Prepare an oil change reminder sticker for the windshield of the vehicle. Refer to the owner's manual or other source for information regarding oil change intervals. What is the recommended oil change interval for this vehicle?	☐
f. Reset the maintenance indicator, if applicable. Research the procedure for resetting this indicator. List the procedure for resetting the maintenance indicator.	☐

Non-Task-Specific Evaluations:	**Step Completed**
1. Tools and equipment were used as directed and returned in good working order.	☐
2. Complied with all general and task-specific safety standards, including proper use of any personal protective equipment.	☐
3. Completed the task in an appropriate time frame (Recommendation: 1.5 or 2 times flat rate).	☐
4. Left the work space clean and orderly.	☐
5. Cared for customer property and returned it undamaged.	☐

Student signature _______________________________ Date _______________________________

Comments:

Have your supervisor/instructor verify satisfactory completion of this procedure, any observations found, and any necessary action(s) recommended.

Evaluation Instructions: The scoring box below is intended to act as a guide for both student and instructor. Each criterion listed will help students understand what is expected from them and evaluators articulate the success at a particular task. The scoring is set up to allow a second attempt at each task (see the "Test" and "Retest" columns). Scoring is designed to reward students for correct completion of the task. Points are lost for failure to complete the employability requirements (see "Non-Task-Specific" criteria). When all criteria are evaluated, tally the points for a total at the bottom of each column.

Tasksheet Scoring

	Test		Retest	
Evaluation Items	**Pass**	**Fail**	**Pass**	**Fail**
Task-Specific Evaluation	**(1 pt)**	**(0 pts)**	**(1 pt)**	**(0 pts)**
Research the oil change procedure, oil type, and capacity.				
Change the engine oil using the proper procedure.				
Change the engine oil filter using the proper procedure.				
Select the correct oil type and fill to the correct level.				
Reset the maintenance indicator.				
Non-Task-Specific Evaluation	**(0 pts)**	**(−1 pt)**	**(0 pts)**	**(−1 pt)**
Student successfully completed at least three of the non-task-specific steps.				
Student successfully completed all five of the non-task-specific steps.				
Total Score: <total # of points / 4 = %>				

Supervisor:

Supervisor/instructor signature _________________________ Date _________________

Comments:

Retest supervisor/instructor signature _________________________ Date _________________

Comments:

CDX Tasksheet Number: A1015

Student/Intern Information

Name ___________________________________ Date ____________ Class ___________________________

Vehicle, Customer, and Service Information

Vehicle used for this activity:

Year ________________ Make ___________________________ Model ___________________________

Odometer ___________________________ VIN ___________________________________

Materials Required

- Blank work order
- Vehicle with available service history records
- Depending on the type of concern, special diagnostic tools may be required. See your supervisor/instructor for instructions to identify what tools may be required.
- Personal protective equipment

Task-Specific Safety Considerations

- Comply with personal and environmental safety practices associated with clothing; eye protection; hand tools; power equipment; proper ventilation; and the handling, storage, and disposal of chemicals/materials in accordance with local, state, and federal safety and environmental regulations.

▶ **TASK** Identify components of the lubrication and cooling systems.

MLR
A1C6

Time off__________________

Time on__________________

Total time__________________

NOTE: Use this sheet, along with a cutaway or picture, as a guide for locating various components of the lubrication and cooling systems.

Student Instructions: Read through the entire procedure prior to starting. Prepare your workspace and any tools or parts that may be needed to complete the task. When directed by your instructor, begin the procedure to complete the task and check the box as each step is finished. Track your time on this procedure for later comparison to the standard completion time (i.e., "flat rate" or customer pay time).

Procedure:	Step Completed
1. Identify and locate the following components of the lubrication system. Describe the path that oil travels through the lubrication system.	☐
a. Oil pump pickup and screen	☐

b. Oil pump	☐
c. Oil pan	☐
d. Oil filter	☐
e. Main oil galleries	☐
f. Crankshaft oil galleries	☐
g. Camshaft oil galleries	☐
h. Lifter and pushrod oil galleries	☐
2. Describe how main bearing clearance affects oil pressure in an engine.	☐
3. Explain how maximum oil pressure is controlled.	☐
4. Identify and locate the following components of a cooling system. Describe the path that coolant flows throughout the system.	☐
a. Radiator	☐
b. Upper and lower radiator hose	☐
c. Water pump	☐
d. Engine block water jackets	☐
e. Cylinder head water jackets	☐
f. Thermostat	☐
g. Radiator cap	☐
h. Heater core and hoses	☐
5. Explain how the temperature is controlled by the cooling system. What controls minimum and maximum coolant temperature?	☐

Non-Task-Specific Evaluations:	Step Completed
1. Tools and equipment were used as directed and returned in good working order.	☐
2. Complied with all general and task-specific safety standards, including proper use of any personal protective equipment.	☐
3. Completed the task in an appropriate time frame (Recommendation: 1.5 or 2 times flat rate).	☐
4. Left the work space clean and orderly.	☐
5. Cared for customer property and returned it undamaged.	☐

Student signature _______________________________ Date _______________________________

Comments:

Have your supervisor/instructor verify satisfactory completion of this procedure, any observations found, and any necessary action(s) recommended.

Evaluation Instructions: The scoring box below is intended to act as a guide for both student and instructor. Each criterion listed will help students understand what is expected from them and evaluators articulate the success at a particular task. The scoring is set up to allow a second attempt at each task (see the "Test" and "Retest" columns). Scoring is designed to reward students for correct completion of the task. Points are lost for failure to complete the employability requirements (see "Non-Task-Specific" criteria). When all the criteria are evaluated, tally the points for a total at the bottom of each column.

Tasksheet Scoring

	Test		Retest	
Evaluation Items	**Pass**	**Fail**	**Pass**	**Fail**
Task-Specific Evaluation	**(1 pt)**	**(0 pts)**	**(1 pt)**	**(0 pts)**
Identify and label lubrication system components.				
Describe main bearings, effects on oil pressure.				
Describe what controls maximum oil pressure.				
Identify and label cooling system components and explain how temperature is controlled in a cooling system.				
Non-Task-Specific Evaluation	**(0 pts)**	**(−1 pt)**	**(0 pts)**	**(−1 pt)**
Student successfully completed at least three of the non-task-specific steps.				
Student successfully completed all five of the non-task-specific steps.				
Total Score: <total # of points / 4 = %>				

Supervisor:

Supervisor/instructor signature _________________________________ Date _________________________

Comments:

Retest supervisor/instructor signature _________________________ Date _________________________

Comments:

2A. Automatic Transmission and Transaxle: General

Learning Objective/Task	CDX Tasksheet Number	ASE Foundation Reference Number; Priority Level
• Research vehicle service information including fluid type, vehicle service history, service precautions, and technical service bulletins.	A2001	A2A1, P-1
• Check fluid level in a transmission or a transaxle equipped with a dip-stick.	A2002	A2A2, P-1
• Check fluid level in a transmission or a transaxle not equipped with a dip-stick.	A2003	A2A3, P-1
• Check transmission fluid condition; check for leaks.	A2004	A2A4, P-2
• Identify drivetrain components and configuration.	A2005	A2A5, P-1

Materials Required

- Blank work order
- Vehicle with available service history records
- Depending on the type of concern, special diagnostic/hand tools may be required. See your supervisor/instructor for instructions to identify what tools may be required.
- Service information database
- Personal protective equipment

Safety Considerations

- When running any vehicles in the shop, make sure you use the shop's exhaust ventilation system to discharge all exhaust gas safely outside.
- Lifting equipment, such as vehicle jacks and stands, vehicle hoists, and engine hoists, are important tools that increase productivity and make the job easier. However, they can also cause severe injury or death if used improperly. Make sure you follow the manufacturer's operation procedures. Also, make sure you have your supervisor's/instructor's permission to use any particular type of lifting equipment.
- Comply with personal and environmental safety practices associated with clothing; eye protection; hand tools; power equipment; proper ventilation; and the handling, storage, and disposal of chemicals/materials in accordance with local, state, and federal safety and environmental regulations.
- If you need to start the vehicle, you should ensure that the parking brake is firmly applied; if necessary, use wheel chocks to prevent the vehicle from moving when the vehicle is started to verify the completion of these tasks.
- Extreme caution must be exercised when working around rotating components.

CDX Tasksheet Number: A2001

Student/Intern Information

Name ________________________________ Date ____________ Class ________________________

Vehicle, Customer, and Service Information

Vehicle used for this activity:

Year ______________ Make ________________________ Model ______________________

Odometer ________________________ VIN ________________________________

Materials Required

- Blank work order
- Vehicle with available service history records
- Depending on the type of concern, special diagnostic/hand tools may be required. See your supervisor/instructor for instructions to identify what tools may be required.
- Service information database
- Personal protective equipment

Task-Specific Safety Considerations

- Shop rules and procedures are critical to your safety. Please give these your utmost attention.
- Comply with personal and environmental safety practices associated with clothing; eye protection; hand tools; power equipment; proper ventilation; and the handling, storage, and disposal of chemicals/materials in accordance with local, state, and federal safety and environmental regulations.

Time off____________

Time on____________

Total time____________

▶ **TASK** Research vehicle service information, including fluid type, vehicle service history, service precautions, and technical service bulletins.

MLR
A2A1

Student Instructions: Read through the entire procedure prior to starting. Prepare your workspace and any tools or parts that may be needed to complete the task. When directed by your instructor, begin the procedure to complete the task and check the box as each step is finished. Track your time on this procedure for later comparison to the standard completion time (i.e., "flat rate" or customer pay time).

Procedure:	Step Completed
1. Research vehicle service information.	
a. List the transmission fluid type used in your specific vehicle.	☐

b. List the full transmission fluid capacity.	☐
2. Research vehicle service history.	
a. Determine (if possible) whether all routine maintenance has been completed.	☐
b. Determine outstanding/completed recall work and list any service precautions below.	☐
c. Determine (if possible) any major repairs in the vehicle service history and list below.	☐
3. Determine TSBs issued for specific vehicle.	
Identify one transmission-related TSB and list the TSB number, vehicle concern, and repair procedure below.	☐

Non-Task-Specific Evaluations:	Step Completed
1. Tools and equipment were used as directed and returned in good working order.	☐
2. Complied with all general and task-specific safety standards, including proper use of any personal protective equipment.	☐
3. Completed the task in an appropriate time frame (Recommendation: 1.5 or 2 times flat rate).	☐
4. Left the work space clean and orderly.	☐
5. Cared for customer property and returned it undamaged.	☐

Student signature _______________________________ Date _______________________________

Comments:

Have your supervisor/instructor verify satisfactory completion of this procedure, any observations found, and any necessary action(s) recommended.

Evaluation Instructions: The scoring box below is intended to act as a guide for both student and instructor. Each criterion listed will help students understand what is expected from them and evaluators articulate success at a particular task. The scoring is set up to allow a second attempt at each task (see the "Test" and "Retest" columns). Scoring is designed to reward students for correct completion of the task. Points are lost for failure to complete the employability requirements (see "Non-Task-Specific" criteria). When all the criteria are evaluated, tally the points for a total at the bottom of each column.

Tasksheet Scoring

	Test		Retest	
Evaluation Items	**Pass**	**Fail**	**Pass**	**Fail**
Task-Specific Evaluation	**(1 pt)**	**(O pts)**	**(1 pt)**	**(O pts)**
Research transmission fluid type and capacity.				
Research vehicle service history and any completed/outstanding recalls.				
Identify service precautions.				
Identify transmission-related TSBs.				
Non-Task-Specific Evaluation	**(O pts)**	**(–1 pt)**	**(O pts)**	**(–1 pt)**
Student successfully completed at least three of the non-task-specific steps.				
Student successfully completed all five of the non-task-specific steps.				
Total Score: <total # of points / 4 = %>				

Supervisor:

Supervisor/instructor signature _______________________________ Date _______________

Comments:

Retest supervisor/instructor signature _______________________________ Date _______________

Comments:

CDX Tasksheet Number: A2002

Student/Intern Information

Name _________________________________ Date ___________ Class _____________________

Vehicle, Customer, and Service Information

Vehicle used for this activity:

Year _______________ Make _____________________________ Model _____________________

Odometer _________________________________ VIN _________________________________

Materials Required

- Blank work order
- Vehicle with available service history records
- Depending on the type of concern, special diagnostic/hand tools may be required. See your supervisor/instructor for instructions to identify what tools may be required.
- Service information database
- Personal protective equipment
- Shop rags

Task-Specific Safety Considerations

- Comply with personal and environmental safety practices associated with clothing; eye protection; hand tools; power equipment; proper ventilation; and the handling, storage, and disposal of chemicals/materials in accordance with local, state, and federal safety and environmental regulations.
- If you need to start the vehicle, you should ensure that the parking brake is firmly applied; if necessary, use wheel chocks to prevent the vehicle from moving when the vehicle is started to verify the completion of these tasks.
- When running any vehicles in the shop, make sure you use the shop's exhaust ventilation system to discharge all exhaust gas safely outside.

Time off_____________

Time on_____________

Total time_____________

▶ **TASK** Check fluid level in a transmission or a transaxle equipped with a dip-stick.

**MLR
A2A2**

Student Instructions: Read through the entire procedure prior to starting. Prepare your workspace and any tools or parts that may be needed to complete the task. When directed by your instructor, begin the procedure to complete the task and check the box as each step is finished. Track your time on this procedure for later comparison to the standard completion time (i.e., "flat rate" or customer pay time).

Procedure:	Step Completed
1. Research the procedure for checking automatic transmission fluid level in your specific vehicle.	
a. List or print and attach to this sheet the proper procedure for checking the fluid level in your assigned vehicle.	☐
b. Is the vehicle running for this task? What result would you expect if the vehicle is not running?	☐
c. What gear selection is used when checking the transmission fluid level in your vehicle?	☐
d. What temperature must the vehicle be to get an accurate reading? How does temperature affect fluid level?	☐
2. Check automatic transmission fluid level.	
a. With the vehicle off, locate the transmission dipstick. Remove the dipstick, take a sample of the fluid, and note the fluid level. Describe the fluid condition and level. List any further recommendations.	☐
b. Following the proper procedure, start the vehicle and place the shifter in the designated position. Measure the fluid level in the vehicle. Describe the fluid level and the condition of the fluid.	☐
c. Explain what signs like milky/discolored fluid might indicate. What would dark black, burnt-smelling fluid indicate?	☐

<table>
<tr><td>d. If the fluid level is low, check the vehicle for signs of leaking. If no external leaks are found, explain other places where fluid may be leaking.</td><td>☐</td></tr>
</table>

Non-Task-Specific Evaluations:	Step Completed
1. Tools and equipment were used as directed and returned in good working order.	☐
2. Complied with all general and task-specific safety standards, including proper use of any personal protective equipment.	☐
3. Completed the task in an appropriate time frame (Recommendation: 1.5 or 2 times flat rate).	☐
4. Left the work space clean and orderly.	☐
5. Cared for customer property and returned it undamaged.	☐

Student signature _______________________ Date _______________________

Comments:

Have your supervisor/instructor verify satisfactory completion of this procedure, any observations found, and any necessary action(s) recommended.

Evaluation Instructions: The scoring box below is intended to act as a guide for both student and instructor. Each criterion listed will help students understand what is expected from them and evaluators articulate success at a particular task. The scoring is set up to allow a second attempt at each task (see the "Test" and "Retest" columns). Scoring is designed to reward students for correct completion of the task. Points are lost for failure to complete the employability requirements (see "Non-Task-Specific" criteria). When all the criteria are evaluated, tally the points for a total at the bottom of each column.

Tasksheet Scoring

	Test		Retest	
Evaluation Items	**Pass**	**Fail**	**Pass**	**Fail**
Task-Specific Evaluation	**(1 pt)**	**(0 pts)**	**(1 pt)**	**(0 pts)**
Research the procedure for checking the transmission fluid level.				
Identify gear selection and transmission fluid temperature when checking fluid level.				
Describe the fluid level and condition.				
Explain signs of bad transmission fluid.				
Non-Task-Specific Evaluation	**(0 pts)**	**(–1 pt)**	**(0 pts)**	**(–1 pt)**
Student successfully completed at least three of the non-task-specific steps.				
Student successfully completed all five of the non-task-specific steps.				
Total Score: <total # of points / 4 = %>				

Supervisor:

Supervisor/instructor signature ________________________________ Date ________________________

Comments:

Retest supervisor/instructor signature ________________________ Date ________________

Comments:

CDX Tasksheet Number: A2003

Student/Intern Information

Name _________________________ Date __________ Class _________________________

Vehicle, Customer, and Service Information

Vehicle used for this activity:

Year ______________ Make _________________________ Model _________________________

Odometer _________________________ VIN _________________________

Materials Required

- Blank work order
- Vehicle with available service history records
- Depending on the type of concern, special diagnostic/hand tools may be required. See your supervisor/instructor for instructions to identify what tools may be required.
- Service information database
- Personal protective equipment
- Shop rags
- Scan tool
- General hand tools
- Vehicle-specific dipstick (if applicable). Some vehicles require a special dipstick to be used that is manufacturer specific. Refer to your instructor for further details.

Task-Specific Safety Considerations

- Comply with personal and environmental safety practices associated with clothing; eye protection; hand tools; power equipment; proper ventilation; and the handling, storage, and disposal of chemicals/materials in accordance with local, state, and federal safety and environmental regulations.
- If you need to start the vehicle, you should ensure that the parking brake is firmly applied; if necessary, use wheel chocks to prevent the vehicle from moving when the vehicle is started to verify the completion of these tasks.
- When running any vehicles in the shop, make sure you use the shop's exhaust ventilation system to discharge all exhaust gas safely outside.
- Lifting equipment, such as vehicle jacks and stands, vehicle hoists, and engine hoists, are important tools that increase productivity and make the job easier. However, they can also cause severe injury or death if used improperly. Make sure you follow the manufacturer's operation procedures. Also, make sure you have your supervisor's/instructor's permission to use any particular type of lifting equipment.

▶ **TASK** Check fluid level in a transmission or a transaxle **not** equipped with a dip-stick.

MLR
A2A3

Student Instructions: Read through the entire procedure prior to starting. Prepare your work space and any tools or parts that may be needed to complete the task. When directed by your instructor, begin the procedure to complete the task and check the box as each step is finished. Track your time on this procedure for later comparison to the standard completion time (i.e., "flat rate" or customer pay time).

Procedure:	Step Completed
1. Research the procedure for checking automatic transmission fluid level in your specific vehicle.	
a. List or print and attach to this sheet the proper procedure for checking fluid level in your assigned vehicle.	☐
b. Is the vehicle running for this task? What result would you expect if the vehicle is not running?	☐
c. What gear selection is used when checking transmission fluid level in your vehicle?	☐
d. What temperature must the vehicle be to get an accurate reading? How does temperature affect fluid level?	☐
e. List any special tools that are required for this task.	☐
2. Check automatic transmission fluid level.	
a. Following the proper procedure, prepare the vehicle for transmission fluid level and condition inspection. *Tip:* You may be instructed to use a scan tool to monitor temperature, use specialty tools such as a vehicle specific dipstick, or follow a procedure to bring the vehicle to a certain temperature range. Follow instructions carefully, as results may vary.	☐

b. What temperature range should the transmission fluid be prior to checking the level?	☐
c. Explain your results. Were the level and condition of the fluid okay?	☐
d. Be sure to replace the drain plug/overflow plug gaskets and to clean any spilled fluid (if applicable).	☐

Non-Task-Specific Evaluations:	Step Completed
1. Tools and equipment were used as directed and returned in good working order.	☐
2. Complied with all general and task-specific safety standards, including proper use of any personal protective equipment.	☐
3. Completed the task in an appropriate time frame (Recommendation: 1.5 or 2 times flat rate).	☐
4. Left the work space clean and orderly.	☐
5. Cared for customer property and returned it undamaged.	☐

Student signature _________________________ Date _______________________

Comments:

Have your supervisor/instructor verify satisfactory completion of this procedure, any observations found, and any necessary action(s) recommended.

Evaluation Instructions: The scoring box below is intended to act as a guide for both student and instructor. Each criterion listed will help students understand what is expected from them and evaluators articulate success at a particular task. The scoring is set up to allow a second attempt at each task (see the "Test" and "Retest" columns). Scoring is designed to reward students for correct completion of the task. Points are lost for failure to complete the employability requirements (see "Non-Task-Specific" criteria). When all the criteria are evaluated, tally the points for a total at the bottom of each column.

Tasksheet Scoring

	Test		Retest	
Evaluation Items	**Pass**	**Fail**	**Pass**	**Fail**
Task-Specific Evaluation	**(1 pt)**	**(0 pts)**	**(1 pt)**	**(0 pts)**
Research procedure for checking transmission fluid level.				
Identify gear selection and transmission fluid temperature when checking fluid level.				
Describe fluid level.				
Describe fluid condition.				
Non-Task-Specific Evaluation	**(0 pts)**	**(−1 pt)**	**(0 pts)**	**(−1 pt)**
Student successfully completed at least three of the non-task-specific steps.				
Student successfully completed all five of the non-task-specific steps.				
Total Score: <total # of points / 4 = %>				

Supervisor:

Supervisor/instructor signature _______________________________ Date _______________________

Comments:

Retest supervisor/instructor signature _______________________________ Date _______________________

Comments:

CDX Tasksheet Number: A2004

Student/Intern Information

Name ________________________________ Date ____________ Class ________________________

Vehicle, Customer, and Service Information

Vehicle used for this activity:

Year ______________ Make ______________________ Model ____________________________

Odometer ________________________ VIN ____________________________________

Task-Specific Safety Considerations

- Shop rules and procedures are critical to your safety. Please give these your utmost attention.
- Comply with personal and environmental safety practices associated with clothing; eye protection; hand tools; power equipment; proper ventilation; and the handling, storage, and disposal of chemicals/materials in accordance with local, state, and federal safety and environmental regulations.
- Lifting equipment, such as vehicle jacks and stands, vehicle hoists, and engine hoists, are important tools that increase productivity and make the job easier. However, they can also cause severe injury or death if used improperly. Make sure you follow the manufacturer's operation procedures. Also make sure you have your supervisor's/instructor's permission to use any particular type of lifting equipment.

Time off________________

Time on________________

Total time________________

▶ TASK Check transmission fluid condition; check for leaks.

MLR
A2A4

Student Instructions: Read through the entire procedure prior to starting. Prepare your work space and any tools or parts that may be needed to complete the task. When directed by your instructor, begin the procedure to complete the task and check the box as each step is finished. Track your time on this procedure for later comparison to the standard completion time (i.e., "flat rate" or customer pay time).

Procedure:	Step Completed
1. Research the procedure for checking automatic transmission fluid level and condition in your specific vehicle.	
a. List or print and attach to this sheet the proper procedure for checking fluid level in your assigned vehicle.	☐
2. Check automatic transmission fluid level and condition.	
a. Describe the fluid level in your vehicle.	☐
b. Describe the condition of the transmission fluid in your vehicle. What color is the fluid? Can you feel grit if you rub the fluid between your fingers? Does the fluid have a particularly burnt smell to it? Can you see any metal flake in the fluid? Explain what these signs may suggest.	☐
3. Check for transmission fluid leaks. Inspect the following components and list any necessary action required.	
a. Transmission pan	☐
b. Bell housing area	☐
c. Transmission cooler/cooler lines	☐
d. Tailshaft/extension housing area	☐

e. Axle shaft/seal (if applicable)	☐
f. Inspect coolant overflow bottle	☐
g. Transmission vent tube	☐

Non-Task-Specific Evaluations:	Step Completed
1. Tools and equipment were used as directed and returned in good working order.	☐
2. Complied with all general and task-specific safety standards, including proper use of any personal protective equipment.	☐
3. Completed the task in an appropriate time frame (Recommendation: 1.5 or 2 times flat rate).	☐
4. Left the work space clean and orderly.	☐
5. Cared for customer property and returned it undamaged.	☐

Student signature _______________________________ Date _______________________________

Comments:

Have your supervisor/instructor verify satisfactory completion of this procedure, any observations found, and any necessary action(s) recommended.

Evaluation Instructions: The scoring box below is intended to act as a guide for both student and instructor. Each criterion listed will help students understand what is expected from them and evaluators articulate success at a particular task. The scoring is set up to allow a second attempt at each task (see the "Test" and "Retest" columns). Scoring is designed to reward students for correct completion of the task. Points are lost for failure to complete the employability requirements (see "Non-Task-Specific" criteria). When all the criteria are evaluated, tally the points for a total at the bottom of each column.

Tasksheet Scoring

		Test		Retest	
Evaluation Items	Pass	Fail	Pass	Fail
Task-Specific Evaluation	**(1 pt)**	**(0 pts)**	**(1 pt)**	**(0 pts)**
Research the procedure for checking transmission fluid level.				
Check transmission fluid level following the correct procedure.				
Describe the transmission fluid condition.				
Perform a fluid leak check.				
Non-Task-Specific Evaluation	**(0 pts)**	**(−1 pt)**	**(0 pts)**	**(−1 pt)**
Student successfully completed at least three of the non-task-specific steps.				
Student successfully completed all five of the non-task-specific steps.				
Total Score: <total # of points / 4 = %>				

Supervisor:

Supervisor/instructor signature _________________________________ Date _____________________

Comments:

Retest supervisor/instructor signature _________________________ Date _____________________

Comments:

CDX Tasksheet Number: A2005

Student/Intern Information

Name _________________________________ Date ___________ Class _______________________

Vehicle, Customer, and Service Information

Vehicle used for this activity:

Year _____________ Make _____________________ Model _____________________

Odometer _____________________ VIN _____________________

Materials Required

- Blank work order
- Vehicle with available service history records
- Depending on the type of concern, special diagnostic/hand tools may be required. See your supervisor/instructor for instructions to identify what tools may be required.
- Service information database
- Personal protective equipment

Task-Specific Safety Considerations

- Lifting equipment, such as vehicle jacks and stands, vehicle hoists, and engine hoists, are important tools that increase productivity and make the job easier. However, they can also cause severe injury or death if used improperly. Make sure you follow the manufacturer's operation procedures. Also, make sure you have your supervisor's/instructor's permission to use any particular type of lifting equipment.
- Comply with personal and environmental safety practices associated with clothing; eye protection; hand tools; power equipment; proper ventilation; and the handling, storage, and disposal of chemicals/materials in accordance with local, state, and federal safety and environmental regulations.

▶ **TASK** Identify drivetrain components and configuration.

MLR
A2A5

Student Instructions: Read through the entire procedure prior to starting. Prepare your workspace and any tools or parts that may be needed to complete the task. When directed by your instructor, begin the procedure to complete the task and check the box as each step is finished. Track your time on this procedure for later comparison to the standard completion time (i.e., "flat rate" or customer pay time).

Procedure:	Step Completed
1. Identify your vehicle's drivetrain configuration. Describe the layout of the drivetrain components on your specific vehicle. Answer the following:	
a. Is the vehicle front-wheel drive?	☐
b. Is the vehicle rear-wheel drive?	☐
c. Is the vehicle four-wheel drive?	☐
d. Is the vehicle all-wheel drive?	☐
e. If the vehicle is four/all-wheel drive, is a manual selection of four-wheel drive required, or is it done automatically?	☐
f. Is the engine and transmission longitudinally mounted? Is it mounted transversely?	☐
g. Describe the difference between an all-wheel drive vehicle and a four-wheel drive vehicle.	☐
2. Locate the following drivetrain components. Some options may not be available on your vehicle.	
a. Transmission/transaxle	☐
b. CV halfshaft and/or axle	☐
c. Transfer case. Explain the purpose of a transfer case.	☐
d. Front driveshaft	☐

e. Rear driveshaft	☐
f. Universal joint	☐
g. Transmission mount	☐
h. Differential	☐
i. Transmission dipstick/drain plug/fill plug	☐

Non-Task-Specific Evaluations:	Step Completed
1. Tools and equipment were used as directed and returned in good working order.	☐
2. Complied with all general and task-specific safety standards, including proper use of any personal protective equipment.	☐
3. Completed the task in an appropriate time frame (Recommendation: 1.5 or 2 times flat rate).	☐
4. Left the work space clean and orderly.	☐
5. Cared for customer property and returned it undamaged.	☐

Student signature _______________________________ Date _______________________________

Comments:

Have your supervisor/instructor verify satisfactory completion of this procedure, any observations found, and any necessary action(s) recommended.

Evaluation Instructions: The scoring box below is intended to act as a guide for both student and instructor. Each criterion listed will help students understand what is expected from them and evaluators articulate success at a particular task. The scoring is setup to allow a second attempt at each task (see the "Test" and "Retest" columns). Scoring is designed to reward students for correct completion of the task. Points are lost for failure to complete the employability requirements (see "Non-Task-Specific" criteria). When all the criteria are evaluated, tally the points for a total at the bottom of each column.

Tasksheet Scoring

	Test		Retest	
Evaluation Items	**Pass**	**Fail**	**Pass**	**Fail**
Task-Specific Evaluation	**(1 pt)**	**(O pts)**	**(1 pt)**	**(O pts)**
Identify the drivetrain configuration and locate various components.				
Identify the difference between a transversely and longitudinally mounted engine and transmission.				
Describe the difference between an all-wheel drive vehicle and a four-wheel drive vehicle.				
Identify purpose and function of a transfer case.				
Non-Task-Specific Evaluation	**(O pts)**	**(−1 pt)**	**(O pts)**	**(−1 pt)**
Student successfully completed at least three of the non-task–specific steps.				
Student successfully completed all five of the non-task-specific steps.				
Total Score: <total # of points / 4 = %>				

Supervisor:

Supervisor/instructor signature _______________________________ Date _______________________

Comments:

Retest supervisor/instructor signature _______________________________ Date _______________

Comments:

Learning Objective/Task	CDX Tasksheet Number	ASE Foundation Reference Number; Priority Level
• Inspect, adjust, and/or replace external manual valve shift linkage, transmission range sensor/switch, and/or park/neutral position switch.	A2006	A2B1, P-2
• Inspect for leakage at external seals, gaskets, and bushings.	A2007	A2B2, P-1
• Inspect, replace, and/or align power train mounts.	A2008	A2B3, P-2
• Drain and replace fluid and filter(s); use proper fluid type per manufacturer specification.	A2009	A2B4, P-1

Materials Required

- Blank work order
- Vehicle with available service history records
- Depending on the type of concern, special diagnostic/hand tools may be required. See your supervisor/instructor for instructions to identify what tools may be required.
- Service information database
- Personal protective equipment
- Vehicle lifting equipment

Safety Considerations

- When running any vehicles in the shop, make sure you use the shop's exhaust ventilation system to discharge all exhaust gas safely outside.
- Lifting equipment, such as vehicle jacks and stands, vehicle hoists, and engine hoists, are important tools that increase productivity and make the job easier. However, they can also cause severe injury or death if used improperly. Make sure you follow the manufacturer's operation procedures. Also, make sure you have your supervisor's/instructor's permission to use any particular type of lifting equipment.
- Comply with personal and environmental safety practices associated with clothing; eye protection; hand tools; power equipment; proper ventilation; and the handling, storage, and disposal of chemicals/materials in accordance with local, state, and federal safety and environmental regulations.
- If you need to start the vehicle, you should ensure that the parking brake is firmly applied; if necessary, use wheel chocks to prevent the vehicle from moving when the vehicle is started to verify the completion of these tasks.
- Extreme caution must be exercised when working around rotating components.

CDX Tasksheet Number: A2006

Student/Intern Information

Name _________________________________ Date ____________ Class _________________________

Vehicle, Customer, and Service Information

Vehicle used for this activity:

Year ______________ Make ____________________________ Model _________________________

Odometer _________________________________ VIN ____________________________________

Materials Required

- Blank work order
- Vehicle with available service history records
- Depending on the type of concern, special diagnostic/hand tools may be required. See your supervisor/instructor for instructions to identify what tools may be required.
- Service information database
- Personal protective equipment
- Vehicle lifting equipment
- General hand tools

Task-Specific Safety Considerations

- Lifting equipment, such as vehicle jacks and stands, vehicle hoists, and engine hoists, are important tools that increase productivity and make the job easier. However, they can also cause severe injury or death if used improperly. Make sure you follow the manufacturer's operation procedures. Also, make sure you have your supervisor's/instructor's permission to use any particular type of lifting equipment.
- Comply with personal and environmental safety practices associated with clothing; eye protection; hand tools; power equipment; proper ventilation; and the handling, storage, and disposal of chemicals/materials in accordance with local, state, and federal safety and environmental regulations.

▶ TASK Inspect, adjust, and/or replace external manual valve shift linkage, transmission range sensor/switch, and/or park/neutral position switch.

MLR
A2B1

Time off_______________

Time on_______________

Student Instructions: Read through the entire procedure prior to starting. Prepare your work space and any tools or parts that may be needed to complete the task. When directed by your instructor, begin the procedure to complete the task and check the box as each step is finished. Track your time on this procedure for later comparison to the standard completion time (i.e., "flat rate" or customer pay time).

Total time_______________

Procedure:	Step Completed
1. Research shift linkage adjustments on your vehicle.	
a. List or print and attach to this sheet the procedure for inspecting and adjusting the shift linkage for your vehicle.	☐
b. Describe how driver gear selection is monitored in your vehicle. Does the vehicle have a transmission range sensor? If so, where is the range sensor located?	☐
c. Describe what could happen to the way a vehicle runs and drives if the shift linkage is not adjusted correctly. Describe signs of a misadjusted shift linkage and any damage that may occur to the vehicle if this happens.	☐
2. Inspect the condition of the shift linkage, range sensor, and electrical connections.	
a. Locate the shift linkage. Describe the condition of the linkage. Are there any broken pieces or parts of the linkage that appear to be binding?	☐
b. Locate the transmission range sensor. Describe the condition of the sensor and its electrical connections. Is the sensor free from rust and corrosion? Is all wiring intact and plugged in properly?	☐
3. Adjust shift linkage and range sensor.	
a. Following the proper procedure, adjust the shift linkage and range sensor so that it operates and gives a proper signal. A scan tool may be necessary to ensure accurate adjustments.	☐
b. Verify the transmission shifts through the gear ranges correctly. Verify proper transmission engagements in reverse, drive, and any manual gear selections.	☐
c. Verify that the engine starts in both park and neutral, and only in these two selections.	☐

Non-Task-Specific Evaluations:	Step Completed
1. Tools and equipment were used as directed and returned in good working order.	☐
2. Complied with all general and task-specific safety standards, including proper use of any personal protective equipment.	☐
3. Completed the task in an appropriate time frame (Recommendation: 1.5 or 2 times flat rate).	☐
4. Left the work space clean and orderly.	☐
5. Cared for customer property and returned it undamaged.	☐

Student signature _______________________________ Date _______________________________

Comments:

Have your supervisor/instructor verify satisfactory completion of this procedure, any observations found, and any necessary action(s) recommended.

Evaluation Instructions: The scoring box below is intended to act as a guide for both student and instructor. Each criterion listed will help students understand what is expected from them and evaluators articulate success at a particular task. The scoring is set up to allow a second attempt at each task (see the "Test" and "Retest" columns). Scoring is designed to reward students for correct completion of the task. Points are lost for failure to complete the employability requirements (see "Non-Task-Specific" criteria). When all the criteria are evaluated, tally the points for a total at the bottom of each column.

Tasksheet Scoring

	Test		Retest	
Evaluation Items	Pass	Fail	Pass	Fail
Task-Specific Evaluation	**(1 pt)**	**(0 pts)**	**(1 pt)**	**(0 pts)**
Research shift linkage adjustments on the vehicle.				
Describe faults that may occur from misadjusted shift.				
Perform shift linkage adjustment.				
Verify correct operation of shift linkage. Ensure that the vehicle starts in park and neutral.				
Non-Task-Specific Evaluation	**(0 pts)**	**(−1 pt)**	**(0 pts)**	**(−1 pt)**
Student successfully completed at least three of the non-task-specific steps.				
Student successfully completed all five of the non-task-specific steps.				
Total Score: <total # of points / 4 = %>				

Supervisor:

Supervisor/instructor signature ______________________________ Date __________________

Comments:

Retest supervisor/instructor signature ______________________________ Date __________________

Comments:

CDX Tasksheet Number: A2007

Student/Intern Information

Name _________________________________ Date _____________ Class _________________________

Vehicle, Customer, and Service Information

Vehicle used for this activity:

Year _______________ Make _______________________ Model _____________________________

Odometer _________________________________ VIN ____

Materials Required

- Blank work order
- Vehicle with available service history records
- Depending on the type of concern, special diagnostic/hand tools may be required. See your supervisor/instructor for instructions to identify what tools may be required.
- Service information database
- Personal protective equipment
- Vehicle lifting equipment

Task-Specific Safety Considerations

- Lifting equipment, such as vehicle jacks and stands, vehicle hoists, and engine hoists, are important tools that increase productivity and make the job easier. However, they can also cause severe injury or death if used improperly. Make sure you follow the manufacturer's operation procedures. Also, make sure you have your supervisor's/instructor's permission to use any particular type of lifting equipment.
- Comply with personal and environmental safety practices associated with clothing; eye protection; hand tools; power equipment; proper ventilation; and the handling, storage, and disposal of chemicals/materials in accordance with local, state, and federal safety and environmental regulations

▶ TASK Inspect for leakage at external seals, gaskets, and bushings.

MLR
A2B2

Time off_____________

Time on_____________

Student Instructions: Read through the entire procedure prior to starting. Prepare your work space and any tools or parts that may be needed to complete the task. When directed by your instructor, begin the procedure to complete the task and check the box as each step is finished. Track your time on this procedure for later comparison to the standard completion time (i.e., "flat rate" or customer pay time).

Total time_____________

Procedure:	Step Completed
1. Research the leak inspection procedure. Are there any special tools that are recommended for performing a leak inspection on your specific vehicle?	☐
2. Inspect the transmission/transaxle for leaks at the following areas. List any action required.	
a. Transmission pan	☐
b. Transmission drain plug	☐
c. CV axle seals	☐
d. Output shaft seal	☐
e. Transmission bell housing	☐

Non-Task-Specific Evaluations:	**Step Completed**
1. Tools and equipment were used as directed and returned in good working order.	☐
2. Complied with all general and task-specific safety standards, including proper use of any personal protective equipment.	☐
3. Completed the task in an appropriate time frame (Recommendation: 1.5 or 2 times flat rate).	☐
4. Left the work space clean and orderly.	☐
5. Cared for customer property and returned it undamaged.	☐

Student signature _______________________________ Date _______________________________

Comments:

Have your supervisor/instructor verify satisfactory completion of this procedure, any observations found, and any necessary action(s) recommended.

Evaluation Instructions: The scoring box below is intended to act as a guide for both student and instructor. Each criterion listed will help students understand what is expected from them and evaluators articulate success at a particular task. The scoring is set up to allow a second attempt at each task (see the "Test" and "Retest" columns). Scoring is designed to reward students for correct completion of the task. Points are lost for failure to complete the employability requirements (see "Non-Task-Specific" criteria). When all the criteria are evaluated, tally the points for a total at the bottom of each column.

Taksheet Scoring

	Test		Retest	
Evaluation Items	**Pass**	**Fail**	**Pass**	**Fail**
Task-Specific Evaluation	**(1 pt)**	**(0 pts)**	**(1 pt)**	**(0 pts)**
Research the transmission leak inspection procedure and any special tools that may be needed.				
Inspect the transmission for leaks in all identified areas.				
Determine the repair procedure of any leaks identified.				
Verify the transmission fluid level.				
Non-Task-Specific Evaluation	**(0 pts)**	**(−1 pt)**	**(0 pts)**	**(−1 pt)**
Student successfully completed at least three of the non-task-specific steps.				
Student successfully completed all five of the non-task-specific steps.				
Total Score: <total # of points / 4 = %>				

Supervisor:

Supervisor/instructor signature ___________________________________ Date _______________________

Comments:

Retest supervisor/instructor signature ___________________________________ Date _______________________

Comments:

CDX Tasksheet Number: A2008

Student/Intern Information

Name _________________________________ Date ____________ Class _________________________

Vehicle, Customer, and Service Information

Vehicle used for this activity:

Year ________________ Make _________________________________ Model _____________________________

Odometer _________________________________ VIN _________________________________

Materials Required
- Blank work order
- Vehicle with available service history records
- Depending on the type of concern, special diagnostic/hand tools may be required. See your supervisor/instructor for instructions to identify what tools may be required.
- Service information database
- Personal protective equipment
- Vehicle lifting equipment
- General hand tools

Task-Specific Safety Considerations
- Lifting equipment, such as vehicle jacks and stands, vehicle hoists, and engine hoists, are important tools that increase productivity and make the job easier. However, they can also cause severe injury or death if used improperly. Make sure you follow the manufacturer's operation procedures. Also, make sure you have your supervisor's/instructor's permission to use any particular type of lifting equipment.
- Comply with personal and environmental safety practices associated with clothing; eye protection; hand tools; power equipment; proper ventilation; and the handling, storage, and disposal of chemicals/materials in accordance with local, state, and federal safety and environmental regulations.

▶ TASK Inspect, replace, and/or align power train mounts.

MLR
A2B3

Time off____________

Time on____________

Student Instructions: Read through the entire procedure prior to starting. Prepare your work space and any tools or parts that may be needed to complete the task. When directed by your instructor, begin the procedure to complete the task and check the box as each step is finished. Track your time on this procedure for later comparison to the standard completion time (i.e., "flat rate" or customer pay time).

Total time____________

Procedure:	Step Completed
1. Inspect the condition of all power train mounts. Look for broken or loose components and torn rubber mounts on the following components. Note any action required.	
a. Engine mounts	☐
b. Transmission/transaxle mounts	☐
c. Driveshaft support mounts	☐
d. Differential mounts	☐
e. Sub frame mounts	☐
2. Remove, replace, and align drivetrain mounts.	
a. Research the procedure for replacing your specific drivetrain mount. List or print and attach to this sheet the procedure for replacing your specific drivetrain mount.	☐
b. Following the proper procedure, remove the damaged or worn mount. Inspect the surrounding area for damage caused by a loose/worn mount. Note any action required.	☐

<table>
<tr><td>c. Compare the new part to the old mount that you removed. Be sure that it is the correct part before you install it. Install the new drivetrain mount following proper procedures. Are there any special precautions that you need to follow according to the service information you researched?</td><td>☐</td></tr>
</table>

Non-Task-Specific Evaluations:	**Step Completed**
1. Tools and equipment were used as directed and returned in good working order.	☐
2. Complied with all general and task-specific safety standards, including proper use of any personal protective equipment.	☐
3. Completed the task in an appropriate time frame (Recommendation: 1.5 or 2 times flat rate).	☐
4. Left the work space clean and orderly.	☐
5. Cared for customer property and returned it undamaged.	☐

Student signature ________________________ Date ________________________

Comments:

Have your supervisor/instructor verify satisfactory completion of this procedure, any observations found, and any necessary action(s) recommended.

Evaluation Instructions: The scoring box below is intended to act as a guide for both student and instructor. Each criterion listed will help students understand what is expected from them and evaluators articulate success at a particular task. The scoring is set up to allow a second attempt at each task (see the "Test" and "Retest" columns). Scoring is designed to reward students for correct completion of the task. Points are lost for failure to complete the employability requirements (see "Non-Task-Specific" criteria). When all the criteria are evaluated, tally the points for a total at the bottom of each column.

Tasksheet Scoring

	Test		Retest	
Evaluation Items	**Pass**	**Fail**	**Pass**	**Fail**
Task-Specific Evaluation	**(1 pt)**	**(0 pts)**	**(1 pt)**	**(0 pts)**
Inspect the condition of all drivetrain mounts.				
Research the procedure for replacing drivetrain mounts.				
Remove the old drivetrain mounts and inspect the surrounding area for damage.				
Install drivetrain mounts using the proper procedure.				
Non-Task-Specific Evaluation	**(0 pts)**	**(−1 pt)**	**(0 pts)**	**(−1 pt)**
Student successfully completed at least three of the non-task-specific steps.				
Student successfully completed all five of the non-task-specific steps.				
Total Score: <total # of points / 4 = %>				

CDX Tasksheet Number: A2009

Student/Intern Information

Name _________________________________ Date ____________ Class _____________________

Vehicle, Customer, and Service Information

Vehicle used for this activity:

Year ______________ Make _________________________ Model _____________________

Odometer _______________________________ VIN _______________________________

Materials Required

- Blank work order
- Vehicle with available service history records
- Depending on the type of concern, special diagnostic/hand tools may be required. See your supervisor/instructor for instructions to identify what tools may be required.
- Service information database
- Personal protective equipment
- Vehicle lifting equipment
- Fluid catch pan
- Transmission fluid
- Transmission filter and gasket/RTV
- Shop rags
- General hand tools
- Scan tool (if applicable)

Task-Specific Safety Considerations

- When running any vehicles in the shop, make sure you use the shop's exhaust ventilation system to discharge all exhaust gas safely outside.
- Lifting equipment, such as vehicle jacks and stands, vehicle hoists, and engine hoists, are important tools that increase productivity and make the job easier. However, they can also cause severe injury or death if used improperly. Make sure you follow the manufacturer's operation procedures. Also, make sure you have your supervisor's/instructor's permission to use any particular type of lifting equipment.
- Comply with personal and environmental safety practices associated with clothing; eye protection; hand tools; power equipment; proper ventilation; and the handling, storage, and disposal of chemicals/materials in accordance with local, state, and federal safety and environmental regulations.
- If you need to start the vehicle, you should ensure that the parking brake is firmly applied; if necessary, use wheel chocks to prevent the vehicle from moving when the vehicle is started to verify the completion of these tasks.

▶ TASK Drain and replace fluid and filter(s); use proper fluid type per manufacturer specification.

MLR
A2B4

Student Instructions: Read through the entire procedure prior to starting. Prepare your work space and any tools or parts that may be needed to complete the task. When directed by your instructor, begin the procedure to complete the task and check the box as each step is finished. Track your time on this procedure for later comparison to the standard completion time (i.e., "flat rate" or customer pay time).

Procedure:	Step Completed
1. Research the procedure for transmission fluid filter replacement.	
a. List or print and attach to this sheet the proper procedure for replacing the transmission fluid filter.	☐
b. List the transmission fluid capacity for a drain and fill with filter replacement.	☐
c. List the type of transmission fluid needed for your specific vehicle.	☐
d. Explain how to check the transmission fluid level for your specific vehicle.	☐
e. Does your vehicle require a transmission pan gasket? Explain what type of gasket is used.	☐
2. Remove the transmission filter.	
a. Following proper procedures, lift the vehicle and prepare to service the transmission. Have a fluid catch pan nearby.	☐
b. Does the transmission pan have a drain plug? If the vehicle is equipped with a drain plug, drain the old fluid. Consult your instructor for instructions, as you may be asked to reuse the transmission fluid.	☐

c. If no drain plug is available, follow the procedure and begin to remove the transmission pan bolts. It may be helpful to leave two bolts halfway in and tilt the transmission pan downward toward your fluid catch pan. Drain as much old fluid as you can before removing the remaining bolts.	☐
d. Inspect the transmission pan. Pans are normally equipped with a magnet to collect debris. A very small amount of black material in the pan/magnet may be considered normal wear. Look for large metallic pieces or for the fluid to have a metallic/shimmer appearance, as this may be a sign of premature wear or failure of an internal component. Consult a repair manual for further clarification.	☐
e. Remove the old filter. Filters may be held in place by bolts or simply snapped into place. Follow your procedure and remove the old filter.	☐
3. Prepare the transmission for reassembly.	
a. Remove the old gasket from the pan and transmission case. It is critical to have a surface free of fluid and old gasket material for proper sealing of the new pan gasket.	☐
b. Clean the transmission pan using the approved methods and solvents. Be sure to clean the magnet inside the pan.	☐
c. Clean the transmission case. Remove old fluid and gasket material. It may be necessary to allow the old fluid to drip from the transmission for a while.	☐
d. Install the new filter. If O-rings or seals are used for the filter, be sure to replace old ones with new ones. Apply a small amount of fresh transmission fluid to any seal before installation. What could happen if a filter seal were torn, rolled, or not installed correctly?	☐
4. Reassemble the transmission and add new fluid.	
a. Install the transmission pan following the proper procedure. It is helpful to start ALL bolts before tightening any single one. Torque bolts to specification. What type of gasket is being used upon reassembly? What torque specification are you using for your transmission bolts?	☐
b. Lower the vehicle. Following proper procedure, add new transmission fluid to the vehicle. Depending on the vehicle, only a small amount of fluid should be added before starting the vehicle. Why would only a small amount of fluid be added first and not the entire amount?	☐

c. Start the vehicle. Top off fluid to the correct level. Allow the vehicle to reach operating temperature and ensure that fluid level is correct. Explain what procedure you are following for checking fluid level.	☐
d. Clean residual fluid from all surfaces. Inspect your work. Lift the vehicle once more and inspect the transmission pan gasket for signs of leakage.	☐
5. Perform the transmission relearn procedure.	
a. Some vehicles require a relearn procedure after servicing the transmission. Research your specific vehicle. Does it require a relearn procedure? If so, explain why this procedure is necessary. List or print and attach to this sheet the method for performing a transmission relearn.	☐
b. If applicable to your vehicle, install a scan tool and begin communication with your vehicle. Follow prompts to begin communication.	☐
c. Under transmission functional tests, select the transmission relearn procedure. Follow prompts on the screen to complete the relearn procedure. Note: Many scan tools have information stored in various sections of the tool. Procedures may vary according to the tool you are using.	☐

Non-Task-Specific Evaluations:	Step Completed
1. Tools and equipment were used as directed and returned in good working order.	☐
2. Complied with all general and task-specific safety standards, including proper use of any personal protective equipment.	☐
3. Completed the task in an appropriate time frame (Recommendation: 1.5 or 2 times flat rate).	☐
4. Left the work space clean and orderly.	☐
5. Cared for customer property and returned it undamaged.	☐

Student signature ___________________________ Date ___________________________

Comments:

Have your supervisor/instructor verify satisfactory completion of this procedure, any observations found, and any necessary action(s) recommended.

Evaluation Instructions: The scoring box below is intended to act as a guide for both student and instructor. Each criterion listed will help students understand what is expected from them and evaluators articulate success at a particular task. The scoring is set up to allow a second attempt at each task (see the "Test" and "Retest" columns). Scoring is designed to reward students for correct completion of the task. Points are lost for failure to complete the employability requirements (see "Non-Task-Specific" criteria). When all the criteria are evaluated, tally the points for a total at the bottom of each column.

Tasksheet Scoring

	Test		Retest	
Evaluation Items	**Pass**	**Fail**	**Pass**	**Fail**
Task-Specific Evaluation	**(1 pt)**	**(0 pts)**	**(1 pt)**	**(0 pts)**
Research the procedure for transmission fluid filter replacement.				
Remove the transmission filter.				
Clean the pan of old fluid and clutch material. Clean the magnet. Clean the transmission case.				
Install the filter and new gasket and reassemble the pan. Add the correct transmission fluid type and amount. Perform the transmission relearn procedure, if applicable.				
Non-Task-Specific Evaluation	**(0 pts)**	**(−1 pt)**	**(0 pts)**	**(−1 pt)**
Student successfully completed at least three of the non-task-specific steps.				
Student successfully completed all five of the non-task-specific steps.				
Total Score: <total # of points / 4 = %>				

Supervisor:

Supervisor/instructor signature ________________________________ Date ________________

Comments:

Retest supervisor/instructor signature ________________________________ Date ________________

Comments:

2C. Off-Vehicle Transmission and Transaxle

Learning Objective/Task	CDX Tasksheet Number	ASE Foundation Reference Number; Priority Level
• Describe the operational characteristics of a continuously variable transmission (CVT).	A2010	A2C1, P-3
• Describe the operational characteristics of a hybrid vehicle drivetrain.	A2011	A2C2, P-3

Materials Required

- Blank work order
- Vehicle with available service history records
- Depending on the type of concern, special diagnostic/hand tools may be required. See your supervisor/instructor for instructions to identify what tools may be required.
- Service information database

Safety Considerations

- Comply with personal and environmental safety practices associated with clothing; eye protection; hand tools; power equipment; proper ventilation; and the handling, storage, and disposal of chemicals/materials in accordance with local, state, and federal safety and environmental regulations.

CDX Tasksheet Number: A2010

Student/Intern Information

Name _________________________________ Date ____________ Class _________________________

Vehicle, Customer, and Service Information

Vehicle used for this activity:

Year ________________ Make _________________________ Model _________________________

Odometer _________________________ VIN _________________________________

Materials Required

- Service information database
- Textbook

Task-Specific Safety Considerations

- Comply with personal and environmental safety practices associated with clothing; eye protection; hand tools; power equipment; proper ventilation; and the handling, storage, and disposal of chemicals/materials in accordance with local, state, and federal safety and environmental regulations.

Time off______________

Time on______________

Total time______________

▶ TASK Describe the operational characteristics of a continuously variable transmission (CVT).

MLR
A2C1

Student Instructions: Read through the entire procedure prior to starting. Prepare your work space and any tools or parts that may be needed to complete the task. When directed by your instructor, begin the procedure to complete the task and check the box as each step is finished. Track your time on this procedure for later comparison to the standard completion time (i.e., "flat rate" or customer pay time).

Procedure:	Step Completed
1. Research CVT design, operation, and necessity.	
a. Explain why the CVT was developed for use in passenger vehicles.	☐

b. What makes a CVT different from an automatic transmission of other designs? Explain how the internal components work to make various gear ratios.	☐
c. What is the main thing that a driver will notice when operating a vehicle equipped with a CVT transmission?	☐
d. Explain maintenance service procedures. Is a CVT serviceable like a "normal" automatic transmission?	☐

Non-Task-Specific Evaluations:	Step Completed
1. Tools and equipment were used as directed and returned in good working order.	☐
2. Complied with all general and task-specific safety standards, including proper use of any personal protective equipment.	☐
3. Completed the task in an appropriate time frame (Recommendation: 1.5 or 2 times flat rate).	☐
4. Left the work space clean and orderly.	☐
5. Cared for customer property and returned it undamaged.	☐

Student signature _________________________ Date _________________________

Comments:

Have your supervisor/instructor verify satisfactory completion of this procedure, any observations found, and any necessary action(s) recommended.

Evaluation Instructions: The scoring box below is intended to act as a guide for both student and instructor. Each criterion listed will help students understand what is expected from them and evaluators articulate success at a particular task. The scoring is set up to allow a second attempt at each task (see the "Test" and "Retest" columns). Scoring is designed to reward students for correct completion of the task. Points are lost for failure to complete the employability requirements (see "Non-Task-Specific" criteria). When all the criteria are evaluated, tally the points for a total at the bottom of each column.

Tasksheet Scoring

	Test		Retest	
Evaluation Items	**Pass**	**Fail**	**Pass**	**Fail**
Task-Specific Evaluation	**(1 pt)**	**(O pts)**	**(1 pt)**	**(O pts)**
Explain the need for CVT development.				
Determine differences between internal components of a CVT and other automatic transmissions.				
Describe operational characteristics and maintenance procedures of a CVT.				
Describe how various gear ratios are accomplished inside a CVT.				
Non-Task-Specific Evaluation	**(O pts)**	**(−1 pt)**	**(O pts)**	**(−1 pt)**
Student successfully completed at least three of the non-task-specific steps.				
Student successfully completed all five of the non-task-specific steps.				
Total Score: <total # of points / 4 = %>				

Supervisor:

Supervisor/instructor signature _________________________________ Date _________________

Comments:

Retest supervisor/instructor signature _________________________________ Date _________________

Comments:

CDX Tasksheet Number: A2O11

Student/Intern Information

Name _________________________ Date __________ Class _________________

Vehicle, Customer, and Service Information

Vehicle used for this activity:

Year ____________ Make ____________________ Model _______________________

Odometer ______________________ VIN _________________________________

Task-Specific Safety Considerations
- Comply with personal and environmental safety practices associated with clothing; eye protection; hand tools; power equipment; proper ventilation; and the handling, storage, and disposal of chemicals/materials in accordance with local, state, and federal safety and environmental regulations.

▶ **TASK** Describe the operational characteristics of a hybrid vehicle drivetrain.

MLR
A2C2

Time off___________

Time on___________

Total time___________

Student Instructions: Read through the entire procedure prior to starting. Prepare your work space and any tools or parts that may be needed to complete the task. When directed by your instructor, begin the procedure to complete the task and check the box as each step is finished. Track your time on this procedure for later comparison to the standard completion time (i.e., "flat rate" or customer pay time).

Procedure:	Step Completed
1. Research hybrid vehicle design, operation, and necessity.	
a. Explain what a hybrid vehicle is. What does *hybrid* mean? Explain why the production of hybrid vehicles has become so popular. What are the benefits of driving a hybrid vehicle?	☐

b. What are the two common propulsion drive types used in a hybrid vehicle?	☐
c. Explain what regenerative braking is and how it works.	☐
d. Explain the safety measures you must take before servicing a hybrid vehicle.	☐
e. What voltages are used on a typical gas-electric hybrid vehicle?	☐
f. Explain what it might be like to drive a hybrid vehicle. What will a driver notice as they are driving this vehicle through a city, in stop-and-go traffic?	☐
g. In a gas-electric hybrid vehicle, explain how electric motors are used to propel the vehicle and how motors are used in stopping the vehicle.	☐

Non-Task-Specific Evaluations:	**Step Completed**
1. Tools and equipment were used as directed and returned in good working order.	☐
2. Complied with all general and task-specific safety standards, including proper use of any personal protective equipment.	☐
3. Completed the task in an appropriate time frame (Recommendation: 1.5 or 2 times flat rate).	☐
4. Left the work space clean and orderly.	☐
5. Cared for customer property and returned it undamaged.	☐

Student signature _________________________ Date _________________________

Comments:

Have your supervisor/instructor verify satisfactory completion of this procedure, any observations found, and any necessary action(s) recommended.

Evaluation Instructions: The scoring box below is intended to act as a guide for both student and instructor. Each criterion listed will help students understand what is expected from them and evaluators articulate success at a particular task. The scoring is set up to allow a second attempt at each task (see the "Test" and "Retest" columns). Scoring is designed to reward students for correct completion of the task. Points are lost for failure to complete the employability requirements (see "Non-Task-Specific" criteria). When all the criteria are evaluated, tally the points for a total at the bottom of each column.

Tasksheet Scoring

	Test		Retest	
Evaluation Items	**Pass**	**Fail**	**Pass**	**Fail**
Task-Specific Evaluation	**(1 pt)**	**(O pts)**	**(1 pt)**	**(O pts)**
Explain the need for hybrid vehicle development. Describe what a hybrid is.				
Explain regenerative braking.				
Explain operational characteristics of a hybrid vehicle.				
Explain propulsion types and how electric motors work in conjunction with a gasoline engine.				
Non-Task-Specific Evaluation	**(O pts)**	**(−1 pt)**	**(O pts)**	**(−1 pt)**
Student successfully completed at least three of the non-task-specific steps.				
Student successfully completed all five of the non-task-specific steps.				
Total Score: <total # of points / 4 = %>				

Supervisor:

Supervisor/instructor signature ________________________________ Date ________________

Comments:

Retest supervisor/instructor signature ________________________________ Date ________________

Comments:

Learning Objective/Task	CDX Tasksheet Number	ASE Education Foundation Reference Number; Priority Level
• Research vehicle service information, including fluid type, vehicle service history, service precautions, and technical service bulletins.	A3001	A3A1, P-1
• Drain and refill manual transmission/transaxle and final drive unit; use proper fluid type per manufacturer specification.	A3002	A3A2, P-1
• Check fluid condition; check for leaks.	A3003	A3A3, P-2
• Identify manual drive train and axle components and configuration.	A3004	A3A4, P-1

Materials Required

- Blank work order
- Vehicle with available service history records
- Depending on the type of concern, special diagnostic/hand tools may be required. See your supervisor/instructor for instructions to identify what tools may be required.
- Service information database
- Personal protective equipment

Safety Considerations

- When running any vehicles in the shop, make sure you use the shop's exhaust ventilation system to discharge all exhaust gas safely outside.
- Lifting equipment such as vehicle jacks and stands, vehicle hoists, and engine hoists are important tools that increase productivity and make the job easier. However, they can also cause severe injury or death if used improperly. Make sure you follow the manufacturer's operation procedures. Also, make sure you have your supervisor's/instructor's permission to use any particular type of lifting equipment.
- Comply with personal and environmental safety practices associated with clothing; eye protection; hand tools; power equipment; proper ventilation; and the handling, storage, and disposal of chemicals/materials in accordance with local, state, and federal safety and environmental regulations.
- If you need to start the vehicle, you should ensure that the parking brake is firmly applied; if necessary, use wheel chocks to prevent the vehicle from moving when the vehicle is started to verify the completion of these tasks.
- Extreme caution must be exercised when working around rotating components.

CDX Tasksheet Number: A3001

Student/Intern Information

Name _________________________________ Date ____________ Class _____________________________

Vehicle, Customer, and Service Information

Vehicle used for this activity:

Year _______________ Make _________________________________ Model ______________________________

Odometer _________________________________ VIN ____

Materials Required

- Blank work order
- Vehicle with available service history records
- Depending on the type of concern, special diagnostic/hand tools may be required. See your supervisor/instructor for instructions to identify what tools may be required.
- Service information database

Task-Specific Safety Considerations

- Comply with personal and environmental safety practices associated with clothing; eye protection; hand tools; power equipment; proper ventilation; and the handling, storage, and disposal of chemicals/materials in accordance with local, state, and federal safety and environmental regulations.

Time off_____________

Time on_____________

Total time_____________

▶ **TASK** Research vehicle service information including fluid type, vehicle service history, service precautions, and technical service bulletins.

MLR
A3A1

Student Instructions: Read through the entire procedure prior to starting. Prepare your workspace and any tools or parts that may be needed to complete the task. When directed by your instructor, begin the procedure to complete the task and check the box as each step is finished. Track your time on this procedure for later comparison to the standard completion time (i.e., "flat rate" or customer pay time).

Procedure:	Step Completed
1. Research vehicle service information.	
a. List the transmission fluid type used in your specific vehicle.	☐

b. List the full transmission fluid capacity.	☐
2. Research the vehicle's service history.	
a. Determine routine maintenance that has been completed.	☐
b. Determine outstanding/completed recall work and list below.	☐
c. Determine any history of major repair and list below.	☐
d. Identify service precautions related to any maintenance or recall related task and list below.	☐
3. Determine technical service bulletins (TSBs) issued for the specific vehicle.	
a. Identify one transmission-related TSB and list the TSB number, vehicle concern, and repair procedure.	☐
b. Identify any service precautions related to the manual transmission in your vehicle and list below.	☐

Non-Task-Specific Evaluations:	**Step Completed**
1. Tools and equipment were used as directed and returned in good working order.	☐
2. Complied with all general and task-specific safety standards, including proper use of any personal protective equipment.	☐
3. Completed the task in an appropriate time frame (Recommendation: 1.5 or 2 times flat rate).	☐
4. Left the work space clean and orderly.	☐
5. Cared for customer property and returned it undamaged.	☐

Student signature _________________________________ Date _________________________________

Comments:

Have your supervisor/instructor verify satisfactory completion of this procedure, any observations found, and any necessary action(s) recommended.

Evaluation Instructions: The scoring box below is intended to act as a guide for both student and instructor. Each criterion listed will help students understand what is expected from them and evaluators articulate success at a particular task. The scoring is set up to allow a second attempt at each task (see the "Test" and "Retest" columns). Scoring is designed to reward students for correct completion of the task. Points are lost for failure to complete the employability requirements (see "Non-Task-Specific" criteria). When all criteria are evaluated, tally the points for a total at the bottom of each column.

Tasksheet Scoring

		Test		Retest	
Evaluation Items	**Pass**	**Fail**	**Pass**	**Fail**	
Task-Specific Evaluation	**(1 pt)**	**(0 pts)**	**(1 pt)**	**(0 pts)**	
Research transmission fluid type and capacity and vehicle service history.					
Identify service precautions.					
Identify completed/outstanding recalls.					
Identify transmission-related TSBs.					
Non-Task-Specific Evaluation	**(0 pts)**	**(−1 pt)**	**(0 pts)**	**(−1 pt)**	
Student successfully completed at least three of the non-task-specific steps.					
Student successfully completed all five of the non-task-specific steps.					
Total Score: <total # of points / 4 = %>					

Supervisor:

Supervisor/instructor signature _______________________ Date _______________

Comments:

Retest supervisor/instructor signature _______________________ Date _______________

Comments:

CDX Tasksheet Number: A3002

Student/Intern Information

Name _________________________________ Date ____________ Class _________________________

Vehicle, Customer, and Service Information

Vehicle used for this activity:

Year _______________ Make _____________________________ Model _________________________

Odometer _________________________________ VIN _________________________________

Materials Required

- Blank work order
- Vehicle with available service history records
- Depending on the type of concern, special diagnostic/hand tools may be required. See your supervisor/instructor for instructions to identify what tools may be required.
- Service information database
- Personal protective equipment
- Transmission fluid
- Fluid catch pan
- General hand tools

Task-Specific Safety Considerations

- Lifting equipment such as vehicle jacks and stands, vehicle hoists, and engine hoists are important tools that increase productivity and make the job easier. However, they can also cause severe injury or death if used improperly. Make sure you follow the manufacturer's operation procedures. Also, make sure you have your supervisor's/instructor's permission to use any particular type of lifting equipment.
- Comply with personal and environmental safety practices associated with clothing; eye protection; hand tools; power equipment; proper ventilation; and the handling, storage, and disposal of chemicals/materials in accordance with local, state, and federal safety and environmental regulations.

▶ **TASK** Drain and refill manual transmission/transaxle and final drive unit; use proper fluid type per manufacturer specification.

MLR
A3A2

Time off____________

Time on____________

Total time____________

Student Instructions: Read through the entire procedure prior to starting. Prepare your workspace and any tools or parts that may be needed to complete the task. When directed by your instructor, begin the procedure to complete the task and check the box as each step is finished. Track your time on this procedure for later comparison to the standard completion time (i.e., "flat rate" or customer pay time).

Procedure:	Step Completed
1. Research fluid type and capacity for your specific vehicle.	
a. What type of fluid is used when servicing the manual transmission in your specific vehicle?	☐
b. What is the fluid capacity for a drain and refill of the manual transmission?	☐
c. What type of fluid is used when servicing the final drive unit on your specific vehicle?	☐
d. What is the fluid capacity for a drain and refill of the final drive unit?	☐
2. Perform a drain and refill of the manual transmission in your specific vehicle.	
a. Prepare the vehicle for service. Safely lift the vehicle so that you can work comfortably underneath. Have a fluid catch pan nearby. Locate the drain plug on the transmission. Before removing the drain plug, it is a good idea to locate and remove the fill plug. Explain why this may be a good idea. Inspect the fluid level prior to draining. Explain what type of tool you will use to remove the drain plug.	☐
b. Drain the old fluid from the transmission. Explain the condition of the fluid. Inspect the magnetic drain plug for signs of excessive debris.	☐
c. Once all fluid has drained, install the drain plug. Apply sealer to the plug if necessary. Torque the drain plug to specification. What torque specification did you use?	☐

d. Using the correct fluid, fill the transmission until the transmission fluid level is correct. Explain how you can check fluid level in a manual transmission. How you know when the transmission is full?	☐
e. Install the fill plug. Apply sealer to the plug if necessary. Torque the fill plug to specification. What torque specification did you use?	☐
3. Perform a drain and refill of the final drive unit on your vehicle.	
a. Have a fluid catch pan nearby. Locate the drain plug on the final drive unit. Explain how the final drive unit is drained. Is there a plug that is removed? Is there a cover that must be removed?	☐
b. Before draining the unit, inspect the level. Is the fluid level in the final drive unit sufficient? Drain the fluid from the final drive unit. Inspect the condition of the fluid. Explain any action that may be required.	☐
c. Install a drain plug or cover. If a cover has been removed, be sure to clean all sealing surfaces thoroughly using approved methods and solvents. Apply sealer as needed.	☐
d. Fill the final drive unit using the proper fluid to the correct level. How do you know when the fluid level is correct? Are there any special additives that you used when filling the final drive unit?	☐
e. Inspect ALL work and verify that no leaks are present. Lower the vehicle and prepare the vehicle for delivery back to the customer.	☐

Non-Task-Specific Evaluations:	**Step Completed**
1. Tools and equipment were used as directed and returned in good working order.	☐
2. Complied with all general and task-specific safety standards, including proper use of any personal protective equipment.	☐
3. Completed the task in an appropriate time frame (Recommendation: 1.5 or 2 times flat rate).	☐
4. Left the work space clean and orderly.	☐
5. Cared for customer property and returned it undamaged.	☐

Student signature _______________________________ Date _______________________________

Comments:

Have your supervisor/instructor verify satisfactory completion of this procedure, any observations found, and any necessary action(s) recommended.

Evaluation Instructions: The scoring box below is intended to act as a guide for both student and instructor. Each criterion listed will help students understand what is expected from them and evaluators articulate success at a particular task. The scoring is set up to allow a second attempt at each task (see the "Test" and "Retest" columns). Scoring is designed to reward students for correct completion of the task. Points are lost for failure to complete the employability requirements (see "Non-Task-Specific" criteria). When all criteria are evaluated, tally the points for a total at the bottom of each column.

Tasksheet Scoring

	Test		Retest	
Evaluation Items	**Pass**	**Fail**	**Pass**	**Fail**
Task-Specific Evaluation	**(1 pt)**	**(0 pts)**	**(1 pt)**	**(0 pts)**
Research the transmission and final drive unit fluid type and capacity. Inspect and describe the fluid condition.				
Perform the drain and fill procedure on the manual transmission.				
Perform the drain and refill procedure on the final drive unit.				
Select the correct type of fluid and fill both the transmission and final drive unit to the correct level.				
Non-Task-Specific Evaluation	**(0 pts)**	**(−1 pt)**	**(0 pts)**	**(−1 pt)**
Student successfully completed at least three of the non-task-specific steps.				
Student successfully completed all five of the non-task-specific steps.				
Total Score: <total # of points / 4 = %>				

Supervisor:

Supervisor/instructor signature _______________________________ Date _______________________

Comments:

Retest supervisor/instructor signature _______________________________ Date _______________________

Comments:

CDX Tasksheet Number: A3003

Student/Intern Information

Name ________________________________ Date ____________ Class ________________________

Vehicle, Customer, and Service Information

Vehicle used for this activity:

Year ______________ Make ________________________ Model ________________________

Odometer ________________________ VIN ________________________________

> **Materials Required**
> - Blank work order
> - Vehicle with available service history records
> - Depending on the type of concern, special diagnostic/hand tools may be required. See your supervisor/instructor for instructions to identify what tools may be required.
> - Service information database
> - Personal protective equipment
> - Vehicle lifting equipment
> - General hand tools

Task-Specific Safety Considerations

- Lifting equipment such as vehicle jacks and stands, vehicle hoists, and engine hoists are important tools that increase productivity and make the job easier. However, they can also cause severe injury or death if used improperly. Make sure you follow the manufacturer's operation procedures. Also, make sure you have your supervisor's/instructor's permission to use any particular type of lifting equipment.
- Comply with personal and environmental safety practices associated with clothing; eye protection; hand tools; power equipment; proper ventilation; and the handling, storage, and disposal of chemicals/materials in accordance with local, state, and federal safety and environmental regulations.

▶ TASK Check fluid condition; check for leaks.

MLR
A3A3

Time off________________

Time on________________

Total time________________

Student Instructions: Read through the entire procedure prior to starting. Prepare your workspace and any tools or parts that may be needed to complete the task. When directed by your instructor, begin the procedure to complete the task and check the box as each step is finished. Track your time on this procedure for later comparison to the standard completion time (i.e., "flat rate" or customer pay time).

Procedure:	Step Completed
1. Check fluid condition.	
a. Explain what fluid in bad condition would look like to you.	☐
b. Using proper procedure, inspect the manual transmission fluid level and condition in your vehicle. Explain how you are checking the fluid condition.	☐
c. Explain your diagnosis of the fluid condition.	☐
2. Check for transmission fluid leaks at the following areas. Note any further action necessary. Refer to your instructor for the recommended procedure for repairing leaking components.	
a. Bell housing	☐
b. Slave cylinder	☐
c. Clutch master cylinder and lines	☐
d. Transmission case	☐
e. Output shaft seal	☐
f. CV axle seals (if applicable)	☐

Non-Task-Specific Evaluations:	Step Completed
1. Tools and equipment were used as directed and returned in good working order.	☐
2. Complied with all general and task-specific safety standards, including proper use of any personal protective equipment.	☐
3. Completed the task in an appropriate time frame (Recommendation: 1.5 or 2 times flat rate).	☐
4. Left the work space clean and orderly.	☐
5. Cared for customer property and returned it undamaged.	☐

Student signature _________________________ Date _________________________

Comments:

Have your supervisor/instructor verify satisfactory completion of this procedure, any observations found, and any necessary action(s) recommended.

Evaluation Instructions: The scoring box below is intended to act as a guide for both student and instructor. Each criterion listed will help students understand what is expected from them and evaluators articulate success at a particular task. The scoring is set up to allow a second attempt at each task (see the "Test" and "Retest" columns). Scoring is designed to reward students for correct completion of the task. Points are lost for failure to complete the employability requirements (see "Non-Task-Specific" criteria). When all criteria are evaluated, tally the points for a total at the bottom of each column.

Taksheet Scoring

	Test		Retest	
Evaluation Items	**Pass**	**Fail**	**Pass**	**Fail**
Task-Specific Evaluation	**(1 pt)**	**(0 pts)**	**(1 pt)**	**(0 pts)**
Inspect the transmission fluid level and condition. Identify the characteristics of a bad fluid and the signs of internal transmission damage.				
Describe the condition of the fluid in your transmission.				
Locate individual manual transmission components and inspect for leaks.				
Repair leaks as designated by your instructor.				
Non-Task-Specific Evaluation	**(0 pts)**	**(−1 pt)**	**(0 pts)**	**(−1 pt)**
Student successfully completed at least three of the non-task-specific steps.				
Student successfully completed all five of the non-task-specific steps.				
Total Score: <total # of points / 4 = %>				

Supervisor:

Supervisor/instructor signature ________________________________ Date ________________

Comments:

Retest supervisor/instructor signature ________________________________ Date ________________

Comments:

CDX Tasksheet Number: A3004

Student/Intern Information

Name _________________________________ Date _____________ Class _________________________

Vehicle, Customer, and Service Information

Vehicle used for this activity:

Year _______________ Make _____________________________ Model _____________________________

Odometer _________________________________ VIN _________________________________

Materials Required
- Blank work order
- Vehicle with available service history records
- Depending on the type of concern, special diagnostic/hand tools may be required. See your supervisor/instructor for instructions to identify what tools may be required.
- Service information database
- Personal protective equipment

Task-Specific Safety Considerations
- Lifting equipment such as vehicle jacks and stands, vehicle hoists, and engine hoists are important tools that increase productivity and make the job easier. However, they can also cause severe injury or death if used improperly. Make sure you follow the manufacturer's operation procedures. Also, make sure you have your supervisor's/instructor's permission to use any particular type of lifting equipment.
- Comply with personal and environmental safety practices associated with clothing; eye protection; hand tools; power equipment; proper ventilation; and the handling, storage, and disposal of chemicals/materials in accordance with local, state, and federal safety and environmental regulations.

▶ **TASK** Identify manual drivetrain and axle components and configuration.

MLR
A3A4

Time off_______________

Time on_______________

Total time_______________

Student Instructions: Read through the entire procedure prior to starting. Prepare your workspace and any tools or parts that may be needed to complete the task. When directed by your instructor, begin the procedure to complete the task and check the box as each step is finished. Track your time on this procedure for later comparison to the standard completion time (i.e., "flat rate" or customer pay time).

Procedure:	Step Completed
1. Identify your vehicle's drivetrain configuration. Describe the layout of the drivetrain components on your specific vehicle. Answer the following:	
a. Is the vehicle front-wheel drive?	☐

b. Is the vehicle rear-wheel drive?	☐
c. Is the vehicle four-wheel drive?	☐
d. Is the vehicle all-wheel drive?	☐
e. If the vehicle is four-/all-wheel drive, is a manual selection of four-wheel drive required, or is it done automatically?	☐
f. Are the engine and transmission longitudinally mounted? Are they mounted transversely?	☐
2. Locate the following drivetrain components. Some options may not be available on your vehicle.	
a. Differential	☐
b. Transmission/transaxle	☐
c. Transfer case. Explain the purpose of a transfer case.	☐
d. Front driveshaft	☐
e. Rear driveshaft	☐
f. Universal joint	☐
g. CV axle	☐
h. Transmission mount(s)	☐
i. Transmission dipstick/fill plug/drain plug	☐

Non-Task-Specific Evaluations:	Step Completed
1. Tools and equipment were used as directed and returned in good working order.	☐
2. Complied with all general and task-specific safety standards, including proper use of any personal protective equipment.	☐
3. Completed the task in an appropriate time frame (Recommendation: 1.5 or 2 times flat rate).	☐
4. Left the work space clean and orderly.	☐
5. Cared for customer property and returned it undamaged.	☐

Student signature _________________________ Date _____________________

Comments:

Have your supervisor/instructor verify satisfactory completion of this procedure, any observations found,

and any necessary action(s) recommended.

Tasksheet Scoring

Evaluation Items	Test		Retest	
	Pass	Fail	Pass	Fail
Task-Specific Evaluation	**(1 pt)**	**(0 pts)**	**(1 pt)**	**(0 pts)**
Identify the drivetrain configuration and describe the difference between transversely and longitudinally mounted engines and transmissions.				
Locate various drivetrain components.				
Describe the difference between an all-wheel drive vehicle and a four-wheel drive vehicle.				
Identify the purpose and function of a transfer case.				
Non-Task-Specific Evaluation	**(0 pts)**	**(−1 pt)**	**(0 pts)**	**(−1 pt)**
Student successfully completed at least three of the non-task-specific steps.				
Student successfully completed all five of the non-task-specific steps.				
Total Score: <total # of points / 4 = %>				

Supervisor:

Supervisor/instructor signature _______________________ Date _______________

Comments:

Retest supervisor/instructor signature _______________________ Date _______________

Comments:

Learning Objective/Task	CDX Tasksheet Number	ASE Education Foundation Reference Number; Priority Level
• Check and adjust clutch master cylinder fluid level; use proper fluid type per manufacturer specification.	A3005	A3B1, P-1
• Check for hydraulic system leaks.	A3006	A3B2, P-1

Materials Required

- Blank work order
- Vehicle with available service history records
- Depending on the type of concern, special diagnostic/hand tools may be required. See your supervisor/instructor for instructions to identify what tools may be required.
- Service information database
- Personal protective equipment

Safety Considerations

- When running any vehicles in the shop, make sure you use the shop's exhaust ventilation system to discharge all exhaust gas safely outside.
- Lifting equipment such as vehicle jacks and stands, vehicle hoists, and engine hoists are important tools that increase productivity and make the job easier. However, they can also cause severe injury or death if used improperly. Make sure you follow the manufacturer's operation procedures. Also, make sure you have your supervisor's/instructor's permission to use any particular type of lifting equipment.
- Comply with personal and environmental safety practices associated with clothing; eye protection; hand tools; power equipment; proper ventilation; and the handling, storage, and disposal of chemicals/materials in accordance with local, state, and federal safety and environmental regulations.
- If you need to start the vehicle, you should ensure that the parking brake is firmly applied; if necessary, use wheel chocks to prevent the vehicle from moving when the vehicle is started to verify the completion of these tasks.
- Extreme caution must be exercised when working around rotating components.

CDX Tasksheet Number: A3005

Student/Intern Information

Name _______________________________ Date ____________ Class _______________________

Vehicle, Customer, and Service Information

Vehicle used for this activity:

Year ______________ Make __________________________ Model _____________________________

Odometer _________________________ VIN ____

Task-Specific Safety Considerations

- Comply with personal and environmental safety practices associated with clothing; eye protection; hand tools; power equipment; proper ventilation; and the handling, storage, and disposal of chemicals/materials in accordance with local, state, and federal safety and environmental regulations.
- **Caution:** The fluid used in a clutch master cylinder in many vehicles may damage painted surfaces, plastic, and glass. Use caution when servicing a customer's vehicle. Always use fender covers and take necessary measures to ensure no damage to the vehicle will occur.

Time off______________

Time on______________

Total time____________

▶ TASK Check and adjust clutch master cylinder fluid level; use proper fluid type per manufacturer specification.

MLR
A3B1

Student Instructions: Read through the entire procedure prior to starting. Prepare your workspace and any tools or parts that may be needed to complete the task. When directed by your instructor, begin the procedure to complete the task and check the box as each step is finished. Track your time on this procedure for later comparison to the standard completion time (i.e., "flat rate" or customer pay time).

<table>
<tr><td>Procedure:</td><td>Step Completed</td></tr>
<tr><td>1. Research the clutch master cylinder fluid type.</td><td></td></tr>
<tr><td>   a. Explain what type of fluid is used in the clutch master cylinder.</td><td>☐</td></tr>
<tr><td>2. Inspect the clutch master cylinder fluid and note the level and condition.</td><td></td></tr>
<tr><td>   a. Explain the location of the clutch master cylinder.</td><td>☐</td></tr>
<tr><td>   b. How will you know if the level is correct in the master cylinder?</td><td>☐</td></tr>
<tr><td>   c. Inspect the fluid level. Add fluid if necessary.</td><td>☐</td></tr>
<tr><td>   d. Explain what symptoms a customer may experience if the clutch master cylinder fluid level is low or empty.</td><td>☐</td></tr>
</table>

<table>
<tr><td>Non-Task-Specific Evaluations:</td><td>Step Completed</td></tr>
<tr><td>1. Tools and equipment were used as directed and returned in good working order.</td><td>☐</td></tr>
<tr><td>2. Complied with all general and task-specific safety standards, including proper use of any personal protective equipment.</td><td>☐</td></tr>
<tr><td>3. Completed the task in an appropriate time frame (Recommendation: 1.5 or 2 times flat rate).</td><td>☐</td></tr>
<tr><td>4. Left the work space clean and orderly.</td><td>☐</td></tr>
<tr><td>5. Cared for customer property and returned it undamaged.</td><td>☐</td></tr>
</table>

Student signature _______________________ Date _______________________

Comments:

Have your supervisor/instructor verify satisfactory completion of this procedure, any observations found, and any necessary action(s) recommended.

Evaluation Instructions: The scoring box below is intended to act as a guide for both student and instructor. Each criterion listed will help students understand what is expected from them and evaluators articulate success at a particular task. The scoring is set up to allow a second attempt at each task (see the "Test" and "Retest" columns). Scoring is designed to reward students for correct completion of the task. Points are lost for failure to complete the employability requirements (see "Non-Task-Specific" criteria). When all criteria are evaluated, tally the points for a total at the bottom of each column.

Tasksheet Scoring

	Test		Retest	
Evaluation Items	**Pass**	**Fail**	**Pass**	**Fail**
Task-Specific Evaluation	**(1 pt)**	**(0 pts)**	**(1 pt)**	**(0 pts)**
Research the clutch master cylinder fluid type and identify the location of the clutch master cylinder.				
Describe the fluid level markings on the clutch master cylinder.				
Top off the fluid level.				
Describe the characteristics/symptoms of low fluid level in a clutch master cylinder.				
Non-Task-Specific Evaluation	**(0 pts)**	**(−1 pt)**	**(0 pts)**	**(−1 pt)**
Student successfully completed at least three of the non-task-specific steps.				
Student successfully completed all five of the non-task-specific steps.				
Total Score: <total # of points / 4 = %>				

Supervisor:

Supervisor/instructor signature _______________________________ Date ___________________

Comments:

Retest supervisor/instructor signature _______________________ Date ___________________

Comments:

CDX Tasksheet Number: A3006

Student/Intern Information

Name _________________________ Date __________ Class _______________________

Vehicle, Customer, and Service Information

Vehicle used for this activity:

Year ______________ Make _______________________ Model ___________________

Odometer _____________________ VIN _______________________________________

Materials Required

- Blank work order
- Vehicle with available service history records
- Depending on the type of concern, special diagnostic/hand tools may be required. See your supervisor/instructor for instructions to identify what tools may be required.
- Service information database
- Personal protective equipment
- Vehicle lifting equipment

Task-Specific Safety Considerations

- Lifting equipment such as vehicle jacks and stands, vehicle hoists, and engine hoists are important tools that increase productivity and make the job easier. However, they can also cause severe injury or death if used improperly. Make sure you follow the manufacturer's operation procedures. Also, make sure you have your supervisor's/instructor's permission to use any particular type of lifting equipment.
- Comply with personal and environmental safety practices associated with clothing; eye protection; hand tools; power equipment; proper ventilation; and the handling, storage, and disposal of chemicals/materials in accordance with local, state, and federal safety and environmental regulations.

▶ **TASK** Check for hydraulic system leaks.

MLR
A3B2

Time off________________

Time on________________

Total time________________

Student Instructions: Read through the entire procedure prior to starting. Prepare your workspace and any tools or parts that may be needed to complete the task. When directed by your instructor, begin the procedure to complete the task and check the box as each step is finished. Track your time on this procedure for later comparison to the standard completion time (i.e., "flat rate" or customer pay time).

Procedure:	Step Completed
1. Inspect the clutch pedal feel.	
a. Do NOT start the vehicle. Ensure the parking brake is set. Pump the clutch pedal. The pedal should have slight resistance on the downward stroke. The pedal should return as you release pressure. Does the pedal feel normal? Explain what is wrong if the pedal would stay down on the floor and not return.	☐
2. Inspect the clutch master cylinder fluid level.	☐
a. Begin leak inspection with a fluid level check. Is the fluid level correct?	☐
3. Inspect for hydraulic system leaks at the following areas. Note any leaks found.	
a. Clutch master cylinder	☐
b. Master cylinder hydraulic line	☐
c. Slave cylinder	☐

Non-Task-Specific Evaluations:	Step Completed
1. Tools and equipment were used as directed and returned in good working order.	☐
2. Complied with all general and task-specific safety standards, including proper use of any personal protective equipment.	☐
3. Completed the task in an appropriate time frame (Recommendation: 1.5 or 2 times flat rate).	☐
4. Left the work space clean and orderly.	☐
5. Cared for customer property and returned it undamaged.	☐

Student signature _________________________ Date _________________________

Comments:

Have your supervisor/instructor verify satisfactory completion of this procedure, any observations found, and any necessary action(s) recommended.

Evaluation Instructions: The scoring box below is intended to act as a guide for both student and instructor. Each criterion listed will help students understand what is expected from them and evaluators articulate success at a particular task. The scoring is set up to allow a second attempt at each task (see the "Test" and "Retest" columns). Scoring is designed to reward students for correct completion of the task. Points are lost for failure to complete the employability requirements (see "Non-Task-Specific" criteria). When all criteria are evaluated, tally the points for a total at the bottom of each column.

Tasksheet Scoring

	Test		Retest	
Evaluation Items	**Pass**	**Fail**	**Pass**	**Fail**
Task-Specific Evaluation	**(1 pt)**	**(0 pts)**	**(1 pt)**	**(0 pts)**
Inspect the clutch pedal feel and explain the symptoms a customer may describe with faults in the clutch hydraulic system.				
Inspect clutch master cylinder fluid level.				
Identify individual components of the clutch hydraulic system.				
Inspect all components for signs of leaking.				
Non-Task-Specific Evaluation	**(0 pts)**	**(−1 pt)**	**(0 pts)**	**(−1 pt)**
Student successfully completed at least three of the non-task-specific steps.				
Student successfully completed all five of the non-task-specific steps.				
Total Score: <total # of points / 4 = %>				

Supervisor:

Supervisor/instructor signature ___________________________ Date ___________________

Comments:

Retest supervisor/instructor signature ___________________________ Date ___________________

Comments:

Learning Objective/Task	CDX Tasksheet Number	ASE Education Foundation Reference Number; Priority Level
• Describe the operational characteristics of an electronically controlled manual transmission/transaxle.	A3007	A3C1, P-2

Materials Required

- Blank work order
- Vehicle with available service history records
- Depending on the type of concern, special diagnostic/hand tools may be required. See your supervisor/instructor for instructions to identify what tools may be required.
- Service information database
- Personal protective equipment

Safety Considerations

- When running any vehicles in the shop, make sure you use the shop's exhaust ventilation system to discharge all exhaust gas safely outside.
- Lifting equipment such as vehicle jacks and stands, vehicle hoists, and engine hoists are important tools that increase productivity and make the job easier. However, they can also cause severe injury or death if used improperly. Make sure you follow the manufacturer's operation procedures. Also, make sure you have your supervisor's/instructor's permission to use any particular type of lifting equipment.
- Comply with personal and environmental safety practices associated with clothing; eye protection; hand tools; power equipment; proper ventilation; and the handling, storage, and disposal of chemicals/materials in accordance with local, state, and federal safety and environmental regulations.
- If you need to start the vehicle, you should ensure that the parking brake is firmly applied; if necessary, use wheel chocks to prevent the vehicle from moving when the vehicle is started to verify the completion of these tasks.
- Extreme caution must be exercised when working around rotating components.

CDX Tasksheet Number: A3007

Student/Intern Information

Name _________________________________ Date _____________ Class _________________________

Vehicle, Customer, and Service Information

Vehicle used for this activity:

Year _______________ Make _________________________ Model _________________________

Odometer _________________________ VIN _________________________________

Materials Required
- Blank work order
- Vehicle with available service history records
- Depending on the type of concern, special diagnostic/hand tools may be required. See your supervisor/instructor for instructions to identify what tools may be required.
- Service information database

Task-Specific Safety Considerations
- Comply with personal and environmental safety practices associated with clothing; eye protection; hand tools; power equipment; proper ventilation; and the handling, storage, and disposal of chemicals/materials in accordance with local, state, and federal safety and environmental regulations.

▶ TASK Describe the operational characteristics of an electronically controlled manual transmission/transaxle.

MLR
A3C1

Time off______________

Time on______________

Total time______________

Student Instructions: Read through the entire procedure prior to starting. Prepare your workspace and any tools or parts that may be needed to complete the task. When directed by your instructor, begin the procedure to complete the task and check the box as each step is finished. Track your time on this procedure for later comparison to the standard completion time (i.e., "flat rate" or customer pay time).

Procedure:	Step Completed
1. Describe the purpose and function of the following electronically controlled manual transmission components:	
a. Gear change lever	☐
b. Transmission control unit	☐

c. Gear shift unit	☐
d. Clutch actuator	☐
2. Explain the different operating modes available in an electronically controlled manual transmission.	
3. Describe how the transmission control unit can override driver input. Why would this be necessary?	

Non-Task-Specific Evaluations:	Step Completed
1. Tools and equipment were used as directed and returned in good working order.	☐
2. Complied with all general and task-specific safety standards, including proper use of any personal protective equipment.	☐
3. Completed the task in an appropriate time frame (Recommendation: 1.5 or 2 times flat rate).	☐
4. Left the work space clean and orderly.	☐
5. Cared for customer property and returned it undamaged.	☐

Student signature _________________________ Date _________________________

Comments:

Have your supervisor/instructor verify satisfactory completion of this procedure, any observations found, and any necessary action(s) recommended.

Evaluation Instructions: The scoring box below is intended to act as a guide for both student and instructor. Each criterion listed will help students understand what is expected from them and evaluators articulate success at a particular task. The scoring is set up to allow a second attempt at each task (see the "Test" and "Retest" columns). Scoring is designed to reward students for correct completion of the task. Points are lost for failure to complete the employability requirements (see "Non-Task-Specific" criteria). When all criteria are evaluated, tally the points for a total at the bottom of each column.

Tasksheet Scoring

	Test		Retest	
Evaluation Items	**Pass**	**Fail**	**Pass**	**Fail**
Task-Specific Evaluation	**(1 pt)**	**(0 pts)**	**(1 pt)**	**(0 pts)**
Describe the purpose and function of the gear change lever and the transmission control unit.				
Describe the purpose and function of the gear shift unit and the clutch actuator.				
Describe operating modes of an electronically-controlled manual transmission.				
Describe transmission control unit operation.				
Non-Task-Specific Evaluation	**(0 pts)**	**(−1 pt)**	**(0 pts)**	**(−1 pt)**
Student successfully completed at least three of the non-task-specific steps.				
Student successfully completed all five of the non-task-specific steps.				
Total Score: <total # of points / 4 = %>				

Supervisor:

Supervisor/instructor signature _________________________ Date _________________

Comments:

Retest supervisor/instructor signature _________________________ Date _________________

Comments:

3D. Drive Shaft, Half Shafts, Universal Joints, and Constant-Velocity (CV) Joints (Front-, Rear-, All-, and Four-wheel Drive)

Learning Objective/Task	CDX Tasksheet Number	ASE Education Foundation Reference Number; Priority Level
• Inspect, remove, and/or replace bearings, hubs, and seals.	A3008	A3D1, P-2
• Inspect, service, and/or replace shafts, yokes, boots, and universal/CV joints.	A3009	A3D2, P-2
• Inspect locking hubs.	A3010	A3D3, P-3
• Check for leaks at drive assembly and transfer case seals; check vents; check fluid level; use proper fluid type per manufacturer specification.	A3011	A3D4, P-2

Materials Required

- Blank work order
- Vehicle with available service history records
- Depending on the type of concern, special diagnostic/hand tools may be required. See your supervisor/instructor for instructions to identify what tools may be required.
- Service information database
- Personal protective equipment

Safety Considerations

- When running any vehicles in the shop, make sure you use the shop's exhaust ventilation system to discharge all exhaust gas safely outside.
- Lifting equipment such as vehicle jacks and stands, vehicle hoists, and engine hoists are important tools that increase productivity and make the job easier. However, they can also cause severe injury or death if used improperly. Make sure you follow the manufacturer's operation procedures. Also, make sure you have your supervisor's/instructor's permission to use any particular type of lifting equipment.
- Comply with personal and environmental safety practices associated with clothing; eye protection; hand tools; power equipment; proper ventilation; and the handling, storage, and disposal of chemicals/materials in accordance with local, state, and federal safety and environmental regulations.
- If you need to start the vehicle, you should ensure that the parking brake is firmly applied; if necessary, use wheel chocks to prevent the vehicle from moving when the vehicle is started to verify the completion of these tasks.
- Extreme caution must be exercised when working around rotating components.

CDX Tasksheet Number: A3008

Student/Intern Information

Name _________________________________ Date ____________ Class _________________________

Vehicle, Customer, and Service Information

Vehicle used for this activity:

Year _______________ Make _________________________ Model _____________________________

Odometer _________________________________ VIN _______________________________________

Materials Required

- Blank work order
- Vehicle with available service history records
- Depending on the type of concern, special diagnostic/hand tools may be required. See your supervisor/instructor for instructions to identify what tools may be required.
- Service information database
- Personal protective equipment
- Hydraulic press
- General hand tools
- Seal remover/installer tool
- Snap ring pliers
- Wheel bearing grease
- Chassis ear

Task-Specific Safety Considerations

- Lifting equipment such as vehicle jacks and stands, vehicle hoists, and engine hoists are important tools that increase productivity and make the job easier. However, they can also cause severe injury or death if used improperly. Make sure you follow the manufacturer's operation procedures. Also, make sure you have your supervisor's/instructor's permission to use any particular type of lifting equipment.
- Comply with personal and environmental safety practices associated with clothing; eye protection; hand tools; power equipment; proper ventilation; and the handling, storage, and disposal of chemicals/materials in accordance with local, state, and federal safety and environmental regulations.
- **Caution:** Diagnosis of this fault may require test-driving the vehicle on the school grounds. Attempt this task only with full permission from your instructor and follow all the guidelines exactly.

▶ **TASK** Inspect, remove, and/or replace bearings, hubs, and seals. **MLR**
A3D1

Student Instructions: Read through the entire procedure prior to starting. Prepare your workspace and any tools or parts that may be needed to complete the task. When directed by your instructor, begin the procedure to complete the task and check the box as each step is finished. Track your time on this procedure for later comparison to the standard completion time (i.e., "flat rate" or customer pay time).

Procedure:	Step Completed
1. Research front wheel bearing removal.	
a. List or print and attach to this sheet the procedure for removing a front wheel bearing from your specific vehicle.	☐
b. Are there any service precautions when performing this task?	☐
c. Does your specific vehicle use a sealed hub bearing assembly?	☐
d. Explain how a sealed hub/bearing assembly is different from a serviceable bearing.	☐
2. Inspect all wheel bearings. Note: There are many methods for diagnosing a faulty wheel bearing. Refer to your instructor for a preferred method of diagnosis.	
a. Test-drive the vehicle to listen for excess wheel bearing noise. Have your instructor drive you or follow EXACT procedure when test-driving a vehicle. List any results below.	☐
b. Lift the vehicle. Using proper procedure, inspect all wheel bearings for looseness. Describe your results.	☐

c. Describe what tools and methods you used to diagnose a faulty bearing.	☐
3. Remove a sealed hub bearing (if applicable).	
a. Following proper procedure, remove the sealed hub bearing from your vehicle.	☐
b. How many bolts are used to hold the bearing in place?	☐
c. Explain the procedure you used to remove the bearing. What types of tools did you use? Did you encounter any difficulties removing the bearing?	☐
d. Does the vehicle have an antilock brake system (ABS)? Did the procedure require you to remove a wheel speed sensor?	☐
e. Thoroughly clean the bearing mounting surface. This surface should be free of rust and corrosion.	☐
4. Install a sealed hub bearing.	
a. Following proper procedure, install the new bearing assembly into the vehicle. Visually inspect the new bearing for signs of damage and be sure that it is the correct part prior to installation. Be sure that the bolt holes are the same, the wheel stud pattern is correct, and, if an axle is installed through the bearing, be sure the splines are the same.	☐
5. List the torque specification, if applicable, for the following:	
a. Sealed hub bearing bolts	☐
b. Axle nut	☐

c. Brake caliper bracket bolts	☐
d. Brake caliper bolts	☐
e. Lug nuts	☐
6. Test-drive the vehicle to verify that you have correctly identified the worn bearing and have repaired it. Many times, other noises can occur after a repair such as wheel bearing replacement. It is common to have brake backing plate noise. Ensure that there are no noises present after your repair.	☐
7. Remove a serviceable wheel bearing (if applicable).	
a. Following proper procedure, remove the serviceable wheel bearing from your vehicle.	☐
b. Using a seal remover tool, remove the inner grease seal from the backside of the rotor. Next, remove the inner bearing.	☐
c. Clean the bearing and hub using the correct methods and solvents.	☐
d. Inspect the bearing rollers, cage, and races for signs of damage. Damage includes pitting, corrosion, spalling, brinelling, or signs of overloading and overheating. Explain the results of your visual inspection. Are you able to reuse the bearing?	☐
8. Install a serviceable wheel bearing.	
a. Apply grease to the wheel bearing. Explain the method you used to pack the bearing full of grease.	☐
b. Install a new inner grease seal.	☐
c. Clean and inspect the spindle before bearing installation. List any action required.	☐

<table>
<tr><td>d. Following procedure, install the serviceable bearing and rotor assembly.</td><td>☐</td></tr>
<tr><td>e. Torque the adjusting nut to the proper preload. What torque specification did you use? Why is a preload necessary?</td><td>☐</td></tr>
<tr><td>f. Finish installation. Torque lug nuts to specification. What torque specification did you use?</td><td>☐</td></tr>
</table>

Non-Task-Specific Evaluations:	Step Completed
1. Tools and equipment were used as directed and returned in good working order.	☐
2. Complied with all general and task-specific safety standards, including proper use of any personal protective equipment.	☐
3. Completed the task in an appropriate time frame (Recommendation: 1.5 or 2 times flat rate).	☐
4. Left the work space clean and orderly.	☐
5. Cared for customer property and returned it undamaged.	☐

Student signature _______________________ Date _______________________

Comments:

Have your supervisor/instructor verify satisfactory completion of this procedure, any observations found, and any necessary action(s) recommended.

Evaluation Instructions: The scoring box below is intended to act as a guide for both student and instructor. Each criterion listed will help students understand what is expected from them and evaluators articulate success at a particular task. The scoring is set up to allow a second attempt at each task (see the "Test" and "Retest" columns). Scoring is designed to reward students for correct completion of the task. Points are lost for failure to complete the employability requirements (see "Non-Task-Specific" criteria). When all criteria are evaluated, tally the points for a total at the bottom of each column.

Tasksheet Scoring

	Test		Retest	
Evaluation Items	Pass	Fail	Pass	Fail
Task-Specific Evaluation	(1 pt)	(0 pts)	(1 pt)	(0 pts)
Research wheel bearing removal. Explain how a sealed hub/bearing assembly is different from a serviceable bearing. Identify bearing type specific to the vehicle you are working on.				
Inspect all bearings.				
Describe what tools and methods can be used to diagnose a faulty wheel bearing.				
Remove and replace a wheel bearing.				
Non-Task-Specific Evaluation	(0 pts)	(−1 pt)	(0 pts)	(−1 pt)
Student successfully completed at least three of the non-task-specific steps.				
Student successfully completed all five of the non-task-specific steps.				
Total Score: <total # of points / 4 = %>				

Supervisor:

Supervisor/instructor signature _________________________________ Date _______________________

Comments:

Retest supervisor/instructor signature _________________________________ Date _______________________

Comments:

CDX Tasksheet Number: A3009

Student/Intern Information

Name _________________________________ Date ___________ Class _____________________

Vehicle, Customer, and Service Information

Vehicle used for this activity:

Year _______________ Make ___________________________ Model ___________________________

Odometer _________________________________ VIN _________________________________

Materials Required

- Blank work order
- Vehicle with available service history records
- Depending on the type of concern, special diagnostic/hand tools may be required. See your supervisor/instructor for instructions to identify what tools may be required.
- Service information database
- Personal protective equipment
- General hand tools
- Universal joint press tool
- Snap ring pliers
- Constant-velocity (CV) boot pliers

Task-Specific Safety Considerations

- Lifting equipment such as vehicle jacks and stands, vehicle hoists, and engine hoists are important tools that increase productivity and make the job easier. However, they can also cause severe injury or death if used improperly. Make sure you follow the manufacturer's operation procedures. Also, make sure you have your supervisor's/instructor's permission to use any particular type of lifting equipment.
- Comply with personal and environmental safety practices associated with clothing; eye protection; hand tools; power equipment; proper ventilation; and the handling, storage, and disposal of chemicals/materials in accordance with local, state, and federal safety and environmental regulations.

▶ TASK Inspect, service, and/or replace shafts, yokes, boots, and universal/CV joints.

MLR
A3D2

Time off____________

Time on____________

Total time____________

Student Instructions: Read through the entire procedure prior to starting. Prepare your workspace and any tools or parts that may be needed to complete the task. When directed by your instructor, begin the procedure to complete the task and check the box as each step is finished. Track your time on this procedure for later comparison to the standard completion time (i.e., "flat rate" or customer pay time).

Procedure:	Step Completed
1. Research the drive shaft and/or CV axle removal procedure.	
a. List or print and attach to this sheet the procedure for removing the driveshaft.	☐
2. Inspect components for looseness, damage, or leaks.	
a. Lift the vehicle and inspect the driveshaft for damage or signs of missing balance weights. **Tip:** Before lifting the vehicle, place the transmission in neutral. This allows the driveshaft to spin and also allows better access to mounting bolts.	☐
b. Explain the signs a customer may notice if a driveshaft is missing its balance weights.	☐
c. Twist the driveshaft back and forth near the universal joint(s). Lift up and down and inspect for any looseness in the joint. List any action required.	☐
d. Check the center support bearing (if equipped). What kind of noise would a center support bearing make if it was defective? When would you hear the noise?	☐
e. Inspect the CV axles for looseness, damage, or torn and leaking axle boots. List any action required.	☐
f. What kind of noise would a defective CV axle make? When would you hear the noise?	☐
3. Following proper procedure, remove the drive shaft and/or CV axle. **Tip:** Use a paint stick to mark the front and rear universal joint flanges of the driveshaft before removal. Explain why it is important to install the driveshaft in the same way that it was removed.	☐

a. Once the drive shaft has been removed, prepare to remove the universal joint. **Tip:** Sometimes it helps to soak the retaining clips of the U-joint in some sort of penetrant to help with disassembly. Remove all retaining clips. Some U-joints are retained by a molded plastic that may have to be melted away. Refer to your instructor for further information.	☐
b. Using the U-joint tool, press the bearing cups from the yolk, one side at a time.	☐
c. Once the U-joint is removed, clean and inspect the driveshaft yolk. The retaining clip grooves must be cleaned thoroughly. Inspect for signs of damage or pitting on the yolk itself.	☐
d. Once the CV axle has been removed, prepare to remove the axle boots. Clamp the axle shaft in a vise with soft jaws.	☐
e. Remove the boot clamps and boots. Refer to your instructions and disassemble the splined end of the axle. You may have to remove grease and locate a retaining clip.	☐
f. Clean and inspect the axle joint cage, race, and bearings for signs of wear. List any action required. What type of CV joint is being used in your vehicle?	☐
4. Reassemble and install the drive shaft and CV axle.	
a. Following proper procedure, install the new U joint into the yolk. Ensure that no needle bearings are lost during the assembly procedure.	☐
b. Install the drive shaft in the vehicle. Be sure to use your painted marks on the flanges for correct alignment. Torque the fasteners to specification. What torque specification did you use?	☐
c. If a grease fitting is being used in the new joint, grease the joint.	☐
d. Following proper procedure, reassemble the CV axle joint. Apply the proper amount of grease to the joint and inside the new CV boot. Using the special boot clamp pliers, install the boot clamps.	☐
e. Install the CV axle into the vehicle. Tighten all fasteners to specification. What is the torque specification for the axle nut?	☐

Non-Task-Specific Evaluations:	Step Completed
1. Tools and equipment were used as directed and returned in good working order.	☐
2. Complied with all general and task-specific safety standards, including proper use of any personal protective equipment.	☐
3. Completed the task in an appropriate time frame (Recommendation: 1.5 or 2 times flat rate).	☐
4. Left the work space clean and orderly.	☐
5. Cared for customer property and returned it undamaged.	☐

Student signature _________________________ Date _____________________

Comments:

Have your supervisor/instructor verify satisfactory completion of this procedure, any observations found, and any necessary action(s) recommended.

Evaluation Instructions: The scoring box below is intended to act as a guide for both student and instructor. Each criterion listed will help students understand what is expected from them and evaluators articulate success at a particular task. The scoring is set up to allow a second attempt at each task (see the "Test" and "Retest" columns). Scoring is designed to reward students for correct completion of the task. Points are lost for failure to complete the employability requirements (see "Non-Task-Specific" criteria). When all criteria are evaluated, tally the points for a total at the bottom of each column.

Tasksheet Scoring

	Test		Retest	
Evaluation Items	**Pass**	**Fail**	**Pass**	**Fail**
Task-Specific Evaluation	**(1 pt)**	**(O pts)**	**(1 pt)**	**(O pts)**
Research the procedure for removing the drive shaft and CV axle. Inspect shafts, yolks, and CV axles for looseness, damage, or leaks.				
Remove the CV axle and replace the axle boot.				
Remove the driveshaft and replace the universal joint.				
Torque all fasteners to specification.				
Non-Task-Specific Evaluation	**(O pts)**	**(−1 pt)**	**(O pts)**	**(−1 pt)**
Student successfully completed at least three of the non-task-specific steps.				
Student successfully completed all five of the non-task-specific steps.				
Total Score: <total # of points / 4 = %>				

Supervisor:

Supervisor/instructor signature _________________________ Date _______________

Comments:

Retest supervisor/instructor signature _________________________ Date _______________

Comments:

CDX Tasksheet Number: A3010

Student/Intern Information

Name _________________________________ Date ___________ Class _______________________

Vehicle, Customer, and Service Information

Vehicle used for this activity:

Year _______________ Make _______________________ Model ___________________________

Odometer _______________________ VIN _______________________________

Materials Required

- Blank work order
- Vehicle with available service history records
- Depending on the type of concern, special diagnostic/hand tools may be required. See your supervisor/instructor for instructions to identify what tools may be required.
- Service information database
- Personal protective equipment
- Four-wheel drive vehicle with locking hubs
- Wheel bearing grease
- Snap ring pliers
- Hub socket

Task-Specific Safety Considerations

- Lifting equipment such as vehicle jacks and stands, vehicle hoists, and engine hoists are important tools that increase productivity and make the job easier. However, they can also cause severe injury or death if used improperly. Make sure you follow the manufacturer's operation procedures. Also, make sure you have your supervisor's/instructor's permission to use any particular type of lifting equipment.
- Comply with personal and environmental safety practices associated with clothing; eye protection; hand tools; power equipment; proper ventilation; and the handling, storage, and disposal of chemicals/materials in accordance with local, state, and federal safety and environmental regulations.

Time off___________

Time on___________

Total time___________

▶ **TASK** Inspect locking hubs.

MLR
A3D3

Student Instructions: Read through the entire procedure prior to starting. Prepare your workspace and any tools or parts that may be needed to complete the task. When directed by your instructor, begin the procedure to complete the task and check the box as each step is finished. Track your time on this procedure for later comparison to the standard completion time (i.e., "flat rate" or customer pay time).

Procedure:	Step Completed
1. Inspect the locking hubs.	
a. Begin by identifying the type of locking hubs on the vehicle. Are they manual or automatic and how can you tell? Describe how an automatic locking hub works.	☐
b. Begin inspection by lifting the vehicle. Locate the switch on the locking hubs. What positions are available on the locking hub?	☐
c. Move the selector switch to the free position. What do you notice about how the wheel rotates? Does it spin the axle along with the wheel?	☐
d. Move the selector switch to the locked position. What do you notice about how the wheel rotates now? Is there more resistance when the wheel is turning? Does the axle now spin with the wheel?	☐
e. With the hubs remaining in the locked position, locate the front driveshaft. Spin the driveshaft by hand. What is happening to the front wheels?	☐
2. Disassemble the locking hubs.	
a. Research the procedure for removal and disassembly of the locking hubs on your vehicle. Are there any special service tools or precautions needed for the task?	☐
b. Following procedure, remove the wheels, brakes, and hub assemblies. Be sure to keep all washers and rings in order as they come apart.	☐

c. Once the hub is removed, clean the old grease from the hub assembly. Inspect for abnormal wear. List any action required.	☐
3. Assemble the locking hubs.	
a. Repack or replace bearings as needed.	☐
b. Following proper procedure, reassemble the hub components. Be sure to follow directions carefully when installing bearings and applying the correct amount of preload. What specifications did you use when tightening the bearing and locknut?	☐
c. Install the remaining components. Verify correct operation of the locking hubs and four-wheel drive system.	☐

Non-Task-Specific Evaluations:	Step Completed
1. Tools and equipment were used as directed and returned in good working order.	☐
2. Complied with all general and task-specific safety standards, including proper use of any personal protective equipment.	☐
3. Completed the task in an appropriate time frame (Recommendation: 1.5 or 2 times flat rate).	☐
4. Left the work space clean and orderly.	☐
5. Cared for customer property and returned it undamaged.	☐

Student signature ________________________________ Date ________________________________

Comments:

Have your supervisor/instructor verify satisfactory completion of this procedure, any observations found, and any necessary action(s) recommended.

Tasksheet Scoring

	Test		Retest	
Evaluation Items	**Pass**	**Fail**	**Pass**	**Fail**
Task-Specific Evaluation	**(1 pt)**	**(0 pts)**	**(1 pt)**	**(0 pts)**
Identify manual or locking hubs on your vehicle.				
Inspect the working condition of locking hubs.				
Research the procedure for removing and servicing the hub assembly. Remove the locking hub assembly.				
Clean, inspect, and install/replace the locking hub assembly.				
Non-Task-Specific Evaluation	**(0 pts)**	**(−1 pt)**	**(0 pts)**	**(−1 pt)**
Student successfully completed at least three of the non-task-specific steps.				
Student successfully completed all five of the non-task-specific steps.				
Total Score: <total # of points / 4 = %>				

Supervisor:

Supervisor/instructor signature _______________________ Date _______________

Comments:

Retest supervisor/instructor signature _______________________ Date _______________

Comments:

CDX Tasksheet Number: A3011

Student/Intern Information

Name _________________________________ Date _____________ Class _____________________

Vehicle, Customer, and Service Information

Vehicle used for this activity:

Year _______________ Make ____________________________ Model ____________________________

Odometer _________________________________ VIN ____

Materials Required

- Blank work order
- Vehicle with available service history records
- Depending on the type of concern, special diagnostic/hand tools may be required. See your supervisor/instructor for instructions to identify what tools may be required.
- Service information database
- Personal protective equipment
- General hand tools
- Transfer case fluid
- Fluid catch pan

Task-Specific Safety Considerations

- Lifting equipment such as vehicle jacks and stands, vehicle hoists, and engine hoists are important tools that increase productivity and make the job easier. However, they can also cause severe injury or death if used improperly. Make sure you follow the manufacturer's operation procedures. Also, make sure you have your supervisor's/instructor's permission to use any particular type of lifting equipment.
- Comply with personal and environmental safety practices associated with clothing; eye protection; hand tools; power equipment; proper ventilation; and the handling, storage, and disposal of chemicals/materials in accordance with local, state, and federal safety and environmental regulations.

Time off______________

Time on______________

Total time______________

▶ TASK Check for leaks at drive assembly and transfer case seals; check vents; check fluid level; use proper fluid type per manufacturer specification.

**MLR
A3D4**

Student Instructions: Read through the entire procedure prior to starting. Prepare your workspace and any tools or parts that may be needed to complete the task. When directed by your instructor, begin the procedure to complete the task and check the box as each step is finished. Track your time on this procedure for later comparison to the standard completion time (i.e., "flat rate" or customer pay time).

Procedure:	Step Completed
1. Research transfer case fluid level check and fluid type and capacity.	
a. How do you check the fluid level in the transfer case?	☐
b. What type of fluid is used in the transfer case?	☐
c. What is the full fluid capacity of the transfer case?	☐
2. Inspect the transfer case for fluid leaks at the following areas. List any action required.	
a. Transfer case front seal	☐
b. Transfer case rear seal	☐
c. Transfer case vent	☐
d. Transfer case shift linkage seals	☐
e. Transfer case halves, drain, and fill plug	☐

3. Inspect the transfer case fluid level. Top off the fluid.	
a. Following procedure, locate the check plug in the transfer case. Using a fluid catch pan, remove the check plug. What is the level of fluid? How do you know if the level is full?	☐
b. If needed, top off the transfer case using the correct type of fluid. Install the fill/check plug. Ensure that no leaks are present at the plug.	☐

Non-Task-Specific Evaluations:	Step Completed
1. Tools and equipment were used as directed and returned in good working order.	☐
2. Complied with all general and task-specific safety standards, including proper use of any personal protective equipment.	☐
3. Completed the task in an appropriate time frame (Recommendation: 1.5 or 2 times flat rate).	☐
4. Left the work space clean and orderly.	☐
5. Cared for customer property and returned it undamaged.	☐

Student signature ___________________________ Date ___________________________

Comments:

Have your supervisor/instructor verify satisfactory completion of this procedure, any observations found, and any necessary action(s) recommended.

Tasksheet Scoring

Evaluation Items	Test		Retest	
	Pass	**Fail**	**Pass**	**Fail**
Task-Specific Evaluation	**(1 pt)**	**(0 pts)**	**(1 pt)**	**(0 pts)**
Research transfer case fluid type, capacity, and the fluid level inspection procedure.				
Inspect the transfer case for leaks.				
Inspect the transfer case fluid level.				
Top off the transfer case fluid.				
Non-Task-Specific Evaluation	**(0 pts)**	**(−1 pt)**	**(0 pts)**	**(−1 pt)**
Student successfully completed at least three of the non-task-specific steps.				
Student successfully completed all five of the non-task-specific steps.				
Total Score: <total # of points / 4 = %>				

Supervisor:

Supervisor/instructor signature _______________________________ Date _______________

Comments:

Retest supervisor/instructor signature _______________________________ Date _______________

Comments:

3E. Differential Case Assembly

Learning Objective/Task	CDX Tasksheet Number	ASE Education Foundation Reference Number; Priority Level
• Clean and inspect differential case; check for leaks; inspect housing vent.	A3012	A3E1, P-1
• Check and adjust differential case fluid level; use proper fluid type per manufacturer specification.	A3013	A3E2, P-1
• Drain and refill differential housing.	A3014	A3E3, P-1
• Inspect and replace drive axle wheel studs.	A3015	A3E4, P-1

Materials Required

- Blank work order
- Vehicle with available service history records
- Depending on the type of concern, special diagnostic/hand tools may be required. See your supervisor/instructor for instructions to identify what tools may be required.
- Service information database
- Personal protective equipment

Safety Considerations

- When running any vehicles in the shop, make sure you use the shop's exhaust ventilation system to discharge all exhaust gas safely outside.
- Lifting equipment such as vehicle jacks and stands, vehicle hoists, and engine hoists are important tools that increase productivity and make the job easier. However, they can also cause severe injury or death if used improperly. Make sure you follow the manufacturer's operation procedures. Also, make sure you have your supervisor's/instructor's permission to use any particular type of lifting equipment.
- Comply with personal and environmental safety practices associated with clothing; eye protection; hand tools; power equipment; proper ventilation; and the handling, storage, and disposal of chemicals/materials in accordance with local, state, and federal safety and environmental regulations.
- If you need to start the vehicle, you should ensure that the parking brake is firmly applied; if necessary, use wheel chocks to prevent the vehicle from moving when the vehicle is started to verify the completion of these tasks.
- Extreme caution must be exercised when working around rotating components.

CDX Tasksheet Number: A3012

Student/Intern Information

Name ________________________________ Date ____________ Class ________________________

Vehicle, Customer, and Service Information

Vehicle used for this activity:

Year ______________ Make ________________________ Model ________________________

Odometer ________________________ VIN ________________________________

Materials Required

- Blank work order
- Vehicle with available service history records
- Depending on the type of concern, special diagnostic/hand tools may be required. See your supervisor/instructor for instructions to identify what tools may be required.
- Service information database
- Personal protective equipment
- General hand tools
- Wire brush
- Scraper

Task-Specific Safety Considerations

- Lifting equipment such as vehicle jacks and stands, vehicle hoists, and engine hoists are important tools that increase productivity and make the job easier. However, they can also cause severe injury or death if used improperly. Make sure you follow the manufacturer's operation procedures. Also, make sure you have your supervisor's/instructor's permission to use any particular type of lifting equipment.
- Comply with personal and environmental safety practices associated with clothing; eye protection; hand tools; power equipment; proper ventilation; and the handling, storage, and disposal of chemicals/materials in accordance with local, state, and federal safety and environmental regulations.

▶ **TASK** Clean and inspect differential case; check for leaks; inspect housing vent.

MLR
A3E1

Time off___________

Time on___________

Student Instructions: Read through the entire procedure prior to starting. Prepare your workspace and any tools or parts that may be needed to complete the task. When directed by your instructor, begin the procedure to complete the task and check the box as each step is finished. Track your time on this procedure for later comparison to the standard completion time (i.e., "flat rate" or customer pay time).

Total time___________

Procedure:	Step Completed
1. Inspect the differential for leaks and note the results.	
a. Lift the vehicle and locate the differential.	☐
b. Inspect the differential cover for leaks.	☐
c. Inspect the axle seals for leaks.	☐
d. Inspect the differential fluid level and condition.	☐
e. Inspect the differential vent. Explain what may happen if the vent becomes clogged.	☐
2. Remove the differential cover and inspect the differential.	
a. Remove the bolts that hold the differential cover in place.	☐
b. Remove the cover. Have a drain pan in place to collect the old differential fluid. **Tip:** Some covers may be difficult to remove. Use caution when trying to pry the cover loose so that the cover or mating surface is not damaged.	☐
c. Describe the condition of the fluid. Are there any abnormal contaminants in the fluid?	☐
d. Inspect the magnet for signs of debris and note the results.	☐

e. Inspect the ring gear for signs of damage and note the results.	☐
f. Using approved solvents and procedures, clean the housing and differential cover sealing surface.	☐
g. Apply a new gasket/sealer and install the differential cover.	☐
h. Torque the cover bolts to specification. List the specification.	☐
i. Identify and fill the differential with the correct type and amount of fluid.	☐

Non-Task-Specific Evaluations:	Step Completed
1. Tools and equipment were used as directed and returned in good working order.	☐
2. Complied with all general and task-specific safety standards, including proper use of any personal protective equipment.	☐
3. Completed the task in an appropriate time frame (Recommendation: 1.5 or 2 times flat rate).	☐
4. Left the work space clean and orderly.	☐
5. Cared for customer property and returned it undamaged.	☐

Student signature _________________________ Date _________________________

Comments:

Have your supervisor/instructor verify satisfactory completion of this procedure, any observations found, and any necessary action(s) recommended.

Evaluation Instructions: The scoring box below is intended to act as a guide for both student and instructor. Each criterion listed will help students understand what is expected from them and evaluators articulate success at a particular task. The scoring is set up to allow a second attempt at each task (see the "Test" and "Retest" columns). Scoring is designed to reward students for correct completion of the task. Points are lost for failure to complete the employability requirements (see "Non-Task-Specific" criteria). When all criteria are evaluated, tally the points for a total at the bottom of each column.

Tasksheet Scoring

	Test		Retest	
Evaluation Items	**Pass**	**Fail**	**Pass**	**Fail**
Task-Specific Evaluation	**(1 pt)**	**(0 pts)**	**(1 pt)**	**(0 pts)**
Inspect the differential and axle seals for leaks.				
Inspect the differential vent.				
Remove the differential cover and inspect fluid condition, magnet, and differential components.				
Clean the differential and apply the gasket. Install the cover and torque fasteners to specifications.				
Non-Task-Specific Evaluation	**(0 pts)**	**(−1 pt)**	**(0 pts)**	**(−1 pt)**
Student successfully completed at least three of the non-task-specific steps.				
Student successfully completed all five of the non-task-specific steps.				
Total Score: <total # of points / 4 = %>				

Supervisor:

Supervisor/instructor signature _________________________________ Date _____________________

Comments:

Retest supervisor/instructor signature _________________________ Date _____________________

Comments:

CDX Tasksheet Number: A3013

Student/Intern Information

Name _________________________________ Date ____________ Class _________________________

Vehicle, Customer, and Service Information

Vehicle used for this activity:

Year ______________ Make ______________________________ Model ____________________________

Odometer ___________________________________ VIN _______________________________________

Materials Required

- Blank work order
- Vehicle with available service history records
- Depending on the type of concern, special diagnostic/hand tools may be required. See your supervisor/instructor for instructions to identify what tools may be required.
- Service information database
- Personal protective equipment
- General hand tools
- Differential fluid
- Fluid catch pan

Task-Specific Safety Considerations

- Lifting equipment such as vehicle jacks and stands, vehicle hoists, and engine hoists are important tools that increase productivity and make the job easier. However, they can also cause severe injury or death if used improperly. Make sure you follow the manufacturer's operation procedures. Also, make sure you have your supervisor's/instructor's permission to use any particular type of lifting equipment.
- Comply with personal and environmental safety practices associated with clothing; eye protection; hand tools; power equipment; proper ventilation; and the handling, storage, and disposal of chemicals/materials in accordance with local, state, and federal safety and environmental regulations.

▶ TASK Check and adjust differential case fluid level; use proper fluid type per manufacturer specification.

MLR A3E2

Time off____________

Time on____________

Student Instructions: Read through the entire procedure prior to starting. Prepare your workspace and any tools or parts that may be needed to complete the task. When directed by your instructor, begin the procedure to complete the task and check the box as each step is finished. Track your time on this procedure for later comparison to the standard completion time (i.e., "flat rate" or customer pay time).

Total time____________

Procedure:	Step Completed
1. Research the differential fluid level check procedure.	
a. Explain how you would check the differential fluid level.	☐
b. How do you know when the differential is full?	☐
c. What type of fluid is used in the differential?	☐
2. Inspect the differential fluid level.	
a. Lift the vehicle and locate the differential fluid check plug.	☐
b. Using a fluid catch pan, remove the check plug. Inspect the fluid level in the differential. What is the level?	☐
3. Adjust the differential fluid level.	
a. Using the correct fluid, fill the differential to the correct level.	☐
b. Install the check/fill plug and torque to specification. What torque specification did you use?	☐

<table>
<tr><td>Non-Task-Specific Evaluations:</td><td>Step Completed</td></tr>
<tr><td>1. Tools and equipment were used as directed and returned in good working order.</td><td>☐</td></tr>
<tr><td>2. Complied with all general and task-specific safety standards, including proper use of any personal protective equipment.</td><td>☐</td></tr>
<tr><td>3. Completed the task in an appropriate time frame (Recommendation: 1.5 or 2 times flat rate).</td><td>☐</td></tr>
<tr><td>4. Left the work space clean and orderly.</td><td>☐</td></tr>
<tr><td>5. Cared for customer property and returned it undamaged.</td><td>☐</td></tr>
</table>

Student signature _______________________________ Date _______________________________

Comments:

Have your supervisor/instructor verify satisfactory completion of this procedure, any observations found, and any necessary action(s) recommended.

Evaluation Instructions: The scoring box below is intended to act as a guide for both student and instructor. Each criterion listed will help students understand what is expected from them and evaluators articulate success at a particular task. The scoring is set up to allow a second attempt at each task (see the "Test" and "Retest" columns). Scoring is designed to reward students for correct completion of the task. Points are lost for failure to complete the employability requirements (see "Non-Task-Specific" criteria). When all criteria are evaluated, tally the points for a total at the bottom of each column.

Tasksheet Scoring

	Test		Retest	
Evaluation Items	**Pass**	**Fail**	**Pass**	**Fail**
Task-Specific Evaluation	**(1 pt)**	**(0 pts)**	**(1 pt)**	**(0 pts)**
Research differential fluid type, capacity, and fluid level inspection procedure.				
Inspect fluid level.				
Top off the differential using the correct fluid.				
Torque fasteners to the correct specification.				
Non-Task-Specific Evaluation	**(0 pts)**	**(−1 pt)**	**(0 pts)**	**(−1 pt)**
Student successfully completed at least three of the non-task-specific steps.				
Student successfully completed all five of the non-task-specific steps.				
Total Score: <total # of points / 4 = %>				

Supervisor:

Supervisor/instructor signature _________________________ Date _________________

Comments:

Retest supervisor/instructor signature _________________________ Date _________________

Comments:

CDX Tasksheet Number: A3014

Student/Intern Information

Name ___________________________________ Date ____________ Class ___________________________

Vehicle, Customer, and Service Information

Vehicle used for this activity:

Year _______________ Make _______________________________ Model _______________________________

Odometer _________________________________ VIN ____

Materials Required

- Blank work order
- Vehicle with available service history records
- Depending on the type of concern, special diagnostic/hand tools may be required. See your supervisor/instructor for instructions to identify what tools may be required.
- Service information database
- Personal protective equipment
- General hand tools
- Differential fluid
- Wire brush
- Razor blade
- Fluid catch pan
- Differential cover gasket/RTV

Task-Specific Safety Considerations

- Lifting equipment such as vehicle jacks and stands, vehicle hoists, and engine hoists are important tools that increase productivity and make the job easier. However, they can also cause severe injury or death if used improperly. Make sure you follow the manufacturer's operation procedures. Also, make sure you have your supervisor's/instructor's permission to use any particular type of lifting equipment.
- Comply with personal and environmental safety practices associated with clothing; eye protection; hand tools; power equipment; proper ventilation; and the handling, storage, and disposal of chemicals/materials in accordance with local, state, and federal safety and environmental regulations.

▶ TASK Drain and refill differential housing.

MLR
A3E3

Time off____________

Time on____________

Student Instructions: Read through the entire procedure prior to starting. Prepare your workspace and any tools or parts that may be needed to complete the task. When directed by your instructor, begin the procedure to complete the task and check the box as each step is finished. Track your time on this procedure for later comparison to the standard completion time (i.e., "flat rate" or customer pay time).

Total time____________

Procedure:	Step Completed
1. Research the differential drain and fill procedure.	
a. List or print and attach to this sheet the procedure for draining and filling the differential in your vehicle.	☐
b. What type of fluid is used in the differential? How much fluid is needed for a drain and fill procedure?	☐
c. Is there a drain plug for your differential? Does a cover and gasket have to be removed in order to drain the fluid?	☐
2. Drain the differential fluid.	
a. Lift the vehicle and locate the fluid drain or cover that must be removed.	☐
b. Using a fluid catch pan, remove the drain plug or the bolts that hold the cover in place. If a cover is being removed, the old gasket material may make it difficult to remove the cover. Be careful prying the cover off the differential housing. Gouging the housing may create fluid leaks. **Tip:** Before prying the cover off, try using a rubber mallet to lightly tap the cover and break the gasket seal. If you must pry, try prying on the top section of the cover, as fluid does not typically reach the upper half of the differential housing.	☐
c. Allow the fluid to drain. Inspect the drain plug, fluid, and differential housing for signs of abnormal wear.	☐
3. Prepare the differential for reassembly.	
a. If a cover has been removed, it is necessary to remove all old gasket material from the cover and housing. Use a scraper or small wire brush to remove old gasket material. **Caution:** Do your best to eliminate dirt and old gasket material from entering the differential housing.	☐
b. Be sure the differential cover has not been bent during the removal process. You must be sure that all surfaces are flat to ensure a good seal.	☐
c. Using approved solvents, clean all sealing surfaces. Surfaces must be free of oil and other contaminants to make a leak-free seal.	☐

4. Reassemble the differential and add the correct fluid.	
a. Apply a new gasket/RTV to the differential cover. Install all fasteners and torque to specification. What torque specification did you use?	☐
b. Add the correct differential fluid. If an RTV sealer was used, allow ample time for the RTV to cure before adding fluid.	☐
c. Install the fluid fill plug. Clean the surrounding areas and double-check that no leaks are present.	☐

Non-Task-Specific Evaluations:	Step Completed
1. Tools and equipment were used as directed and returned in good working order.	☐
2. Complied with all general and task-specific safety standards, including proper use of any personal protective equipment.	☐
3. Completed the task in an appropriate time frame (Recommendation: 1.5 or 2 times flat rate).	☐
4. Left the work space clean and orderly.	☐
5. Cared for customer property and returned it undamaged.	☐

Student signature ________________________________ Date ________________________

Comments:

Have your supervisor/instructor verify satisfactory completion of this procedure, any observations found,

and any necessary action(s) recommended.

Tasksheet Scoring

Evaluation Items	Test		Retest	
	Pass	**Fail**	**Pass**	**Fail**
Task-Specific Evaluation	**(1 pt)**	**(0 pts)**	**(1 pt)**	**(0 pts)**
Research differential fluid drain/fill procedure. Research fluid type and capacity.				
Remove the cover, drain the fluid, and clean and inspect the differential cover.				
Apply gasket material and install the differential cover.				
Use the correct fluid type and amount. Torque all fasteners to specification.				
Non-Task-Specific Evaluation	**(0 pts)**	**(−1 pt)**	**(0 pts)**	**(−1 pt)**
Student successfully completed at least three of the non-task-specific steps.				
Student successfully completed all five of the non-task-specific steps.				
Total Score: <total # of points / 4 = %>				

Supervisor:

Supervisor/instructor signature _______________________________ Date _______________

Comments:

Retest supervisor/instructor signature _______________________________ Date _______________

Comments:

CDX Tasksheet Number: A3015

Student/Intern Information

Name _________________________________ Date ___________ Class ___________________________

Vehicle, Customer, and Service Information

Vehicle used for this activity:

Year _______________ Make _________________________ Model _________________________

Odometer _________________________ VIN _________________________________

Materials Required

- Blank work order
- Vehicle with available service history records
- Depending on the type of concern, special diagnostic/hand tools may be required. See your supervisor/instructor for instructions to identify what tools may be required.
- Service information database
- Personal protective equipment
- Wheel stud and wheel nut
- General hand tools

Task-Specific Safety Considerations

- Lifting equipment such as vehicle jacks and stands, vehicle hoists, and engine hoists are important tools that increase productivity and make the job easier. However, they can also cause severe injury or death if used improperly. Make sure you follow the manufacturer's operation procedures. Also, make sure you have your supervisor's/instructor's permission to use any particular type of lifting equipment.
- Comply with personal and environmental safety practices associated with clothing; eye protection; hand tools; power equipment; proper ventilation; and the handling, storage, and disposal of chemicals/materials in accordance with local, state, and federal safety and environmental regulations.

▶ **TASK** Inspect and replace drive axle wheel studs.

MLR
A3E4

Time off____________

Time on____________

Student Instructions: Read through the entire procedure prior to starting. Prepare your workspace and any tools or parts that may be needed to complete the task. When directed by your instructor, begin the procedure to complete the task and check the box as each step is finished. Track your time on this procedure for later comparison to the standard completion time (i.e., "flat rate" or customer pay time).

Total time____________

Procedure:	Step Completed
1. Research the procedure for replacing a damaged wheel stud.	
a. Are there any special service tools or precautions for this procedure?	☐
b. What is the torque specification for the lug nuts on your vehicle?	☐
2. Inspect the wheel studs.	
a. Inspect the condition of all wheel studs and nuts. Describe the condition. Are any wheel studs stripped out? Are any broken, rusted, or cracked?	☐
3. Replace the wheel stud and nut.	
a. Following proper procedure, remove the brake assembly on the wheel with the damaged wheel stud.	☐
b. Once the brakes are removed, inspect the hub flange for any damage. List any action required.	☐
c. Use a hammer to remove the old wheel stud. The stud should come out with minimal force.	☐
d. Insert the new wheel stud from the rear of the hub flange. Use the stud installation tool if available. If not, select three or four washers or a spacer to slip over the new wheel stud and install a lug nut. Install a lug nut of the same diameter and thread pitch with the tapered side of the nut facing away from the washers.	☐
e. Ensure the stud is straight as you begin to draw the stud through the flange. Tighten the lug nut until the new stud is fully seated in the backside of the flange. Remove the tool, lug nut, and any washers or spacer.	☐
f. Inspect the new wheel stud for damage that may have occurred during the installation.	☐
g. Assemble the brake components and wheel. Torque all fasteners to specification. What torque specification did you use when tightening the wheel? What pattern did you use when tightening the wheel?	☐

Non-Task-Specific Evaluations:	Step Completed
1. Tools and equipment were used as directed and returned in good working order.	☐
2. Complied with all general and task-specific safety standards, including proper use of any personal protective equipment.	☐
3. Completed the task in an appropriate time frame (Recommendation: 1.5 or 2 times flat rate).	☐
4. Left the work space clean and orderly.	☐
5. Cared for customer property and returned it undamaged.	☐

Student signature _______________________________ Date _______________________________

Comments:

Have your supervisor/instructor verify satisfactory completion of this procedure, any observations found, and any necessary action(s) recommended.

Evaluation Instructions: The scoring box below is intended to act as a guide for both student and instructor. Each criterion listed will help students understand what is expected from them and evaluators articulate success at a particular task. The scoring is set up to allow a second attempt at each task (see the "Test" and "Retest" columns). Scoring is designed to reward students for correct completion of the task. Points are lost for failure to complete the employability requirements (see "Non-Task-Specific" criteria). When all criteria are evaluated, tally the points for a total at the bottom of each column.

Taksheet Scoring

	Test		Retest	
Evaluation Items	**Pass**	**Fail**	**Pass**	**Fail**
Task-Specific Evaluation	**(1 pt)**	**(0 pts)**	**(1 pt)**	**(0 pts)**
Research the procedure for replacing a wheel stud. Research the torque specification.				
Remove a damaged wheel stud.				
Install a new wheel stud.				
Reassemble the vehicle and torque all fasteners to specification.				
Non-Task-Specific Evaluation	**(0 pts)**	**(−1 pt)**	**(0 pts)**	**(−1 pt)**
Student successfully completed at least three of the non-task-specific steps.				
Student successfully completed all five of the non-task-specific steps.				
Total Score: <total # of points / 4 = %>				

Supervisor:

Supervisor/instructor signature _______________________________________ Date _______________________

Comments:

Retest supervisor/instructor signature _______________________________ Date _______________________

Comments:

Learning Objective/Task	CDX Tasksheet Number	ASE Education Foundation Reference Number; Priority Level
• Research vehicle service information, including fluid type, vehicle service history, service precautions, and technical service bulletins.	A4001	A4A1, P-1
• Disable and enable supplemental restraint system (SRS); verify indicator lamp operation.	A4002	A4A2, P-1
• Identify suspension and steering system components and configurations.	A4003	A4A3, P-1

Materials Required

- Blank work order
- Vehicle with available service history records
- Depending on the type of concern, special diagnostic/hand tools may be required. See your supervisor/instructor for instructions to identify what tools may be required.
- Service information database
- Personal protective equipment

Safety Considerations

- When running any vehicles in the shop, make sure you use the shop's exhaust ventilation system to discharge all exhaust gas safely outside.
- Lifting equipment, such as vehicle jacks and stands, vehicle hoists, and engine hoists, are important tools that increase productivity and make the job easier. However, they can also cause severe injury or death if used improperly. Make sure you follow the manufacturer's operation procedures. Also, make sure you have your supervisor's/instructor's permission to use any particular type of lifting equipment.
- Comply with personal and environmental safety practices associated with clothing; eye protection; hand tools; power equipment; proper ventilation; and the handling, storage, and disposal of chemicals/materials in accordance with local, state, and federal safety and environmental regulations.
- If you need to start the vehicle, you should ensure that the parking brake is firmly applied; if necessary, use wheel chocks to prevent the vehicle from moving when the vehicle is started to verify the completion of these tasks.
- Extreme caution must be exercised when working around rotating components.

CDX Tasksheet Number: A4001

Student/Intern Information

Name _________________________________ Date _____________ Class _____________________

Vehicle, Customer, and Service Information

Vehicle used for this activity:

Year _______________ Make _______________________ Model ___________________________

Odometer _________________________ VIN _____________________________________

Task-Specific Safety Considerations

- Comply with personal and environmental safety practices associated with clothing; eye protection; hand tools; power equipment; proper ventilation; and the handling, storage, and disposal of chemicals/materials in accordance with local, state, and federal safety and environmental regulations.

Time off_______________

Time on_______________

Total time_______________

▶ TASK Research vehicle service information including fluid type, vehicle service history, service precautions, and technical service bulletins.

MLR
A4A1

Student Instructions: Read through the entire procedure prior to starting. Prepare your work space and any tools or parts that may be needed to complete the task. When directed by your instructor, begin the procedure to complete the task and check the box as each step is finished. Track your time on this procedure for later comparison to the standard completion time (i.e., "flat rate" or customer pay time).

Procedure:	Step Completed
1. Research the vehicle service information.	
a. What type of fluid is used in the power steering system of your vehicle?	☐

b. What is the full power steering system capacity?	☐
2. Research the vehicle service history.	
a. Determine routine maintenance that has been completed.	☐
b. Determine any completed/outstanding recall work.	☐
c. Determine any major repair in service history.	☐
d. Identify service precautions related to any maintenance- or recall-related task.	☐
3. Determine TSBs issued for the specific vehicle.	
a. Identify one steering/suspension-related TSB and list the TSB number, vehicle concern, and repair procedure.	☐
b. Identify any service precautions related to the steering/suspension system in your vehicle.	☐

Non-Task-Specific Evaluations:	**Step Completed**
1. Tools and equipment were used as directed and returned in good working order.	☐
2. Complied with all general and task-specific safety standards, including proper use of any personal protective equipment.	☐
3. Completed the task in an appropriate time frame (Recommendation: 1.5 or 2 times flat rate).	☐
4. Left the work space clean and orderly.	☐
5. Cared for customer property and returned it undamaged.	☐

Student signature _______________________________ Date _______________________________

Comments:

>

Have your supervisor/instructor verify satisfactory completion of this procedure, any observations found, and any necessary action(s) recommended.

Tasksheet Scoring

Evaluation Items	Test		Retest	
	Pass	**Fail**	**Pass**	**Fail**
Task-Specific Evaluation	**(1 pt)**	**(0 pts)**	**(1 pt)**	**(0 pts)**
Research the power steering fluid type and capacity. Research the vehicle service history.				
Identify any service precautions.				
Identify any completed/outstanding recalls.				
Identify any steering/suspension related TSBs.				
Non-Task-Specific Evaluation	**(0 pts)**	**(−1 pt)**	**(0 pts)**	**(−1 pt)**
Student successfully completed at least three of the non-task-specific steps.				
Student successfully completed all five of the non-task-specific steps.				
Total Score: <total # of points / 4 = %>				

Supervisor:

Supervisor/instructor signature _________________________ Date _______________

Comments:

Retest supervisor/instructor signature _________________________ Date _______________

Comments:

CDX Tasksheet Number: A4002

Student/Intern Information

Name _______________________________ Date ___________ Class _____________________

Vehicle, Customer, and Service Information

Vehicle used for this activity:

Year _______________ Make ___________________________ Model _____________________

Odometer _________________________ VIN ____

Materials Required

- Blank work order
- Vehicle with available service history records
- Depending on the type of concern, special diagnostic/hand tools may be required. See your supervisor/instructor for instructions to identify what tools may be required.
- Service information database
- Personal protective equipment
- General hand tools

Task-Specific Safety Considerations

- Comply with personal and environmental safety practices associated with clothing; eye protection; hand tools; power equipment; proper ventilation; and the handling, storage, and disposal of chemicals/materials in accordance with local, state, and federal safety and environmental regulations.
- **Caution:** Working with airbags in the supplemental restraint systems can cause serious injury or even death if not handled with extreme care. Preventive measures must be taken at all time.

Time off______________

Time on______________

Total time______________

▶ TASK Disable and enable supplemental restraint system (SRS); verify indicator lamp operation.

MLR
A4A2

Student Instructions: Read through the entire procedure prior to starting. Prepare your work space and any tools or parts that may be needed to complete the task. When directed by your instructor, begin the procedure to complete the task and check the box as each step is finished. Track your time on this procedure for later comparison to the standard completion time (i.e., "flat rate" or customer pay time).

Procedure:	Step Completed
1. Research the SRS system in your vehicle.	
a. List the main components of the SRS system in your vehicle.	☐
b. What color is the wiring that leads up to airbag components?	☐
c. According to your service information, how is the SRS system disabled in order to service components in your vehicle?	☐
d. How long must you wait after disabling the SRS system, to service the vehicle?	☐
e. As the driver of the vehicle, are you able to disable the passenger side SRS system manually? If so, how is that done?	☐
2. Verify correct operation of the airbag indicator light on the instrument cluster.	
a. Turn the ignition to the ON position. Verify that the airbag light comes on. Did the light come on and then turn off after 3–5 seconds? Is this a normal condition?	☐
b. Verify that the seatbelt light goes out as you buckle the seat belt.	☐

c. If the airbag light remains on, what does this mean?	☐
3. Disable the SRS system in your vehicle.	
a. Following proper procedure, disable the SRS system. Explain the steps you took to disable the system.	☐
b. If you were to remove the driver airbag, explain how you would carry it and how you would place it on the bench for service. Do not remove the airbag.	☐
4. Enable the SRS system in your vehicle.	
a. Following proper procedure, enable the SRS system. Explain the steps you took to enable the system.	☐
b. After enabling the SRS, verify correct operation by cycling the ignition key once more. Verify that the airbag light comes on momentarily and then goes out, indicating that the self-check is complete.	☐

Non-Task-Specific Evaluations:	Step Completed
1. Tools and equipment were used as directed and returned in good working order.	☐
2. Complied with all general and task-specific safety standards, including proper use of any personal protective equipment.	☐
3. Completed the task in an appropriate time frame (Recommendation: 1.5 or 2 times flat rate).	☐
4. Left the work space clean and orderly.	☐
5. Cared for customer property and returned it undamaged.	☐

Student signature _________________________ Date _________________________

Comments:

Have your supervisor/instructor verify satisfactory completion of this procedure, any observations found, and any necessary action(s) recommended.

Evaluation Instructions: The scoring box below is intended to act as a guide for both student and instructor. Each criterion listed will help students understand what is expected from them and evaluators articulate success at a particular task. The scoring is set up to allow a second attempt at each task (see the "Test" and "Retest" columns). Scoring is designed to reward students for correct completion of the task. Points are lost for failure to complete the employability requirements (see "Non-Task-Specific" criteria). When all criteria are evaluated, tally the points for a total at the bottom of each column.

Tasksheet Scoring

	Test		Retest	
Evaluation Items	Pass	Fail	Pass	Fail
Task-Specific Evaluation	**(1 pt)**	**(0 pts)**	**(1 pt)**	**(0 pts)**
Research the SRS system in your vehicle, then list the SRS components, wiring color, and safety precautions.				
Verify correct operation of the airbag light.				
Disable SRS system.				
Enable SRS system.				
Non-Task-Specific Evaluation	**(0 pts)**	**(−1 pt)**	**(0 pts)**	**(−1 pt)**
Student successfully completed at least three of the non-task-specific steps.				
Student successfully completed all five of the non-task-specific steps.				
Total Score: <total # of points / 4 = %>				

Supervisor:

Supervisor/instructor signature _______________________________ Date _____________________

Comments:

Retest supervisor/instructor signature _______________________ Date _____________________

Comments:

CDX Tasksheet Number: A4003

Student/Intern Information

Name _________________________ Date __________ Class _________________

Vehicle, Customer, and Service Information

Vehicle used for this activity:

Year ____________ Make ____________________ Model ____________________

Odometer ______________________ VIN ___________________________

Materials Required

- Blank work order
- Vehicle with available service history records
- Depending on the type of concern, special diagnostic/hand tools may be required. See your supervisor/instructor for instructions to identify what tools may be required.
- Service information database
- Personal protective equipment

Task-Specific Safety Considerations

- Lifting equipment, such as vehicle jacks and stands, vehicle hoists, and engine hoists, are important tools that increase productivity and make the job easier. However, they can also cause severe injury or death if used improperly. Make sure you follow the manufacturer's operation procedures. Also, make sure you have your supervisor's/instructor's permission to use any particular type of lifting equipment.
- Comply with personal and environmental safety practices associated with clothing; eye protection; hand tools; power equipment; proper ventilation; and the handling, storage, and disposal of chemicals/materials in accordance with local, state, and federal safety and environmental regulations.

▶ TASK Identify suspension and steering system components and configurations.

MLR
A4A3

Time off________

Time on________

Total time________

Student Instructions: Read through the entire procedure prior to starting. Prepare your work space and any tools or parts that may be needed to complete the task. When directed by your instructor, begin the procedure to complete the task and check the box as each step is finished. Track your time on this procedure for later comparison to the standard completion time (i.e., "flat rate" or customer pay time).

Procedure:	Step Completed
1. Identify the steering configuration in your vehicle.	
a. Research the type of steering used in your vehicle. Does your vehicle use a rack and pinion or a parallelogram-style steering configuration?	☐
b. Locate either the rack and pinion or steering gear box.	☐
c. Locate the tie rods on your vehicle. What are the tie-rods attached to?	☐
d. In a parallelogram-style steering configuration, locate the component that transfers the motion from your steering gear box to the tie rod ends. What is this component called?	☐
e. Explain the difference between a rack-and-pinion-type steering system and a parallelogram-type steering system. Which system is made up of more moving parts?	☐
f. What component is attached to the steering knuckle, which allows for the steering knuckle to pivot right and left, as well as tilt inward and outward when cornering?	☐
2. Identify the front suspension configuration.	
a. Research the type of suspension that is used in the front of your vehicle. Does the vehicle use struts? Does the vehicle use an upper and lower control arm? What is the name for the suspension configuration that uses two different length control arms?	☐

b. Is a coil spring used on the front suspension? If so, where is the spring located?	☐
c. Is a shock absorber used in the front suspension design of your vehicle? If no shock absorber is present, which component absorbs road shock?	☐
d. What is the purpose of a sway bar (anti-roll bar)? Does your vehicle's front suspension incorporate a sway bar? How is a sway bar mounted to the frame of a vehicle? How does a sway bar attach to each wheel?	☐

3. Identify the rear suspension configuration.	
a. Does your vehicle incorporate an independent rear suspension design or a solid axle/beam-type suspension? Explain the difference between the two designs.	☐
b. Is a coil spring used in the rear suspension? If so, where is the coil spring located?	☐
c. Is a shock absorber used in the rear suspension design of your vehicle? If no shock absorber is present, which components absorbs road shock?	☐
d. Is a leaf spring used in the rear suspension design of your vehicle? Explain some advantages of using a leaf spring suspension design in a vehicle. What are some disadvantages?	☐
e. Is a sway bar used in the rear suspension design of your vehicle?	☐

f. Explain the purpose of a lateral link. Does your vehicle incorporate this component in its rear suspension design? If not, what supports the wheel/hub assembly from lateral movement?	☐

Non-Task-Specific Evaluations:	Step Completed
1. Tools and equipment were used as directed and returned in good working order.	☐
2. Complied with all general and task-specific safety standards, including proper use of any personal protective equipment.	☐
3. Completed the task in an appropriate time frame (Recommendation: 1.5 or 2 times flat rate).	☐
4. Left the work space clean and orderly.	☐
5. Cared for customer property and returned it undamaged.	☐

Student signature _______________________________ Date _______________________________

Comments:

Have your supervisor/instructor verify satisfactory completion of this procedure, any observations found,

and any necessary action(s) recommended.

Evaluation Instructions: The scoring box below is intended to act as a guide for both student and instructor. Each criterion listed will help students understand what is expected from them and evaluators articulate success at a particular task. The scoring is set up to allow a second attempt at each task (see the "Test" and "Retest" columns). Scoring is designed to reward students for correct completion of the task. Points are lost for failure to complete the employability requirements (see "Non-Task-Specific" criteria). When all criteria are evaluated, tally the points for a total at the bottom of each column.

Tasksheet Scoring

	Test		Retest	
Evaluation Items	**Pass**	**Fail**	**Pass**	**Fail**
Task-Specific Evaluation	**(1 pt)**	**(0 pts)**	**(1 pt)**	**(0 pts)**
Research the steering configuration and front suspension design of your vehicle.				
Research the rear suspension design of your vehicle.				
Explain the differences in steering system designs.				
Explain the differences in suspension system designs.				
Non-Task-Specific Evaluation	**(0 pts)**	**(−1 pt)**	**(0 pts)**	**(−1 pt)**
Student successfully completed at least three of the non-task-specific steps.				
Student successfully completed all five of the non-task-specific steps.				
Total Score: <total # of points / 4 = %>				

Supervisor:

Supervisor/instructor signature _______________________________ Date _______________

Comments:

Retest supervisor/instructor signature _______________________ Date _______________

Comments:

4B. Related Suspension and Steering Service

Learning Objective/Task	CDX Tasksheet Number	ASE Education Foundation Reference Number; Priority Level
• Inspect rack and pinion steering gear inner tie rod ends (sockets) and bellows boots.	A4004	A4B1, P-1
• Inspect power steering fluid level and condition.	A4005	A4B2, P-1
• Flush, fill, and bleed power steering system; use proper fluid type per manufacturer specification.	A4006	A4B3, P-2
• Inspect for power steering fluid leakage.	A4007	A4B4, P-1
• Remove, inspect, replace, and/or adjust power steering pump drive belt.	A4008	A4B5, P-1
• Inspect and replace power steering hoses and fittings.	A4009	A4B6, P-2
• Inspect pitman arm, relay (centerlink/intermediate) rod, idler arm, mountings, and steering linkage damper.	A4010	A4B7, P-1
• Inspect tie rod ends (sockets), tie rod sleeves, and clamps.	A4011	A4B8, P-1
• Inspect upper and lower control arms, bushings, and shafts.	A4012	A4B9, P-1
• Inspect and replace rebound bumpers.	A4013	A4B10, P-1
• Inspect track bar, strut rods/radius arms, and related mounts and bushings.	A4014	A4B11, P-1
• Inspect upper and lower ball joints (with or without wear indicators).	A4015	A4B12, P-1
• Inspect suspension system coil springs and spring insulators (silencers).	A4016	A4B13, P-1
• Inspect suspension system torsion bars and mounts.	A4017	A4B14, P-1
• Inspect and/or replace front/rear stabilizer bar (sway bar) bushings, brackets, and links.	A4018	A4B15, P-1
• Inspect, remove, and/or replace strut cartridge or assembly; inspect mounts and bushings.	A4019	A4B16, P-2
• Inspect front strut bearing and mount.	A4020	A4B17, P-1
• Inspect rear suspension system lateral links/arms (track bars), control (trailing) arms.	A4021	A4B18, P-1
• Inspect rear suspension system leaf spring(s), spring insulators (silencers), shackles, brackets, bushings, center pins/bolts, and mounts.	A4022	A4B19, P-1
• Inspect, remove, and/or replace shock absorbers; inspect mounts and bushings.	A4023	A4B20, P-1
• Inspect electric power steering assist system.	A4024	A4B21, P-2
• Identify hybrid vehicle power steering system electrical circuits and safety precautions.	A4025	A4B22, P-2
• Describe the function of suspension and steering control systems and components (i.e., active suspension, and stability control).	A4026	A4B23, P-3

Safety Considerations

- When running any vehicles in the shop, make sure you use the shop's exhaust ventilation system to discharge all exhaust gas safely outside.
- Lifting equipment, such as vehicle jacks and stands, vehicle hoists, and engine hoists, are important tools that increase productivity and make the job easier. However, they can also cause severe injury or death if used improperly. Make sure you follow the manufacturer's operation procedures. Also, make sure you have your supervisor's/instructor's permission to use any particular type of lifting equipment.
- Comply with personal and environmental safety practices associated with clothing; eye protection; hand tools; power equipment; proper ventilation; and the handling, storage, and disposal of chemicals/materials in accordance with local, state, and federal safety and environmental regulations.
- If you need to start the vehicle, you should ensure that the parking brake is firmly applied; if necessary, use wheel chocks to prevent the vehicle from moving when the vehicle is started to verify the completion of these tasks.
- Extreme caution must be exercised when working around rotating components.

CDX Tasksheet Number: A4004

Student/Intern Information

Name _________________________________ Date ___________ Class _____________________

Vehicle, Customer, and Service Information

Vehicle used for this activity:

Year _______________ Make _________________________ Model _____________________

Odometer _____________________________ VIN _______________________________________

Materials Required

- Blank work order
- Vehicle with available service history records
- Depending on the type of concern, special diagnostic/hand tools may be required. See your supervisor/instructor for instructions to identify what tools may be required.
- Service information database
- Personal protective equipment

Task-Specific Safety Considerations

- Lifting equipment, such as vehicle jacks and stands, vehicle hoists, and engine hoists, are important tools that increase productivity and make the job easier. However, they can also cause severe injury or death if used improperly. Make sure you follow the manufacturer's operation procedures. Also, make sure you have your supervisor's/instructor's permission to use any particular type of lifting equipment.
- Comply with personal and environmental safety practices associated with clothing; eye protection; hand tools; power equipment; proper ventilation; and the handling, storage, and disposal of chemicals/materials in accordance with local, state, and federal safety and environmental regulations.

▶ TASK Inspect rack and pinion steering gear inner tie rod ends (sockets) and bellows boots.

MLR
A4B1

Student Instructions: Read through the entire procedure prior to starting. Prepare your work space and any tools or parts that may be needed to complete the task. When directed by your instructor, begin the procedure to complete the task and check the box as each step is finished. Track your time on this procedure for later comparison to the standard completion time (i.e., "flat rate" or customer pay time).

Time off______________

Time on______________

Total time______________

Procedure:	Step Completed
1. Inspect the rack-and-pinion steering gear for damage, looseness, and leaks.	
a. With the vehicle on the ground, start the vehicle and turn the steering wheel from lock to lock. Is there any abnormal noise, binding, or looseness felt in the steering wheel? **Tip:** If a test drive is required to verify correct steering operation, follow EXACT procedure from your instructor.	☐
b. Lift the vehicle. Locate the rack and pinion. How many bolts mount the rack and pinion to the frame? Are the mounts, fasteners, and bushings in good condition?	☐
c. Locate the power steering lines that are connected at the rack and pinion. Inspect the lines for damage, excess rust, and leaks. Describe your results.	☐
d. Inspect the steering shaft coming from the steering wheel. Inspect this area for power steering fluid leaks.	☐
e. Locate the rubber bellow boots. Inspect for cracks or tears. If fluid is leaking from the rack and pinion inner seals, it may accumulate in the bellow boots. Verify the boots are dry by loosening the boot clamps and inspecting inside.	☐
f. Inspect the inner tie rod ends by lightly moving them back and forth by hand. Can you feel any looseness at the inner tie rods?	☐
2. Based on your inspection, what is your recommendation for service?	

Non-Task-Specific Evaluations:	Step Completed
1. Tools and equipment were used as directed and returned in good working order.	☐
2. Complied with all general and task-specific safety standards, including proper use of any personal protective equipment.	☐
3. Completed the task in an appropriate time frame (Recommendation: 1.5 or 2 times flat rate).	☐
4. Left the work space clean and orderly.	☐
5. Cared for customer property and returned it undamaged.	☐

Student signature _______________________________ Date _______________________________

Comments:

Have your supervisor/instructor verify satisfactory completion of this procedure, any observations found, and any necessary action(s) recommended.

Evaluation Instructions: The scoring box below is intended to act as a guide for both student and instructor. Each criterion listed will help students understand what is expected from them and evaluators articulate success at a particular task. The scoring is set up to allow a second attempt at each task (see the "Test" and "Retest" columns). Scoring is designed to reward students for correct completion of the task. Points are lost for failure to complete the employability requirements (see "Non-Task-Specific" criteria). When all criteria are evaluated, tally the points for a total at the bottom of each column.

Tasksheet Scoring

	Test		Retest	
Evaluation Items	**Pass**	**Fail**	**Pass**	**Fail**
Task-Specific Evaluation	**(1 pt)**	**(O pts)**	**(1 pt)**	**(O pts)**
Verify correct steering operation prior to rack-and-pinion inspection.				
Inspect the rack and pinion mounts, fasteners, and bushings. Inspect power steering lines for damage, excessive rust, or leaks.				
Inspect the inner tie rod ends and bellow boots.				
Make recommendations based on your inspection.				
Non-Task-Specific Evaluation	**(O pts)**	**(−1 pt)**	**(O pts)**	**(−1 pt)**
Student successfully completed at least three of the non-task-specific steps.				
Student successfully completed all five of the non-task-specific steps.				
Total Score: <total # of points / 4 = %>				

Supervisor:

Supervisor/instructor signature ________________________________ Date ________________________

Comments:

Retest supervisor/instructor signature ________________________________ Date ________________________

Comments:

CDX Tasksheet Number: A4005

Student/Intern Information

Name _________________________ Date _________ Class _________________

Vehicle, Customer, and Service Information

Vehicle used for this activity:

Year _____________ Make _______________________ Model _________________

Odometer _______________________ VIN _________________________

Materials Required

- Blank work order
- Vehicle with available service history records
- Depending on the type of concern, special diagnostic/hand tools may be required. See your supervisor/instructor for instructions to identify what tools may be required.
- Service information database
- Personal protective equipment
- Shop rags

Task-Specific Safety Considerations

- When running any vehicles in the shop, make sure you use the shop's exhaust ventilation system to discharge all exhaust gas safely outside.
- Lifting equipment, such as vehicle jacks and stands, vehicle hoists, and engine hoists, are important tools that increase productivity and make the job easier. However, they can also cause severe injury or death if used improperly. Make sure you follow the manufacturer's operation procedures. Also, make sure you have your supervisor's/instructor's permission to use any particular type of lifting equipment.
- Comply with personal and environmental safety practices associated with clothing; eye protection; hand tools; power equipment; proper ventilation; and the handling, storage, and disposal of chemicals/materials in accordance with local, state, and federal safety and environmental regulations. If you need to start the vehicle, you should ensure that the parking brake is firmly applied; if necessary, use wheel chocks to prevent the vehicle from moving when the vehicle is started to verify the completion of these tasks.
- Extreme caution must be exercised when working around rotating components.

▶ TASK Inspect power steering fluid level and condition.

MLR
A4B2

Time off_____________

Time on_____________

Total time_____________

Student Instructions: Read through the entire procedure prior to starting. Prepare your work space and any tools or parts that may be needed to complete the task. When directed by your instructor, begin the procedure to complete the task and check the box as each step is finished. Track your time on this procedure for later comparison to the standard completion time (i.e., "flat rate" or customer pay time).

Procedure:	Step Completed
1. Listen for abnormal power steering pump noise.	
a. Describe what kind of noise a power steering pump may make if the fluid level is low.	☐
b. Start the vehicle. Listen to the power steering pump and the drive belt. Are there any abnormal noises that can be heard at this time? Shut the vehicle off.	☐
2. Inspect the power steering fluid level.	
a. Research the procedure for checking power steering fluid. Should the engine be off or running when performing this check?	☐
b. Where is the power steering dipstick located?	☐
c. Remove the power steering cap. Are there markings for "Cold" and "Hot" fluid level checks?	☐
d. Use a rag and remove the fluid from the dipstick. Re-insert the dipstick to take an accurate measurement. Describe the fluid level.	☐
3. Inspect the power steering fluid condition.	
a. Describe what color new power steering fluid for your vehicle should be.	☐

b. Take a sample of the fluid. Describe its condition. Is it dark and discolored? Does it have a burnt smell? Are there signs of metal in the fluid?	☐
c. Is the fluid aerated (foamy)? Explain what could cause this condition.	☐

Non-Task-Specific Evaluations:	Step Completed
1. Tools and equipment were used as directed and returned in good working order.	☐
2. Complied with all general and task-specific safety standards, including proper use of any personal protective equipment.	☐
3. Completed the task in an appropriate time frame (Recommendation: 1.5 or 2 times flat rate).	☐
4. Left the work space clean and orderly.	☐
5. Cared for customer property and returned it undamaged.	☐

Student signature _______________________________ Date _______________________________

Comments:

Have your supervisor/instructor verify satisfactory completion of this procedure, any observations found, and any necessary action(s) recommended.

Evaluation Instructions: The scoring box below is intended to act as a guide for both student and instructor. Each criterion listed will help students understand what is expected from them and evaluators articulate success at a particular task. The scoring is set up to allow a second attempt at each task (see the "Test" and "Retest" columns). Scoring is designed to reward students for correct completion of the task. Points are lost for failure to complete the employability requirements (see "Non-Task-Specific" criteria). When all criteria are evaluated, tally the points for a total at the bottom of each column.

Tasksheet Scoring

	Test		Retest	
Evaluation Items	**Pass**	**Fail**	**Pass**	**Fail**
Task-Specific Evaluation	**(1 pt)**	**(0 pts)**	**(1 pt)**	**(0 pts)**
Describe abnormal power steering pump noises.				
Research the procedure for checking power steering fluid.				
Locate the power steering fluid dipstick and inspect fluid level.				
Inspect and describe the fluid's condition.				
Non-Task-Specific Evaluation	**(0 pts)**	**(−1 pt)**	**(0 pts)**	**(−1 pt)**
Student successfully completed at least three of the non-task-specific steps.				
Student successfully completed all five of the non-task-specific steps.				
Total Score: <total # of points / 4 = %>				

Supervisor:

Supervisor/instructor signature _________________________ Date _________________

Comments:

Retest supervisor/instructor signature _________________________ Date _________________

Comments:

CDX Tasksheet Number: A4006

Student/Intern Information

Name _________________________________ Date _____________ Class _____________________

Vehicle, Customer, and Service Information

Vehicle used for this activity:

Year _______________ Make _________________________ Model _____________________________

Odometer _________________________________ VIN _____________________________________

Materials Required

- Blank work order
- Vehicle with available service history records
- Depending on the type of concern, special diagnostic/hand tools may be required. See your supervisor/instructor for instructions to identify what tools may be required.
- Service information database
- Personal protective equipment
- Power steering flush equipment
- Power steering fluid
- General hand tools
- Fluid catch pan

Task-Specific Safety Considerations

- Lifting equipment, such as vehicle jacks and stands, vehicle hoists, and engine hoists, are important tools that increase productivity and make the job easier. However, they can also cause severe injury or death if used improperly. Make sure you follow the manufacturer's operation procedures. Also, make sure you have your supervisor's/instructor's permission to use any particular type of lifting equipment.
- Comply with personal and environmental safety practices associated with clothing; eye protection; hand tools; power equipment; proper ventilation; and the handling, storage, and disposal of chemicals/materials in accordance with local, state, and federal safety and environmental regulations.
- When running any vehicles in the shop, make sure you use the shop's exhaust ventilation system to discharge all exhaust gas safely outside.

Time off_____________

Time on_____________

Total time_____________

▶ TASK Flush, fill, and bleed power steering system; use proper fluid type per manufacturer specification.

MLR
A4B3

Student Instructions: Read through the entire procedure prior to starting. Prepare your work space and any tools or parts that may be needed to complete the task. When directed by your instructor, begin the procedure to complete the task and check the box as each step is finished. Track your time on this procedure for later comparison to the standard completion time (i.e., "flat rate" or customer pay time).

Procedure:	Step Completed
1. Research the power steering flush procedure, fluid type, and fluid capacity.	
a. Describe the procedure for performing a power steering flush on your specific vehicle. List instructions and special precautions that may be found on the power steering flush equipment.	☐
b. What type of fluid is used in the power steering system?	☐
c. What is the full system capacity?	☐
2. Flush the power steering system.	
a. Following proper procedure, install the power steering flush machine. You may be directed to tap into a fitting or hose. Ensure that all connections are tight before starting the equipment.	☐
b. Explain what type of cleaner you are using in the flush machine. Are you using a cleaner or are you just flushing the system with new power steering fluid?	☐
c. Continue to flush the system until all the old fluid has been removed from the system and replaced with new fluid. Carefully remove the power steering flush equipment.	☐
d. Verify that all hoses and fittings are secure. Top off the power steering fluid level if needed. Verify that no air has been trapped in the system. Start the vehicle and perform a bleed procedure. Explain how you are bleeding the power steering system.	☐

Non-Task-Specific Evaluations:	Step Completed
1. Tools and equipment were used as directed and returned in good working order.	☐
2. Complied with all general and task-specific safety standards, including proper use of any personal protective equipment.	☐
3. Completed the task in an appropriate time frame (Recommendation: 1.5 or 2 times flat rate).	☐
4. Left the work space clean and orderly.	☐
5. Cared for customer property and returned it undamaged.	☐

Student signature _______________________________ Date _______________________________

Comments:

Have your supervisor/instructor verify satisfactory completion of this procedure, any observations found, and any necessary action(s) recommended.

Evaluation Instructions: The scoring box below is intended to act as a guide for both student and instructor. Each criterion listed will help students understand what is expected from them and evaluators articulate success at a particular task. The scoring is set up to allow a second attempt at each task (see the "Test" and "Retest" columns). Scoring is designed to reward students for correct completion of the task. Points are lost for failure to complete the employability requirements (see "Non-Task-Specific" criteria). When all criteria are evaluated, tally the points for a total at the bottom of each column.

Tasksheet Scoring

	Test		Retest	
Evaluation Items	**Pass**	**Fail**	**Pass**	**Fail**
Task-Specific Evaluation	**(1 pt)**	**(0 pts)**	**(1 pt)**	**(0 pts)**
Research the power steering flush procedure, fluid type, and capacity.				
Install power steering flush equipment and flush entire system.				
Disassemble the flush equipment and verify all hoses and fittings are secure and leak-free.				
List service precautions associated with using flush equipment and cleaners or detergents.				
Non-Task-Specific Evaluation	**(0 pts)**	**(−1 pt)**	**(0 pts)**	**(−1 pt)**
Student successfully completed at least three of the non-task-specific steps.				
Student successfully completed all five of the non-task-specific steps.				
Total Score: <total # of points / 4 = %>				

Supervisor:

Supervisor/instructor signature _________________________ Date _________________

Comments:

Retest supervisor/instructor signature _________________________ Date _________________

Comments:

CDX Tasksheet Number: A4007

Student/Intern Information

Name _________________________________ Date ____________ Class _____________________

Vehicle, Customer, and Service Information

Vehicle used for this activity:

Year ________________ Make _______________________________ Model ______________________________

Odometer _________________________________ VIN _______________________________________

Materials Required

- Blank work order
- Vehicle with available service history records
- Depending on the type of concern, special diagnostic/hand tools may be required. See your supervisor/instructor for instructions to identify what tools may be required.
- Service information database
- Personal protective equipment

Task-Specific Safety Considerations

- Lifting equipment, such as vehicle jacks and stands, vehicle hoists, and engine hoists, are important tools that increase productivity and make the job easier. However, they can also cause severe injury or death if used improperly. Make sure you follow the manufacturer's operation procedures. Also, make sure you have your supervisor's/instructor's permission to use any particular type of lifting equipment.
- Comply with personal and environmental safety practices associated with clothing; eye protection; hand tools; power equipment; proper ventilation; and the handling, storage, and disposal of chemicals/materials in accordance with local, state, and federal safety and environmental regulations.

▶ TASK Inspect for power steering fluid leakage.

MLR
A4B4

Time off___________

Time on___________

Total time___________

Student Instructions: Read through the entire procedure prior to starting. Prepare your work space and any tools or parts that may be needed to complete the task. When directed by your instructor, begin the procedure to complete the task and check the box as each step is finished. Track your time on this procedure for later comparison to the standard completion time (i.e., "flat rate" or customer pay time).

Procedure:	Step Completed
1. Inspect the power steering fluid level.	
a. Start your inspection by taking a reading of initial power steering fluid level. What is the level of fluid?	☐
b. Inspect the reservoir cap. Make sure the cap seal is in good condition and that the cap is securely tightened to the reservoir.	☐
2. Inspect the power steering hoses and fittings for leaks.	
a. Inspect the power steering hoses, beginning at the reservoir and following the hoses to the rack and pinion/steering gear box. List any action required.	☐
b. Inspect all hose clamps and/or crimped connections for leaks.	☐
c. Inspect the power steering cooler if the vehicle comes equipped with one.	☐
d. If leaks are found, describe the color of the fluid. Does it compare to the color of the fluid found in the reservoir? Explain how you would go about repairing any leaks that you may have found.	☐
4. Inspect the rack and pinion/steering gear box.	
a. Inspect the rack and pinion bellows (boots). These are normally clamped to the inner tie rods. Leaks can occur at the inner seals of the rack and pinion and allow fluid to collect inside of the bellows. If a gear box is used, inspect fittings and hoses that are attached to the box. Inspect steering shaft seals for signs of leaks.	☐

Non-Task-Specific Evaluations:	Step Completed
1. Tools and equipment were used as directed and returned in good working order.	☐
2. Complied with all general and task-specific safety standards, including proper use of any personal protective equipment.	☐
3. Completed the task in an appropriate time frame (Recommendation: 1.5 or 2 times flat rate).	☐
4. Left the work space clean and orderly.	☐
5. Cared for customer property and returned it undamaged.	☐

Student signature _______________________________ Date _______________________________

Comments:

Have your supervisor/instructor verify satisfactory completion of this procedure, any observations found, and any necessary action(s) recommended.

Evaluation Instructions: The scoring box below is intended to act as a guide for both student and instructor. Each criterion listed will help students understand what is expected from them and evaluators articulate success at a particular task. The scoring is set up to allow a second attempt at each task (see the "Test" and "Retest" columns). Scoring is designed to reward students for correct completion of the task. Points are lost for failure to complete the employability requirements (see "Non-Task-Specific" criteria). When all criteria are evaluated, tally the points for a total at the bottom of each column.

Tasksheet Scoring

	Test		Retest	
Evaluation Items	**Pass**	**Fail**	**Pass**	**Fail**
Task-Specific Evaluation	**(1 pt)**	**(0 pts)**	**(1 pt)**	**(0 pts)**
Inspect the power steering fluid level, reservoir, and cap condition.				
Inspect the fluid hoses, lines, fittings, and gear box.				
Inspect the power steering fluid cooler.				
Describe the fluid color at source of leak. Describe your recommended repair procedure.				
Non-Task-Specific Evaluation	**(0 pts)**	**(−1 pt)**	**(0 pts)**	**(−1 pt)**
Student successfully completed at least three of the non-task-specific steps.				
Student successfully completed all five of the non-task-specific steps.				
Total Score: <total # of points / 4 = %>				

Supervisor:

Supervisor/instructor signature _________________________ Date _______________

Comments:

Retest supervisor/instructor signature _________________________ Date _______________

Comments:

CDX Tasksheet Number: A4008

Student/Intern Information

Name _________________________________ Date ____________ Class _____________________

Vehicle, Customer, and Service Information

Vehicle used for this activity:

Year _______________ Make _____________________________ Model _____________________

Odometer _________________________________ VIN _____________________________________

Materials Required

- Blank work order
- Vehicle with available service history records
- Depending on the type of concern, special diagnostic tools may be required. See your supervisor/instructor for instructions to identify what tools may be required.
- Service information
- Serpentine belt removal tool set
- Belt tension gauge

Task-Specific Safety Considerations

- When running any vehicles in the shop, make sure you use the shop's exhaust ventilation system to discharge all exhaust gas safely outside.
- Lifting equipment, such as vehicle jacks and stands, vehicle hoists, and engine hoists, are important tools that increase productivity and make the job easier. However, they can also cause severe injury or death if used improperly. Make sure you follow the manufacturer's operation procedures. Also, make sure you have your supervisor's/instructor's permission to use any particular type of lifting equipment.
- Comply with personal and environmental safety practices associated with clothing; eye protection; hand tools; power equipment; proper ventilation; and the handling, storage, and disposal of chemicals/materials in accordance with local, state, and federal safety and environmental regulations.
- If you need to start the vehicle, you should ensure that the parking brake is firmly applied; if necessary, use wheel chocks to prevent the vehicle from moving when the vehicle is started to verify the completion of these tasks.
- Extreme caution must be exercised when working around rotating components.

Time off_______________

Time on_______________

Total time_______________

Student Instructions: Read through the entire procedure prior to starting. Prepare your work space and any tools or parts that may be needed to complete the task. When directed by your instructor, begin the procedure to complete the task and check the box as each step is finished. Track your time on this procedure for later comparison to the standard completion time (i.e., "flat rate" or customer pay time).

Procedure:	Step Completed
1. Locate the belt routing diagram.	
a. Draw or print off and attach to this sheet the belt routing diagram for your specific vehicle.	☐
2. Identify the belt and tensioner type.	
a. Explain what type of belt is used to drive the power steering pump. Is a single belt used to drive all accessories?	☐
b. What type of tensioner is being used for the belt?	☐
3. Using the proper procedure, use the serpentine belt removal tool/hand tools and remove the belt.	
a. Inspect the belt for cracks or fraying. Based on your observation, what is your recommendation?	☐
b. Inspect all applicable engine accessory drive pulleys for binding or looseness. List any further action.	☐
4. Replace the belt.	
a. If an automatic tensioner is being used, install the belt. Check that the belt is installed completely.	☐
b. If a manual tensioner is being used, install the belt to the recommended tension.	☐

c. Use a tension gauge to verify correct installation. List tension specification used for installation.	☐
5. Identify concerns related to improper belt tension.	
a. How would the power steering pump and system be affected if belt tension was too loose?	☐
b. How would the power steering pump and system be affected if belt tension was too tight?	☐

Non-Task-Specific Evaluations:	Step Completed
1. Tools and equipment were used as directed and returned in good working order.	☐
2. Complied with all general and task-specific safety standards, including proper use of any personal protective equipment.	☐
3. Completed the task in an appropriate time frame (Recommendation: 1.5 or 2 times flat rate).	☐
4. Left the work space clean and orderly.	☐
5. Cared for customer property and returned it undamaged.	☐

Student signature _______________________________ Date _______________________________

Comments:

Have your supervisor/instructor verify satisfactory completion of this procedure, any observations found, and any necessary action(s) recommended.

Evaluation Instructions: The scoring box below is intended to act as a guide for both student and instructor. Each criterion listed will help students understand what is expected from them and evaluators articulate success at a particular task. The scoring is set up to allow a second attempt at each task (see the "Test" and "Retest" columns). Scoring is designed to reward students for correct completion of the task. Points are lost for failure to complete the employability requirements (see "Non-Task-Specific" criteria). When all criteria are evaluated, tally the points for a total at the bottom of each column.

Tasksheet Scoring

	Test		Retest	
Evaluation Items	**Pass**	**Fail**	**Pass**	**Fail**
Task-Specific Evaluation	**(1 pt)**	**(0 pts)**	**(1 pt)**	**(0 pts)**
Locate the belt routing diagram and identify the belt and tensioner type.				
Remove and inspect the belt.				
Replace the belt and adjust tension.				
Identify any issues related to improper belt tension.				
Non-Task-Specific Evaluation	**(0 pts)**	**(−1 pt)**	**(0 pts)**	**(−1 pt)**
Student successfully completed at least three of the non-task-specific steps.				
Student successfully completed all five of the non-task-specific steps.				
Total Score: <total # of points / 4 = %>				

Supervisor:

Supervisor/instructor signature _________________________________ Date _____________________

Comments:

Retest supervisor/instructor signature _________________________ Date _____________________

Comments:

CDX Tasksheet Number: A4009

Student/Intern Information

Name _________________________________ Date _____________ Class _____________________

Vehicle, Customer, and Service Information

Vehicle used for this activity:

Year _______________ Make _____________________________ Model _____________________

Odometer _____________________________ VIN _____________________________

> ### Materials Required
> - Blank work order
> - Vehicle with available service history records
> - Depending on the type of concern, special diagnostic/hand tools may be required. See your supervisor/instructor for instructions to identify what tools may be required.
> - Service information database
> - Personal protective equipment
> - General hand tools
> - Fluid catch pan
> - Power steering fluid

Task-Specific Safety Considerations

- When running any vehicles in the shop, make sure you use the shop's exhaust ventilation system to discharge all exhaust gas safely outside.
- Lifting equipment, such as vehicle jacks and stands, vehicle hoists, and engine hoists, are important tools that increase productivity and make the job easier. However, they can also cause severe injury or death if used improperly. Make sure you follow the manufacturer's operation procedures. Also, make sure you have your supervisor's/instructor's permission to use any particular type of lifting equipment.
- Comply with personal and environmental safety practices associated with clothing; eye protection; hand tools; power equipment; proper ventilation; and the handling, storage, and disposal of chemicals/materials in accordance with local, state, and federal safety and environmental regulations.

Time off_______________

Time on_______________

Total time_______________

▶ TASK Inspect and replace power steering hoses and fittings.

MLR
A4B6

Student Instructions: Read through the entire procedure prior to starting. Prepare your work space and any tools or parts that may be needed to complete the task. When directed by your instructor, begin the procedure to complete the task and check the box as each step is finished. Track your time on this procedure for later comparison to the standard completion time (i.e., "flat rate" or customer pay time).

Procedure:	Step Completed
1. Inspect the power steering hoses for leaks.	
a. Locate the hoses attached to the power steering pump. Inspect for leaks at the hose clamps, crimped line connections, or threaded fittings. Follow the power steering lines to the rack and pinion or steering gear box and continue inspection.	☐
b. List any action required.	☐
2. Remove the power steering hoses.	
a. Research the procedure for removing the power steering hoses. Then, research the fluid type and amount. Are there any special tools or precautions needed for the task?	☐
b. Following proper procedure, remove the power steering lines. Use a fluid catch pan to collect power steering fluid as it drips from the hoses. If possible, remove fluid from the power steering reservoir to eliminate extra fluid from spilling out.	☐
c. Determine which line is a pressure line and which one is a return line. How can you tell the difference between the two lines?	☐
d. Inspect the hoses internally. Sometimes, hoses can become clogged or collapse internally and show no signs of damage externally. Using light pressure, use a shop air gun to blow air through the lines. **Caution:** compressed air can be dangerous. Lines will have residual fluid, so place them in a fluid catch pan. Do not point the lines at anyone. Always wear your safety glasses. List any action required.	☐
3. Install the power steering hoses.	
a. Following proper procedure, install the power steering hoses. List any torque specifications that you used or that may have been provided by your service information.	☐

b. Once the hoses are installed, select the correct power steering fluid and fill the reservoir. What type of fluid is used in your vehicle? How much fluid is required to fill the system?	☐
c. Once the reservoir is filled, start the vehicle. Air will be trapped in the system. What kind of noise indicates air in a power steering system? Following procedure, bleed air from the power steering by turning the steering wheel back and forth, lock to lock, several times. It may take some time to bleed all air out of the system. **Tip:** After bleeding the system initially, turn the vehicle off and allow the fluid to settle in the reservoir. You may notice many air bubbles at this time. Once the fluid has settled for 10–12 minutes, bleed the system once more. The system is fully bled when there is no more noise present when turning the wheel.	☐
d. Inspect all hoses and fittings once more to ensure that there are no leaks or kinks in the hoses.	☐

Non-Task-Specific Evaluations:	Step Completed
1. Tools and equipment were used as directed and returned in good working order.	☐
2. Complied with all general and task-specific safety standards, including proper use of any personal protective equipment.	☐
3. Completed the task in an appropriate time frame (Recommendation: 1.5 or 2 times flat rate).	☐
4. Left the work space clean and orderly.	☐
5. Cared for customer property and returned it undamaged.	☐

Student signature _________________________________ Date _________________________________

Comments:

Have your supervisor/instructor verify satisfactory completion of this procedure, any observations found, and any necessary action(s) recommended.

Evaluation Instructions: The scoring box below is intended to act as a guide for both student and instructor. Each criterion listed will help students understand what is expected from them and evaluators articulate success at a particular task. The scoring is set up to allow a second attempt at each task (see the "Test" and "Retest" columns). Scoring is designed to reward students for correct completion of the task. Points are lost for failure to complete the employability requirements (see "Non-Task-Specific" criteria). When all criteria are evaluated, tally the points for a total at the bottom of each column.

Tasksheet Scoring

Evaluation Items	Test		Retest	
	Pass	**Fail**	**Pass**	**Fail**
Task-Specific Evaluation	**(1 pt)**	**(0 pts)**	**(1 pt)**	**(0 pts)**
Research the power steering hose replacement procedure, fluid type, and capacity.				
Inspect the hoses and fittings for signs of leaks.				
Remove and replace the power steering hoses.				
Select the correct fluid type and bleed the power steering system.				
Non-Task-Specific Evaluation	**(0 pts)**	**(1 pt)**	**(0 pts)**	**(−1 pt)**
Student successfully completed at least three of the non-task-specific steps.				
Student successfully completed all five of the non-task-specific steps.				
Total Score: <total # of points / 4 = %>				

Supervisor:

Supervisor/instructor signature _______________________________ Date _______________________

Comments:

Retest supervisor/instructor signature _______________________ Date _______________________

Comments:

CDX Tasksheet Number: A4010

Student/Intern Information

Name ___________________________ Date ___________ Class ___________________

Vehicle, Customer, and Service Information

Vehicle used for this activity:

Year _____________ Make _________________________ Model ___________________

Odometer _____________________________ VIN _________________________________

Task-Specific Safety Considerations

- Lifting equipment, such as vehicle jacks and stands, vehicle hoists, and engine hoists, are important tools that increase productivity and make the job easier. However, they can also cause severe injury or death if used improperly. Make sure you follow the manufacturer's operation procedures. Also, make sure you have your supervisor's/instructor's permission to use any particular type of lifting equipment.
- Comply with personal and environmental safety practices associated with clothing; eye protection; hand tools; power equipment; proper ventilation; and the handling, storage, and disposal of chemicals/materials in accordance with local, state, and federal safety and environmental regulations.

Time off_____________

Time on_____________

Total time_____________

▶ **TASK** Inspect pitman arm, relay (centerlink/intermediate) rod, idler arm, mountings, and steering linkage damper.

MLR
A4B7

Student Instructions: Read through the entire procedure prior to starting. Prepare your work space and any tools or parts that may be needed to complete the task. When directed by your instructor, begin the procedure to complete the task and check the box as each step is finished. Track your time on this procedure for later comparison to the standard completion time (i.e., "flat rate" or customer pay time).

Procedure:	Step Completed
1. Inspect the steering wheel for correct operation.	
a. With the vehicle on the ground, start the vehicle and turn the steering wheel from lock to lock. Is there any abnormal noise, binding, or looseness felt in the steering wheel? Tip: If a test drive is required to verify correct steering operation, follow the EXACT procedure from your instructor.	☐
2. Inspect the pitman arm, relay rod, idler arm, and mounts for correct operation.	
a. With the vehicle on the ground, have a helper gently rock the steering wheel back and forth. As this is happening, inspect the steering shaft, pitman arm, relay rod, and idler arm for looseness. It is normal to have a small amount of freeplay. Inspect for excessive looseness. Describe your results.	☐
b. Lift the vehicle. Perform a thorough inspection of all steering components. Rock each component back and forth by hand. Inspect for damage, binding, or looseness. Describe your results.	☐
c. Do the individual steering components have grease fittings? Does it appear that each joint has the proper amount of grease?	☐
3. Inspect the steering linkage damper.	
a. Describe what the steering may feel like to the customer if the steering linkage damper was worn or damaged.	☐
b. Inspect the steering linkage damper for damage or fluid leaks. Inspect the mount bushings for wear, excessive cracking, or dislodged bushings.	☐
4. Based on your inspection, what are your recommendations?	

Non-Task-Specific Evaluations:	Step Completed
1. Tools and equipment were used as directed and returned in good working order.	☐
2. Complied with all general and task-specific safety standards, including proper use of any personal protective equipment.	☐
3. Completed the task in an appropriate time frame (Recommendation: 1.5 or 2 times flat rate).	☐
4. Left the work space clean and orderly.	☐
5. Cared for customer property and returned it undamaged.	☐

Student signature _________________________________ Date _______________________________

Comments:

Have your supervisor/instructor verify satisfactory completion of this procedure, any observations found, and any necessary action(s) recommended.

Evaluation Instructions: The scoring box below is intended to act as a guide for both student and instructor. Each criterion listed will help students understand what is expected from them and evaluators articulate success at a particular task. The scoring is set up to allow a second attempt at each task (see the "Test" and "Retest" columns). Scoring is designed to reward students for correct completion of the task. Points are lost for failure to complete the employability requirements (see "Non-Task-Specific" criteria). When all criteria are evaluated, tally the points for a total at the bottom of each column.

Tasksheet Scoring

	Test		Retest	
Evaluation Items	**Pass**	**Fail**	**Pass**	**Fail**
Task-Specific Evaluation	**(1 pt)**	**(0 pts)**	**(1 pt)**	**(0 pts)**
Verify correct steering operation prior to inspection.				
Inspect the steering wheel for excessive freeplay.				
Inspect the pitman arm, relay rod, idler arm, and steering damper.				
Make recommendations for service based on your inspection.				
Non-Task-Specific Evaluation	**(0 pts)**	**(−1 pt)**	**(0 pts)**	**(−1 pt)**
Student successfully completed at least three of the non-task-specific steps.				
Student successfully completed all five of the non-task-specific steps.				
Total Score: <total # of points / 4 = %>				

Supervisor:

Supervisor/instructor signature ___________________________ Date ___________________

Comments:

Retest supervisor/instructor signature ___________________________ Date ___________________

Comments:

CDX Tasksheet Number: A4011

Student/Intern Information

Name _________________________ Date _________ Class _________________

Vehicle, Customer, and Service Information

Vehicle used for this activity:

Year _____________ Make _________________ Model _________________

Odometer _____________________ VIN _________________________

Materials Required

- Blank work order
- Vehicle with available service history records
- Depending on the type of concern, special diagnostic/hand tools may be required. See your supervisor/instructor for instructions to identify what tools may be required.
- Service information database
- Personal protective equipment

Task-Specific Safety Considerations

- Lifting equipment such as vehicle jacks and stands, vehicle hoists, and engine hoists are important tools that increase productivity and make the job easier. However, they can also cause severe injury or death if used improperly. Make sure you follow the manufacturer's operation procedures. Also, make sure you have your supervisor's/instructor's permission to use any particular type of lifting equipment.
- Comply with personal and environmental safety practices associated with clothing; eye protection; hand tools; power equipment; proper ventilation; and the handling, storage, and disposal of chemicals/materials in accordance with local, state, and federal safety and environmental regulations.

Time off_____________

Time on_____________

Total time_____________

▶ TASK Inspect tie rod ends (sockets), tie rod sleeves, and clamps.

MLR
A4B8

Student Instructions: Read through the entire procedure prior to starting. Prepare your work space and any tools or parts that may be needed to complete the task. When directed by your instructor, begin the procedure to complete the task and check the box as each step is finished. Track your time on this procedure for later comparison to the standard completion time (i.e., "flat rate" or customer pay time).

Procedure:	Step Completed
1. Verify correct steering wheel operation.	
a. With the vehicle on the ground, start the vehicle and turn the steering wheel from lock to lock. Is there any abnormal noises, binding, or looseness felt in the steering wheel? **Tip:** If a test drive is required to verify correct steering operation, follow EXACT procedure from your instructor.	☐
2. Inspect the tie rod ends, adjusting sleeves, and clamps.	
a. Lift the vehicle. Rock each front wheel from side to side. Your hands should be placed at the 3 and 9 o'clock position when performing this check. The wheel should have no free play. If needed, have a helper rock the wheels while you inspect the tie rod ends for looseness. Describe your results.	☐
b. By hand, rock the tie rod ends up and down, and back and forth. There should be no play in the tie rod end.	☐
c. Inspect the rubber boots on the tie rod end for cracks, tears, or leaking grease.	☐
d. Verify that the tie rod adjusting sleeves, clamps, or lock nuts are tight.	☐
e. Describe how the vehicle may steer if the tie rod ends are worn, loose, or damaged.	☐
3. Based on your inspection, what is your recommendation for service?	☐

Non-Task-Specific Evaluations:	Step Completed
1. Tools and equipment were used as directed and returned in good working order.	☐
2. Complied with all general and task-specific safety standards, including proper use of any personal protective equipment.	☐
3. Completed the task in an appropriate time frame (Recommendation: 1.5 or 2 times flat rate).	☐
4. Left the work space clean and orderly.	☐
5. Cared for customer property and returned it undamaged.	☐

Student signature ___________________________ Date ___________________________

Comments:

Have your supervisor/instructor verify satisfactory completion of this procedure, any observations found, and any necessary action(s) recommended.

Evaluation Instructions: The scoring box below is intended to act as a guide for both student and instructor. Each criterion listed will help students understand what is expected from them and evaluators articulate success at a particular task. The scoring is set up to allow a second attempt at each task (see the "Test" and "Retest" columns). Scoring is designed to reward students for correct completion of the task. Points are lost for failure to complete the employability requirements (see "Non-Task-Specific" criteria). When all criteria are evaluated, tally the points for a total at the bottom of each column.

Tasksheet Scoring

	Test		Retest	
Evaluation Items	**Pass**	**Fail**	**Pass**	**Fail**
Task-Specific Evaluation	**(1 pt)**	**(0 pts)**	**(1 pt)**	**(0 pts)**
Verify correct steering wheel operation.				
Inspect the tie rod end socket for looseness. Inspect the tie rod adjusting sleeves, clamps, and/or locknuts.				
Describe the symptoms a customer may have if tie rod ends are worn, loose, or damaged.				
Make recommendations for service based on your inspection.				
Non-Task-Specific Evaluation	**(0 pts)**	**(−1 pt)**	**(0 pts)**	**(−1 pt)**
Student successfully completed at least three of the non-task-specific steps.				
Student successfully completed all five of the non-task-specific steps.				
Total Score: <total # of points / 4 = %>				

Supervisor:

Supervisor/instructor signature _________________________________ Date _____________________

Comments:

Retest supervisor/instructor signature _________________________ Date _____________________

Comments:

CDX Tasksheet Number: A4012

Student/Intern Information

Name _________________________________ Date ___________ Class _________________________

Vehicle, Customer, and Service Information

Vehicle used for this activity:

Year _______________ Make _____________________________ Model _________________________

Odometer _________________________________ VIN _________________________________

Materials Required
- Blank work order
- Vehicle with available service history records
- Depending on the type of concern, special diagnostic/hand tools may be required. See your supervisor/instructor for instructions to identify what tools may be required.
- Service information database
- Personal protective equipment
- General hand tools
- Pry bar

Task-Specific Safety Considerations
- Lifting equipment, such as vehicle jacks and stands, vehicle hoists, and engine hoists, are important tools that increase productivity and make the job easier. However, they can also cause severe injury or death if used improperly. Make sure you follow the manufacturer's operation procedures. Also, make sure you have your supervisor's/instructor's permission to use any particular type of lifting equipment.
- Comply with personal and environmental safety practices associated with clothing; eye protection; hand tools; power equipment; proper ventilation; and the handling, storage, and disposal of chemicals/materials in accordance with local, state, and federal safety and environmental regulations.

▶ **TASK** Inspect upper and lower control arms, bushings, and shafts.

MLR
A4B9

Time off________________

Time on________________

Total time________________

Student Instructions: Read through the entire procedure prior to starting. Prepare your work space and any tools or parts that may be needed to complete the task. When directed by your instructor, begin the procedure to complete the task and check the box as each step is finished. Track your time on this procedure for later comparison to the standard completion time (i.e., "flat rate" or customer pay time).

Procedure:	Step Completed
1. Research the procedure for inspecting control arms, bushings, and shafts.	
a. Explain how you would inspect the control arms and what you are looking for.	☐
2. Inspect the control arms and bushings for damage, wear, and excessive rust or corrosion.	
a. Lift the vehicle and remove the wheels.	☐
b. Explain what type of suspension configuration your vehicle has. Are there one or two control arms per wheel? Is a spring, shock, or strut being used?	☐
c. Inspect the control arms for signs of damage, rust, or corrosion. List any action required.	☐
d. Inspect the control arm bushings. Look for signs of cracking, tears, or dislodged bushings. Use a pry bar to lightly pry the control arm up and down, back and forth. Is there any excessive movement? Are there signs of metal-to-metal contact between the control arm and where it mounts to the frame?	☐
e. Install the wheels back on the vehicle and torque to specifications. Lower the vehicle.	☐
3. Based on your inspection, what is your recommendation?	

Non-Task-Specific Evaluations:	Step Completed
1. Tools and equipment were used as directed and returned in good working order.	☐
2. Complied with all general and task-specific safety standards, including proper use of any personal protective equipment.	☐
3. Completed the task in an appropriate time frame (Recommendation: 1.5 or 2 times flat rate).	☐
4. Left the work space clean and orderly.	☐
5. Cared for customer property and returned it undamaged.	☐

Student signature _______________________________ Date _______________________________

Comments:

Have your supervisor/instructor verify satisfactory completion of this procedure, any observations found, and any necessary action(s) recommended.

Evaluation Instructions: The scoring box below is intended to act as a guide for both student and instructor. Each criterion listed will help students understand what is expected from them and evaluators articulate success at a particular task. The scoring is set up to allow a second attempt at each task (see the "Test" and "Retest" columns). Scoring is designed to reward students for correct completion of the task. Points are lost for failure to complete the employability requirements (see "Non-Task-Specific" criteria). When all criteria are evaluated, tally the points for a total at the bottom of each column.

Tasksheet Scoring

	Test		Retest	
Evaluation Items	**Pass**	**Fail**	**Pass**	**Fail**
Task-Specific Evaluation	**(1 pt)**	**(0 pts)**	**(1 pt)**	**(0 pts)**
Explain how you would inspect the control arms and what you are looking for. Explain what type of suspension configuration your vehicle has.				
Inspect the control arms for signs of damage, rust, or corrosion.				
Inspect the control arm bushings.				
Make recommendations for service based on your inspection.				
Non-Task-Specific Evaluation	**(0 pts)**	**(−1 pt)**	**(0 pts)**	**(−1 pt)**
Student successfully completed at least three of the non-task-specific steps.				
Student successfully completed all five of the non-task-specific steps.				
Total Score: <total # of points / 4 = %>				

Supervisor:

Supervisor/instructor signature ___________________________ Date ___________________

Comments:

Retest supervisor/instructor signature ___________________________ Date ___________________

Comments:

CDX Tasksheet Number: A4013

Student/Intern Information

Name ___________________________ Date __________ Class ___________________

Vehicle, Customer, and Service Information

Vehicle used for this activity:

Year ______________ Make ____________________________ Model ___________________

Odometer ______________________________ VIN ________________________________

Materials Required

- Blank work order
- Vehicle with available service history records
- Depending on the type of concern, special diagnostic/hand tools may be required. See your supervisor/instructor for instructions to identify what tools may be required.
- Service information database
- Personal protective equipment

Task-Specific Safety Considerations

- Lifting equipment, such as vehicle jacks and stands, vehicle hoists, and engine hoists, are important tools that increase productivity and make the job easier. However, they can also cause severe injury or death if used improperly. Make sure you follow the manufacturer's operation procedures. Also, make sure you have your supervisor's/instructor's permission to use any particular type of lifting equipment.
- Comply with personal and environmental safety practices associated with clothing; eye protection; hand tools; power equipment; proper ventilation; and the handling, storage, and disposal of chemicals/materials in accordance with local, state, and federal safety and environmental regulations.

▶ **TASK** Inspect and replace rebound bumpers.

MLR
A4B10

Time off___________

Time on___________

Total time___________

Student Instructions: Read through the entire procedure prior to starting. Prepare your work space and any tools or parts that may be needed to complete the task. When directed by your instructor, begin the procedure to complete the task and check the box as each step is finished. Track your time on this procedure for later comparison to the standard completion time (i.e., "flat rate" or customer pay time).

Procedure:	Step Completed
1. Explain the purpose of rebound bumpers on a vehicle.	
a. What does a rebound bumper do?	☐
b. Where are rebound bumpers located on a vehicle?	☐
c. Explain what may happen if a rebound bumper is missing.	☐
2. Inspect rebound bumpers on your vehicle.	
a. Lift the vehicle and remove the wheels if necessary.	☐
b. Locate all suspension rebound bumpers. Explain the location of the rebound bumpers. **Tip:** Vehicles that use Macpherson strut assemblies may have a rebound bumper located underneath the rubber boot that protects the strut piston shaft. You should be able to lift the rubber boot to inspect the condition of the rebound bumper.	☐
c. Based on your inspection, what is your recommendation for service?	☐
3. Remove and replace the rebound bumpers.	
a. Following proper procedure, remove the rebound bumpers on your vehicle. Special care must be taken to eliminate the chance of breaking fasteners that hold the bumpers in place, as they are prone to rust and corrosion. **Note:** Consult your teacher for instructions in regards to replacing bumpers that are part of a strut assembly.	☐
b. Replace the old rebound bumpers with new bumpers. Torque fasteners to specification.	☐
c. Install the wheels and torque lug nuts to specification. Lower the vehicle. Jounce the vehicle and verify that no noise is present.	☐

Non-Task-Specific Evaluations:	Step Completed
1. Tools and equipment were used as directed and returned in good working order.	☐
2. Complied with all general and task-specific safety standards, including proper use of any personal protective equipment.	☐
3. Completed the task in an appropriate time frame (Recommendation: 1.5 or 2 times flat rate).	☐
4. Left the work space clean and orderly.	☐
5. Cared for customer property and returned it undamaged.	☐

Student signature _______________________________ Date _______________________________

Comments:

Have your supervisor/instructor verify satisfactory completion of this procedure, any observations found, and any necessary action(s) recommended.

Evaluation Instructions: The scoring box below is intended to act as a guide for both student and instructor. Each criterion listed will help students understand what is expected from them and evaluators articulate success at a particular task. The scoring is set up to allow a second attempt at each task (see the "Test" and "Retest" columns). Scoring is designed to reward students for correct completion of the task. Points are lost for failure to complete the employability requirements (see "Non-Task-Specific" criteria). When all criteria are evaluated, tally the points for a total at the bottom of each column.

Tasksheet Scoring

	Test		Retest	
Evaluation Items	**Pass**	**Fail**	**Pass**	**Fail**
Task-Specific Evaluation	**(1 pt)**	**(0 pts)**	**(1 pt)**	**(0 pts)**
Explain the purpose of rebound bumpers.				
Explain faults that may occur if rebound bumpers are damaged or missing.				
Inspect the rebound bumper location and condition.				
Remove and replace rebound bumpers.				
Non-Task-Specific Evaluation	**(0 pts)**	**(−1 pt)**	**(0 pts)**	**(−1 pt)**
Student successfully completed at least three of the non-task-specific steps.				
Student successfully completed all five of the non-task-specific steps.				
Total Score: <total # of points / 4 = %>				

Supervisor:

Supervisor/instructor signature ________________________________ Date ________________

Comments:

Retest supervisor/instructor signature ________________________ Date ________________

Comments:

CDX Tasksheet Number: A4014

Student/Intern Information

Name _________________________ Date __________ Class _________________________

Vehicle, Customer, and Service Information

Vehicle used for this activity:

Year ______________ Make __________________________ Model _________________________

Odometer __________________________ VIN _________________________

Task-Specific Safety Considerations

- Lifting equipment, such as vehicle jacks and stands, vehicle hoists, and engine hoists, are important tools that increase productivity and make the job easier. However, they can also cause severe injury or death if used improperly. Make sure you follow the manufacturer's operation procedures. Also, make sure you have your supervisor's/instructor's permission to use any particular type of lifting equipment.
- Comply with personal and environmental safety practices associated with clothing; eye protection; hand tools; power equipment; proper ventilation; and the handling, storage, and disposal of chemicals/materials in accordance with local, state, and federal safety and environmental regulations.

Time off_____________

Time on_____________

Total time_____________

▶ TASK Inspect track bar, strut rods/radius arms, and related mounts and bushings.

MLR
A4B11

Student Instructions: Read through the entire procedure prior to starting. Prepare your work space and any tools or parts that may be needed to complete the task. When directed by your instructor, begin the procedure to complete the task and check the box as each step is finished. Track your time on this procedure for later comparison to the standard completion time (i.e., "flat rate" or customer pay time).

Procedure:	Step Completed
1. Describe the purposes and functions of the track bar, strut rod, and radius arm.	
a. Explain the purpose of a strut rod and radius arm. Where are these components located on a vehicle?	☐
b. Explain the purpose of a track bar. Where is this component located on a vehicle?	☐
c. Explain what symptoms a vehicle may have if these components were worn or damaged.	☐
2. Inspect the track bar, strut rods, radius arms, and related mounts and bushings.	
a. Lift the vehicle. Locate the strut rods and radius arms. Using a pry bar or your hand, lightly pry the arms back and forth to check for looseness.	☐
b. Inspect the mount bushings for signs of dry rot, cracking, or dislodged bushings. Inspect the mount areas for signs of metal-on-metal contact between the strut rod and the frame.	☐
c. Locate the track bar. Perform a similar inspection by prying the track bar back and forth, inspecting for looseness.	☐
d. Inspect the track bar mounts for cracks, tears, or dislodged bushings.	☐
3. Based on your inspection, what are your recommendations for service?	

Non-Task-Specific Evaluations:	Step Completed
1. Tools and equipment were used as directed and returned in good working order.	☐
2. Complied with all general and task-specific safety standards, including proper use of any personal protective equipment.	☐
3. Completed the task in an appropriate time frame (Recommendation: 1.5 or 2 times flat rate).	☐
4. Left the work space clean and orderly.	☐
5. Cared for customer property and returned it undamaged.	☐

Student signature _______________________________ Date _______________________________

Comments:

Have your supervisor/instructor verify satisfactory completion of this procedure, any observations found, and any necessary action(s) recommended.

Tasksheet Scoring

Evaluation Items	Test		Retest	
	Pass	**Fail**	**Pass**	**Fail**
Task-Specific Evaluation	**(1 pt)**	**(0 pts)**	**(1 pt)**	**(0 pts)**
Explain the purpose and function of a strut rod and radius arm and a track bar.				
Inspect the strut rod, radius arm, and mount bushings.				
Inspect the track bar and mount bushings.				
Make recommendations for service based on your inspection.				
Non-Task-Specific Evaluation	**(0 pts)**	**(−1 pt)**	**(0 pts)**	**(−1 pt)**
Student successfully completed at least three of the non-task-specific steps.				
Student successfully completed all five of the non-task-specific steps.				
Total Score: <total # of points / 4 = %>				

Supervisor:

Supervisor/instructor signature _____________________________ Date _____________________

Comments:

Retest supervisor/instructor signature _____________________________ Date _____________________

Comments:

CDX Tasksheet Number: A4015

Student/Intern Information

Name _______________________________ Date ____________ Class _______________________

Vehicle, Customer, and Service Information

Vehicle used for this activity:

Year ______________ Make ___________________________ Model ______________________

Odometer ___________________________ VIN _________________________________

Materials Required

- Blank work order
- Vehicle with available service history records
- Depending on the type of concern, special diagnostic/hand tools may be required. See your supervisor/instructor for instructions to identify what tools may be required.
- Service information database
- Personal protective equipment
- Drive-on lift or tall Jack stand
- Pry bar
- Dial indicator

Task-Specific Safety Considerations

- Lifting equipment, such as vehicle jacks and stands, vehicle hoists, and engine hoists, are important tools that increase productivity and make the job easier. However, they can also cause severe injury or death if used improperly. Make sure you follow the manufacturer's operation procedures. Also, make sure you have your supervisor's/instructor's permission to use any particular type of lifting equipment.
- Comply with personal and environmental safety practices associated with clothing; eye protection; hand tools; power equipment; proper ventilation; and the handling, storage, and disposal of chemicals/materials in accordance with local, state, and federal safety and environmental regulations.

▶ TASK Inspect upper and lower ball joints (with or without wear indicators).

MLR
A4B12

Time off________

Time on________

Total time________

Student Instructions: Read through the entire procedure prior to starting. Prepare your work space and any tools or parts that may be needed to complete the task. When directed by your instructor, begin the procedure to complete the task and check the box as each step is finished. Track your time on this procedure for later comparison to the standard completion time (i.e., "flat rate" or customer pay time).

Procedure:	Step Completed
1. Research the ball joint inspection procedure.	
a. Explain how you should check the ball joints in your vehicle.	☐
b. Explain the difference between a load-carrying ball joint and a follower ball joint, and how you would inspect each type.	☐
c. Determine if the ball joints on your vehicle are equipped with wear indicators. Explain how you would know if a ball joint needs replaced if it is equipped with a wear indicator.	☐
d. What is the specification for ball joint movement?	☐
e. How many ball joints are located on each side of the front of your vehicle?	☐
2. Inspect the ball joints.	
a. Lift the vehicle. Rock the wheel up and down, back and forth. Is there any obvious looseness that can be felt? If needed, have a helper perform this procedure while you inspect the ball joints for looseness.	☐
b. Position a tall jack stand in the desired spot to check the ball joints.	☐
c. Mount a dial indicator to a fixed position to measure the amount of movement in the ball joint that you are checking. Using a pry bar, gently pry the control arms up and down and record the measurement on the dial indicator.	☐

d. If you are working with a ball joint that is equipped with a wear indicator, describe its condition.	☐
3. Based on your inspection, what is your recommendation for service?	☐

Non-Task-Specific Evaluations:	Step Completed
1. Tools and equipment were used as directed and returned in good working order.	☐
2. Complied with all general and task-specific safety standards, including proper use of any personal protective equipment.	☐
3. Completed the task in an appropriate time frame (Recommendation: 1.5 or 2 times flat rate).	☐
4. Left the work space clean and orderly.	☐
5. Cared for customer property and returned it undamaged.	☐

Student signature _______________________ Date _______________________

Comments:

Have your supervisor/instructor verify satisfactory completion of this procedure, any observations found, and any necessary action(s) recommended.

Evaluation Instructions: The scoring box below is intended to act as a guide for both student and instructor. Each criterion listed will help students understand what is expected from them and evaluators articulate success at a particular task. The scoring is set up to allow a second attempt at each task (see the "Test" and "Retest" columns). Scoring is designed to reward students for correct completion of the task. Points are lost for failure to complete the employability requirements (see "Non-Task-Specific" criteria). When all criteria are evaluated, tally the points for a total at the bottom of each column.

Taksheet Scoring

	Test		Retest	
Evaluation Items	**Pass**	**Fail**	**Pass**	**Fail**
Task-Specific Evaluation	**(1 pt)**	**(0 pts)**	**(1 pt)**	**(0 pts)**
Research the procedure for checking ball joints and describe what type and how many ball joints your vehicle is equipped with.				
Explain the difference between a load-carrying and a follower ball joint.				
Determine the specification for ball joint movement and how to inspect ball joints equipped with wear indicators. Inspect ball joints for wear.				
Make recommendations for service based on your inspection.				
Non-Task-Specific Evaluation	**(0 pts)**	**(−1 pt)**	**(0 pts)**	**(−1 pt)**
Student successfully completed at least three of the non-task-specific steps.				
Student successfully completed all five of the non-task-specific steps.				
Total Score: <total # of points / 4 = %>				

Supervisor:

Supervisor/instructor signature _________________________________ Date _______________________

Comments:

Retest supervisor/instructor signature _________________________________ Date _______________________

Comments:

CDX Tasksheet Number: A4016

Student/Intern Information

Name _________________________________ Date ___________ Class _____________________

Vehicle, Customer, and Service Information

Vehicle used for this activity:

Year _______________ Make _____________________________ Model _____________________

Odometer _________________________________ VIN _________________________________

Task-Specific Safety Considerations
- Lifting equipment, such as vehicle jacks and stands, vehicle hoists, and engine hoists, are important tools that increase productivity and make the job easier. However, they can also cause severe injury or death if used improperly. Make sure you follow the manufacturer's operation procedures. Also, make sure you have your supervisor's/instructor's permission to use any particular type of lifting equipment.
- Comply with personal and environmental safety practices associated with clothing; eye protection; hand tools; power equipment; proper ventilation; and the handling, storage, and disposal of chemicals/materials in accordance with local, state, and federal safety and environmental regulations.

Time off____________

Time on____________

▶ TASK Inspect suspension system coil springs and spring insulators (silencers).

MLR
A4B13

Total time____________

Student Instructions: Read through the entire procedure prior to starting. Prepare your work space and any tools or parts that may be needed to complete the task. When directed by your instructor, begin the procedure to complete the task and check the box as each step is finished. Track your time on this procedure for later comparison to the standard completion time (i.e., "flat rate" or customer pay time).

Procedure:	Step Completed
1. Research the suspension configuration and the coil spring inspection procedure.	
a. How many coil springs is your vehicle equipped with?	☐
b. Where are the springs located?	☐
c. Explain what you are looking for during an inspection of the coil springs and spring insulators.	☐
d. Explain what may happen if a coil spring is damaged or an insulator is missing.	☐
2. Inspect the coil spring and insulators.	
a. Lift the vehicle and remove the wheels if necessary.	☐
b. Locate the coil springs. Inspect the springs for damage, such as broken coils, excessive rust, or corrosion, or a spring that is not seated properly.	☐
c. Inspect the insulators. Ensure that springs are seated correctly against the insulators. Insulators may be located at both ends of the spring as well as wrapped around a section of coils.	☐
d. Based on your inspection, what is your recommendation for service?	☐
e. Install the wheels. Torque the lug nuts to specification and lower the vehicle.	☐

Non-Task-Specific Evaluations:	**Step Completed**
1. Tools and equipment were used as directed and returned in good working order.	☐
2. Complied with all general and task-specific safety standards, including proper use of any personal protective equipment.	☐
3. Completed the task in an appropriate time frame (Recommendation: 1.5 or 2 times flat rate).	☐
4. Left the work space clean and orderly.	☐
5. Cared for customer property and returned it undamaged.	☐

Student signature _______________________________ Date _________________________________

Comments:

Have your supervisor/instructor verify satisfactory completion of this procedure, any observations found, and any necessary action(s) recommended.

Tasksheet Scoring

	Test		Retest	
Evaluation Items	**Pass**	**Fail**	**Pass**	**Fail**
Task-Specific Evaluation	**(1 pt)**	**(0 pts)**	**(1 pt)**	**(0 pts)**
Research the suspension configuration and coil spring inspection procedure.				
Identify issues related to damaged coil springs and/or insulators.				
Inspect the coil springs and insulators.				
Make recommendations for service based on your inspection.				
Non-Task-Specific Evaluation	**(0 pts)**	**(−1 pt)**	**(0 pts)**	**(−1 pt)**
Student successfully completed at least three of the non-task-specific steps.				
Student successfully completed all five of the non-task-specific steps.				
Total Score: <total # of points / 4 = %>				

Supervisor:

Supervisor/instructor signature _________________________________ Date _____________________

Comments:

Retest supervisor/instructor signature _________________________________ Date _____________________

Comments:

CDX Tasksheet Number: A4O17

Student/Intern Information

Name _________________________________ Date _____________ Class _____________________

Vehicle, Customer, and Service Information

Vehicle used for this activity:

Year _______________ Make _______________________ Model _____________________________

Odometer _____________________________ VIN _______________________________________

Task-Specific Safety Considerations

- Lifting equipment, such as vehicle jacks and stands, vehicle hoists, and engine hoists, are important tools that increase productivity and make the job easier. However, they can also cause severe injury or death if used improperly. Make sure you follow the manufacturer's operation procedures. Also, make sure you have your supervisor's/instructor's permission to use any particular type of lifting equipment.
- Comply with personal and environmental safety practices associated with clothing; eye protection; hand tools; power equipment; proper ventilation; and the handling, storage, and disposal of chemicals/materials in accordance with local, state, and federal safety and environmental regulations.

▶ TASK Inspect suspension system torsion bars and mounts.

MLR
A4B14

Time off________________

Time on________________

Total time________________

Student Instructions: Read through the entire procedure prior to starting. Prepare your work space and any tools or parts that may be needed to complete the task. When directed by your instructor, begin the procedure to complete the task and check the box as each step is finished. Track your time on this procedure for later comparison to the standard completion time (i.e., "flat rate" or customer pay time).

Procedure:	Step Completed
1. Research the torsion bar adjustment and inspection procedures.	
a. Explain how torsion bars affect vehicle ride height.	☐
b. Explain when torsion bars might require adjustment.	☐
c. Explain how ride height adjustments are made with torsion bars.	☐
d. Explain how you would measure vehicle ride height and how you would inspect torsion bars for proper function.	☐
e. Using your service information, find the vehicle ride height specification. What is the specification?	☐
2. Inspect the coil spring and insulators.	
a. Following proper procedure, measure the vehicle ride height and compare it to your specifications. Describe your results.	☐
b. Lift the vehicle and locate the torsion bars. Inspect the bars for signs of damage including bent, broken, or cracked components. Is there excess rust or corrosion present?	☐

	Step Completed
c. Inspect the adjusting screws and describe their condition. Do the adjusting screws look like they are in a condition where an adjustment could be made?	☐
d. Inspect the torsion bar mounts. Explain where the torsion bars mount to. Describe the condition of the mounts.	☐
3. Based on your inspection, what is your recommendation for service?	☐

Non-Task-Specific Evaluations:	Step Completed
1. Tools and equipment were used as directed and returned in good working order.	☐
2. Complied with all general and task-specific safety standards, including proper use of any personal protective equipment.	☐
3. Completed the task in an appropriate time frame (Recommendation: 1.5 or 2 times flat rate).	☐
4. Left the work space clean and orderly.	☐
5. Cared for customer property and returned it undamaged.	☐

Student signature _______________________ Date _______________________

Comments:

Have your supervisor/instructor verify satisfactory completion of this procedure, any observations found, and any necessary action(s) recommended.

Tasksheet Scoring

	Test		Retest	
Evaluation Items	**Pass**	**Fail**	**Pass**	**Fail**
Task-Specific Evaluation	**(1 pt)**	**(0 pts)**	**(1 pt)**	**(0 pts)**
Explain how torsion bars affect ride height and how adjustments can be made.				
Using your service information, find the vehicle ride height specification.				
Inspect the torsion bars for signs of damage, wear, excess rust or corrosion.				
Inspect the torsion bar mounts and adjusting screws.				
Non-Task-Specific Evaluation	**(0 pts)**	**(−1 pt)**	**(0 pts)**	**(−1 pt)**
Student successfully completed at least three of the non-task-specific steps.				
Student successfully completed all five of the non-task-specific steps.				
Total Score: <total # of points / 4 = %>				

Supervisor:

Supervisor/instructor signature _________________________________ Date _____________________

Comments:

Retest supervisor/instructor signature _________________________ Date _____________________

Comments:

CDX Tasksheet Number: A4018

Student/Intern Information

Name _________________________________ Date ____________ Class ___________________________

Vehicle, Customer, and Service Information

Vehicle used for this activity:

Year _______________ Make _______________________________ Model ____________________________

Odometer _________________________________ VIN ____

Materials Required

- Blank work order
- Vehicle with available service history records
- Depending on the type of concern, special diagnostic/hand tools may be required. See your supervisor/instructor for instructions to identify what tools may be required.
- Service information database
- Personal protective equipment
- General hand tools

Task-Specific Safety Considerations

- Lifting equipment, such as vehicle jacks and stands, vehicle hoists, and engine hoists, are important tools that increase productivity and make the job easier. However, they can also cause severe injury or death if used improperly. Make sure you follow the manufacturer's operation procedures. Also, make sure you have your supervisor's/instructor's permission to use any particular type of lifting equipment.
- Comply with personal and environmental safety practices associated with clothing; eye protection; hand tools; power equipment; proper ventilation; and the handling, storage, and disposal of chemicals/materials in accordance with local, state, and federal safety and environmental regulations.

▶ **TASK** Inspect and/or replace front/rear stabilizer bar (sway bar) bushings, brackets, and links.

MLR
A4B15

Time off____________

Time on____________

Total time____________

Student Instructions: Read through the entire procedure prior to starting. Prepare your work space and any tools or parts that may be needed to complete the task. When directed by your instructor, begin the procedure to complete the task and check the box as each step is finished. Track your time on this procedure for later comparison to the standard completion time (i.e., "flat rate" or customer pay time).

Procedure:	Step Completed
1. Inspect the front/rear stabilizer bar, bushings, brackets, and links. Lift the vehicle and perform the following inspections.	
a. Inspect the stabilizer bar for signs of damage, rust/corrosion, or excessive wear. Describe the condition of the stabilizer bar.	☐
b. Inspect the stabilizer bar bushings for signs of cracking. Check to see that the bushing is correctly positioned inside the bracket and that there is no looseness of the bar inside the bushing/bracket. Note your inspection results.	☐
c. Inspect the end links. Describe what they are attached to. Some use a ball and socket design while others are mounted with a stud and bushings. Check for any signs of looseness, worn bushings, or damage. Describe the condition of the links.	☐
2. Research the stabilizer bushing and end link replacement procedure.	
a. List or print off the procedure for replacing the stabilizer bar bushings and end links on your vehicle.	☐
b. Are there any special tools or precautions needed for this task?	☐
3. Replace the stabilizer bar bushings and end links.	
a. Following the proper procedure, remove the stabilizer bar end links.	☐
b. Compare the new links to the old links. Install the new links. Be careful to install spacers and bushings in the correct order (if applicable).	☐
c. Remove the old stabilizer bar bushings by removing the brackets and retaining bolts. A rust penetrant may be needed to help rusted bolts come loose.	☐

d. Compare the new bushings to the old bushings. Many times, there are a variety of bushing/bar size combinations, so be sure that the part is correct for your application. Install the new bushings. **Tip:** It may be advisable to use a specific lubricant on the bushings to eliminate noise from occurring. Refer to your service procedure and/or instructions that come with the new parts.	☐
e. Torque all fasteners to specification. What specification did you use for the end links and bushing brackets?	☐

Non-Task-Specific Evaluations:	Step Completed
1. Tools and equipment were used as directed and returned in good working order.	☐
2. Complied with all general and task-specific safety standards, including proper use of any personal protective equipment.	☐
3. Completed the task in an appropriate time frame (Recommendation: 1.5 or 2 times flat rate).	☐
4. Left the work space clean and orderly.	☐
5. Cared for customer property and returned it undamaged.	☐

Student signature ___________________________ Date ___________________________

Comments:

Have your supervisor/instructor verify satisfactory completion of this procedure, any observations found, and any necessary action(s) recommended.

Tasksheet Scoring

		Test		Retest	
Evaluation Items	**Pass**	**Fail**	**Pass**	**Fail**
Task-Specific Evaluation	**(1 pt)**	**(0 pts)**	**(1 pt)**	**(0 pts)**
Inspect the stabilizer bushings, brackets, and end links.				
Research the procedures for removal and installation of stabilizer bushings and end links, then remove and replace the stabilizer bar bushings.				
Remove and replace the stabilizer bar end links.				
Torque all fasteners to specification.				
Non-Task-Specific Evaluation	**(0 pts)**	**(−1 pt)**	**(0 pts)**	**(−1 pt)**
Student successfully completed at least three of the non-task-specific steps.				
Student successfully completed all five of the non-task-specific steps.				
Total Score: <total # of points / 4 = %>				

Supervisor:

Supervisor/instructor signature _________________________ Date _______________

Comments:

Retest supervisor/instructor signature _________________________ Date _______________

Comments:

CDX Tasksheet Number: A4019

Student/Intern Information

Name _________________________________ Date ____________ Class _____________________

Vehicle, Customer, and Service Information

Vehicle used for this activity:

Year _______________ Make _______________________ Model ___________________________

Odometer _______________________ VIN ______________________________________

Materials Required

- Blank work order
- Vehicle with available service history records
- Depending on the type of concern, special diagnostic/hand tools may be required. See your supervisor/instructor for instructions to identify what tools may be required.
- Service information database
- Personal protective equipment
- General hand tools
- Strut spring compressor

Task-Specific Safety Considerations

- Lifting equipment, such as vehicle jacks and stands, vehicle hoists, and engine hoists, are important tools that increase productivity and make the job easier. However, they can also cause severe injury or death if used improperly. Make sure you follow the manufacturer's operation procedures. Also, make sure you have your supervisor's/instructor's permission to use any particular type of lifting equipment.
- Comply with personal and environmental safety practices associated with clothing; eye protection; hand tools; power equipment; proper ventilation; and the handling, storage, and disposal of chemicals/materials in accordance with local, state, and federal safety and environmental regulations.

<table>
<tr><td>Time off___________</td></tr>
<tr><td>Time on___________</td></tr>
<tr><td>Total time___________</td></tr>
</table>

▶ TASK Inspect, remove, and/or replace strut cartridge or assembly, and inspect mounts and bushings.

MLR
A4B16

Student Instructions: Read through the entire procedure prior to starting. Prepare your work space and any tools or parts that may be needed to complete the task. When directed by your instructor, begin the procedure to complete the task and check the box as each step is finished. Track your time on this procedure for later comparison to the standard completion time (i.e., "flat rate" or customer pay time).

Procedure:	Step Completed
1. Inspect the strut assembly for damage or wear. List any action required at each step of the inspection process.	
a. Jounce the vehicle while it is on the ground. It should rebound and stabilize quickly. If it continues to jounce, this could indicate worn struts.	☐
b. As you jounce the vehicle, listen for any abnormal knocking sounds as this could be related to a worn strut, mount, or bearing. Note the results here.	☐
c. Turn the steering wheel from lock to lock. Listen for abnormal noise coming from the upper strut mount/bearing assembly. **Tip:** It may be helpful to have an assistant turn the wheel while you listen under the hood. It may also be helpful to place your hand on the strut mount, as any binding or worn bearings may be felt as the steering wheel is turned. Note the results here.	☐
d. Lift the vehicle. Inspect the struts and springs (if applicable) for signs of damage, fluid leaks, or excessive rust or corrosion. Note the results here.	☐
e. Inspect for abnormal tire wear (cupping) that could be caused from worn struts. Note the results here.	☐
2. Research strut removal and disassembly procedure.	
a. List or print off the procedure for strut removal and disassembly.	☐
b. Are any special tools or precautions needed for this task?	☐

<table>
<tr><td>c. List all torque specifications associated with this task.</td><td>☐</td></tr>
<tr><td colspan="2">3. Remove and disassemble the strut assembly.</td></tr>
<tr><td>a. Following the proper procedure, remove the strut assembly. **Tip:** It may be helpful to make some sort of identifying mark on the top mount to ensure proper installation during reassembly.</td><td>☐</td></tr>
<tr><td>b. After the strut assembly has been removed, use a spring compressor to compress the coil spring. This relieves tension from the top mount. Remove the mount nut and mount. Keep all items in order as they are removed.</td><td>☐</td></tr>
<tr><td>c. Inspect the coil spring, strut, bearing plate, fasteners, and bump stops. If any components are questionable, it is a good practice to replace them.</td><td>☐</td></tr>
<tr><td colspan="2">4. Reassemble and install the strut assembly.</td></tr>
<tr><td>a. Reassemble the struts assembly using new parts that are needed. Torque all fasteners to specification. What is the specification for the top mount bearing plate nut?</td><td>☐</td></tr>
<tr><td>b. Install the strut in the vehicle. Torque all fasteners to specification. What is the specification for the strut's top and lower mounting bolts?</td><td>☐</td></tr>
<tr><td>c. Should an alignment check be performed after the installation is complete? Explain your answer.</td><td>☐</td></tr>
</table>

Non-Task-Specific Evaluations:	**Step Completed**
1. Tools and equipment were used as directed and returned in good working order.	☐
2. Complied with all general and task-specific safety standards, including proper use of any personal protective equipment.	☐
3. Completed the task in an appropriate time frame (Recommendation: 1.5 or 2 times flat rate).	☐
4. Left the work space clean and orderly.	☐
5. Cared for customer property and returned it undamaged.	☐

Student signature _______________________________ Date _______________________________

Comments:

Have your supervisor/instructor verify satisfactory completion of this procedure, any observations found, and any necessary action(s) recommended.

Tasksheet Scoring

	Test		Retest	
Evaluation Items	Pass	Fail	Pass	Fail
Task-Specific Evaluation	**(1 pt)**	**(0 pts)**	**(1 pt)**	**(0 pts)**
Inspect the strut assembly for damage or wear.				
Remove the strut assembly and disassemble the strut assembly using the spring compressor.				
Reassemble the strut assembly.				
Torque all fasteners to specification.				
Non-Task-Specific Evaluation	**(0 pts)**	**(−1 pt)**	**(0 pts)**	**(−1 pt)**
Student successfully completed at least three of the non-task-specific steps.				
Student successfully completed all five of the non-task-specific steps.				
Total Score: <total # of points / 4 = %>				

Supervisor:

Supervisor/instructor signature ______________________________ Date ___________________

Comments:

Retest supervisor/instructor signature ______________________________ Date ___________________

Comments:

CDX Tasksheet Number: A4020

Student/Intern Information

Name _________________________________ Date ____________ Class _____________________

Vehicle, Customer, and Service Information

Vehicle used for this activity:

Year _______________ Make ___________________________ Model _____________________

Odometer _______________________________ VIN _______________________________

Materials Required

- Blank work order
- Vehicle with available service history records
- Depending on the type of concern, special diagnostic/hand tools may be required. See your supervisor/instructor for instructions to identify what tools may be required.
- Service information database
- Personal protective equipment

Task-Specific Safety Considerations

- Lifting equipment, such as vehicle jacks and stands, vehicle hoists, and engine hoists, are important tools that increase productivity and make the job easier. However, they can also cause severe injury or death if used improperly. Make sure you follow the manufacturer's operation procedures. Also, make sure you have your supervisor's/instructor's permission to use any particular type of lifting equipment.
- Comply with personal and environmental safety practices associated with clothing; eye protection; hand tools; power equipment; proper ventilation; and the handling, storage, and disposal of chemicals/materials in accordance with local, state, and federal safety and environmental regulations.

▶ TASK Inspect front strut bearing and mount.

MLR
A4B17

Time off____________

Time on____________

Student Instructions: Read through the entire procedure prior to starting. Prepare your work space and any tools or parts that may be needed to complete the task. When directed by your instructor, begin the procedure to complete the task and check the box as each step is finished. Track your time on this procedure for later comparison to the standard completion time (i.e., "flat rate" or customer pay time).

Total time____________

Procedure:	Step Completed
1. Research the procedure for inspecting strut bearings and mounts.	
a. Explain what struts use bearings and why.	☐
b. What complaints may a customer have if a strut mount and/or bearing were to go bad?	☐
2. Inspect the strut bearing and/or mount.	
a. Jounce the vehicle while it is on the ground. It should rebound and stabilize quickly. As you jounce the vehicle, listen for any abnormal knocking sounds as this could be related to a worn strut, mount, or bearing.	☐
b. With the vehicle still on the ground, turn the steering wheel from lock to lock. Listen for abnormal noise coming from the upper strut mount/bearing assembly. **Tip:** It may be helpful to have an assistant turn the wheel while you listen under the hood. It may also be helpful to place your hand on the strut mount, as any binding or worn bearings may be felt as the steering wheel is turned.	☐
c. If a test drive is necessary, follow your instructor's procedures EXACTLY. If possible, use a set of chassis ears to pinpoint the location of strut noise.	☐
d. If a thorough inspection of the strut mount and bearing from the top side does not reveal any issues, it may be necessary to lift the vehicle and inspect the strut from the underside. The wheel can be turned lock to lock; meanwhile, listen for any abnormal noises due to shifting of the wheel/strut assembly.	☐
3. Based on your inspection, what is your recommendation for service?	
a. Explain how you would repair a worn strut mount and/or bearing.	☐

Non-Task-Specific Evaluations:	**Step Completed**
1. Tools and equipment were used as directed and returned in good working order.	☐
2. Complied with all general and task-specific safety standards, including proper use of any personal protective equipment.	☐
3. Completed the task in an appropriate time frame (Recommendation: 1.5 or 2 times flat rate).	☐
4. Left the work space clean and orderly.	☐
5. Cared for customer property and returned it undamaged.	☐

Student signature ______________________________ Date ______________________________

Comments:

Have your supervisor/instructor verify satisfactory completion of this procedure, any observations found, and any necessary action(s) recommended.

Evaluation Instructions: The scoring box below is intended to act as a guide for both student and instructor. Each criterion listed will help students understand what is expected from them and evaluators articulate success at a particular task. The scoring is set up to allow a second attempt at each task (see the "Test" and "Retest" columns). Scoring is designed to reward students for correct completion of the task. Points are lost for failure to complete the employability requirements (see "Non-Task-Specific" criteria). When all criteria are evaluated, tally the points for a total at the bottom of each column.

Tasksheet Scoring

	Test		Retest	
Evaluation Items	**Pass**	**Fail**	**Pass**	**Fail**
Task-Specific Evaluation	**(1 pt)**	**(0 pts)**	**(1 pt)**	**(0 pts)**
Research the procedure for inspecting the strut bearings and mounts.				
Explain any issues that may result from worn strut mounts and/or bearings.				
Inspect the strut mounts and bearings.				
Make recommendations for service based on your inspection.				
Non-Task-Specific Evaluation	**(0 pts)**	**(−1 pt)**	**(0 pts)**	**(−1 pt)**
Student successfully completed at least three of the non-task-specific steps.				
Student successfully completed all five of the non-task-specific steps.				
Total Score: <total # of points / 4 = %>				

Supervisor:

Supervisor/instructor signature ______________________________ Date ______________

Comments:

Retest supervisor/instructor signature ______________________________ Date ______________

Comments:

CDX Tasksheet Number: A4021

Student/Intern Information

Name _________________________________ Date ___________ Class _____________________

Vehicle, Customer, and Service Information

Vehicle used for this activity:

Year _______________ Make _______________________ Model _____________________

Odometer _______________________ VIN _________________________________

> **Materials Required**
> - Blank work order
> - Vehicle with available service history records
> - Depending on the type of concern, special diagnostic/hand tools may be required. See your supervisor/instructor for instructions to identify what tools may be required.
> - Service information database
> - Personal protective equipment
> - Pry bar

Task-Specific Safety Considerations

- Lifting equipment, such as vehicle jacks and stands, vehicle hoists, and engine hoists, are important tools that increase productivity and make the job easier. However, they can also cause severe injury or death if used improperly. Make sure you follow the manufacturer's operation procedures. Also, make sure you have your supervisor's/instructor's permission to use any particular type of lifting equipment.
- Comply with personal and environmental safety practices associated with clothing; eye protection; hand tools; power equipment; proper ventilation; and the handling, storage, and disposal of chemicals/materials in accordance with local, state, and federal safety and environmental regulations.

▶ TASK Inspect rear suspension system lateral links/arms (track bars), control (trailing) arms.

MLR
A4B18

Time off________

Time on________

Total time________

Student Instructions: Read through the entire procedure prior to starting. Prepare your work space and any tools or parts that may be needed to complete the task. When directed by your instructor, begin the procedure to complete the task and check the box as each step is finished. Track your time on this procedure for later comparison to the standard completion time (i.e., "flat rate" or customer pay time).

Procedure:	Step Completed
1. Describe the purpose and function of the lateral link and trailing arm.	
a. Explain the purpose and function of a lateral link. Where is it located on the vehicle?	☐
b. Explain the purpose and function of a trailing arm. Where is it located on the vehicle?	☐
c. Explain what symptoms a vehicle may have if these components were worn or damaged.	☐
2. Inspect the rear lateral link and trailing arms and related mounts and bushings.	
a. Lift the vehicle. Locate the lateral link. Using a pry bar, lightly pry the lateral link back and forth and up and down. Is there any excessive movement? Note the results of your inspection.	☐
b. Inspect the condition of the lateral link. Check for signs of damage or excessive rust and corrosion. Note the results below.	☐
c. Inspect the mounting area and bushings for wear such as dry rot, cracking, or missing pieces of rubber. Note the results below.	☐
d. Locate the trailing arms. Perform a similar inspection procedure by lightly prying on the bars to check for excessive movement. Inspect the mounts and bushings for signs of wear. Note the results below.	☐

<table>
<tr><td>3. Based on your inspection, what are your recommendations for service?</td><td></td></tr>
</table>

Non-Task-Specific Evaluations:	**Step Completed**
1. Tools and equipment were used as directed and returned in good working order.	☐
2. Complied with all general and task-specific safety standards, including proper use of any personal protective equipment.	☐
3. Completed the task in an appropriate time frame (Recommendation: 1.5 or 2 times flat rate).	☐
4. Left the work space clean and orderly.	☐
5. Cared for customer property and returned it undamaged.	☐

Student signature ___________________________ Date ___________________________

Comments:

Have your supervisor/instructor verify satisfactory completion of this procedure, any observations found, and any necessary action(s) recommended.

Tasksheet Scoring

	Test		Retest	
Evaluation Items	**Pass**	**Fail**	**Pass**	**Fail**
Task-Specific Evaluation	**(1 pt)**	**(0 pts)**	**(1 pt)**	**(0 pts)**
Describe the purpose and function of lateral links and trailing arms.				
Explain what symptoms a vehicle may have if these components were worn or damaged.				
Inspect the condition of lateral links, trailing arms, all bushings, and all mounts.				
Make recommendations for service based on your inspection.				
Non-Task-Specific Evaluation	**(0 pts)**	**(−1 pt)**	**(0 pts)**	**(−1 pt)**
Student successfully completed at least three of the non-task-specific steps.				
Student successfully completed all five of the non-task-specific steps.				
Total Score: <total # of points / 4 = %>				

Supervisor:

Supervisor/instructor signature _________________________________ Date _________________

Comments:

Retest supervisor/instructor signature _________________________________ Date _________________

Comments:

CDX Tasksheet Number: A4022

Student/Intern Information

Name ___________________________ Date __________ Class _______________

Vehicle, Customer, and Service Information

Vehicle used for this activity:

Year ____________ Make ________________ Model _________________

Odometer ____________________ VIN ___________________

Task-Specific Safety Considerations

- Lifting equipment, such as vehicle jacks and stands, vehicle hoists, and engine hoists, are important tools that increase productivity and make the job easier. However, they can also cause severe injury or death if used improperly. Make sure you follow the manufacturer's operation procedures. Also, make sure you have your supervisor's/instructor's permission to use any particular type of lifting equipment.
- Comply with personal and environmental safety practices associated with clothing; eye protection; hand tools; power equipment; proper ventilation; and the handling, storage, and disposal of chemicals/materials in accordance with local, state, and federal safety and environmental regulations.

▶ **TASK** Inspect rear suspension system leaf spring(s), spring insulators (silencers), shackles, brackets, bushings, center pins/bolts, and mounts.

MLR
A4B19

Time off____________

Time on____________

Total time____________

Student Instructions: Read through the entire procedure prior to starting. Prepare your work space and any tools or parts that may be needed to complete the task. When directed by your instructor, begin the procedure to complete the task and check the box as each step is finished. Track your time on this procedure for later comparison to the standard completion time (i.e., "flat rate" or customer pay time).

Procedure:	Step Completed
1. Inspect the rear suspension equipped with leaf springs.	
a. Lift the vehicle and locate the leaf springs in the rear of the vehicle. How many leaf springs are there per side?	☐
b. Explain how adding or subtracting springs will affect the vehicle's ride and payload capability.	☐
c. Inspect each spring for signs of damage, including cracked or broken springs. Inspect the spring insulators that are located between each leaf.	☐
d. Ensure that the springs are firmly mounted to the rear axle and that the centering pin, U-bolts, and fasteners are secure.	☐
e. Inspect the leaf spring shackles for signs of excessive wear, rust and/or corrosion.	☐
f. Inspect the bushings in the shackles for signs of dry rotting or excess wear and ensure that the bushings are seated correctly.	☐
g. Based on your inspection, what are your recommendations for service?	☐

Non-Task-Specific Evaluations:	Step Completed
1. Tools and equipment were used as directed and returned in good working order.	☐
2. Complied with all general and task-specific safety standards, including proper use of any personal protective equipment.	☐
3. Completed the task in an appropriate time frame (Recommendation: 1.5 or 2 times flat rate).	☐
4. Left the work space clean and orderly.	☐
5. Cared for customer property and returned it undamaged.	☐

Student signature _______________________________ Date _______________________________

Comments:

Have your supervisor/instructor verify satisfactory completion of this procedure, any observations found, and any necessary action(s) recommended.

Evaluation Instructions: The scoring box below is intended to act as a guide for both student and instructor. Each criterion listed will help students understand what is expected from them and evaluators articulate success at a particular task. The scoring is set up to allow a second attempt at each task (see the "Test" and "Retest" columns). Scoring is designed to reward students for correct completion of the task. Points are lost for failure to complete the employability requirements (see "Non-Task-Specific" criteria). When all criteria are evaluated, tally the points for a total at the bottom of each column.

Taksheet Scoring

	Test		Retest	
Evaluation Items	Pass	Fail	Pass	Fail
Task-Specific Evaluation	**(1 pt)**	**(0 pts)**	**(1 pt)**	**(0 pts)**
Lift the vehicle. Locate and identify how many leaf springs are used per side on your specific vehicle.				
Explain how the vehicle's ride and payload capability are affected by the addition or subtraction of leaf springs.				
Inspect the leaf springs, insulators, mounts, centering pins, shackles, and bushings for wear.				
Make recommendations for service based on your inspection.				
Non-Task-Specific Evaluation	**(0 pts)**	**(−1 pt)**	**(0 pts)**	**(−1 pt)**
Student successfully completed at least three of the non-task-specific steps.				
Student successfully completed all five of the non-task-specific steps.				
Total Score: <total # of points / 4 = %>				

Supervisor:

Supervisor/instructor signature _______________________________ Date _______________

Comments:

Retest supervisor/instructor signature _______________________________ Date _______________

Comments:

CDX Tasksheet Number: A4023

Student/Intern Information

Name _________________________________ Date ____________ Class _________________________

Vehicle, Customer, and Service Information

Vehicle used for this activity:

Year ______________ Make _____________________________ Model _________________________

Odometer _________________________________ VIN _________________________________

Materials Required

- Blank work order
- Vehicle with available service history records
- Depending on the type of concern, special diagnostic/hand tools may be required. See your supervisor/instructor for instructions to identify what tools may be required.
- Service information database
- Personal protective equipment
- General hand tools
- Shock absorbers
- Tall jack stand

Task-Specific Safety Considerations

- Lifting equipment, such as vehicle jacks and stands, vehicle hoists, and engine, are important tools that increase productivity and make the job easier. However, they can also cause severe injury or death if used improperly. Make sure you follow the manufacturer's operation procedures. Also, make sure you have your supervisor's/instructor's permission to use any particular type of lifting equipment.
- Comply with personal and environmental safety practices associated with clothing; eye protection; hand tools; power equipment; proper ventilation; and the handling, storage, and disposal of chemicals/materials in accordance with local, state, and federal safety and environmental regulations.

▶ **TASK** Inspect, remove, and/or replace shock absorbers; inspect mounts and bushings.

MLR
A4B20

Time off____________

Time on____________

Student Instructions: Read through the entire procedure prior to starting. Prepare your work space and any tools or parts that may be needed to complete the task. When directed by your instructor, begin the procedure to complete the task and check the box as each step is finished. Track your time on this procedure for later comparison to the standard completion time (i.e., "flat rate" or customer pay time).

Total time____________

Procedure:	Step Completed
1. Inspect the condition and operation of the shock absorber.	
a. Jounce the vehicle on the ground. Watch as the car rebounds. The vehicle should stabilize quickly. If the vehicle continues to jounce itself, this could be an indication that the shocks are worn out. List any action required.	☐
b. Lift the vehicle and inspect the shocks for signs of oil leakage and excessive rust, and inspect the mounts and bushings for signs of deterioration. List any action required.	☐
c. Inspect the tires for signs of irregular wear (cupping) that may be caused by faulty shock absorbers. List any action required.	☐
2. Research the shock absorber replacement procedure.	
a. Is there a recommended inspection/replacement interval according to the maintenance guide for the vehicle? If so, what is the interval?	☐
b. List or print off the procedure for replacing the shock absorbers in your specific vehicle.	☐
c. Are there any service precautions that you must follow when replacing a shock absorber?	☐
3. Remove and replace a shock absorber.	
a. Support the axle or control arm with a tall jack stand before loosening the shock mount bolts.	☐
b. Remove the bolts that hold the shock in place. It may be necessary to lower the tall jack stand to allow the shock to be removed once the bolts are out.	☐

c. Compare the old shock to the new one. Inspect the mounting hardware. If a fastener appears questionable, it is good practice to replace it with a new part.	☐
d. Install the new shock absorber. Depending on the design, it may be necessary to hold the top of the shock where the nut threads on with a wrench to keep the shock piston from rotating in the housing.	☐
e. Torque all fasteners to specification. Which specification did you use?	☐
f. Lower the vehicle. Jounce the vehicle once more to verify your repair.	☐

Non-Task-Specific Evaluations:	Step Completed
1. Tools and equipment were used as directed and returned in good working order.	☐
2. Complied with all general and task-specific safety standards, including proper use of any personal protective equipment.	☐
3. Completed the task in an appropriate time frame (Recommendation: 1.5 or 2 times flat rate).	☐
4. Left the work space clean and orderly.	☐
5. Cared for customer property and returned it undamaged.	☐

Student signature _______________________________ Date _______________________________

Comments:

Have your supervisor/instructor verify satisfactory completion of this procedure, any observations found, and any necessary action(s) recommended.

Evaluation Instructions: The scoring box below is intended to act as a guide for both student and instructor. Each criterion listed will help students understand what is expected from them and evaluators articulate success at a particular task. The scoring is set up to allow a second attempt at each task (see the "Test" and "Retest" columns). Scoring is designed to reward students for correct completion of the task. Points are lost for failure to complete the employability requirements (see "Non-Task-Specific" criteria). When all criteria are evaluated, tally the points for a total at the bottom of each column.

Tasksheet Scoring

	Test		Retest	
Evaluation Items	**Pass**	**Fail**	**Pass**	**Fail**
Task-Specific Evaluation	**(1 pt)**	**(0 pts)**	**(1 pt)**	**(0 pts)**
Inspect the condition and operation of the shock absorber.				
Research the shock absorber removal procedure. Research any service precautions. Remove the shock absorber.				
Replace the shock absorber. Torque the fasteners to specification.				
Verify the installation and correct operation of the shock absorbers.				
Non-Task-Specific Evaluation	**(0 pts)**	**(−1 pt)**	**(0 pts)**	**(−1 pt)**
Student successfully completed at least three of the non-task-specific steps.				
Student successfully completed all five of the non-task-specific steps.				
Total Score: <total # of points / 4 = %>				

Supervisor:

Supervisor/instructor signature _________________________________ Date ____________________

Comments:

Retest supervisor/instructor signature _________________________________ Date ____________________

Comments:

CDX Tasksheet Number: A4024

Student/Intern Information

Name _______________________________ Date ___________ Class _______________________

Vehicle, Customer, and Service Information

Vehicle used for this activity:

Year _____________ Make ___________________________ Model _____________________

Odometer _________________________ VIN _________________________________

Materials Required

- Blank work order
- Vehicle with available service history records
- Depending on the type of concern, special diagnostic/hand tools may be required. See your supervisor/instructor for instructions to identify what tools may be required.
- Service information database
- Personal protective equipment
- Scan tool

Task-Specific Safety Considerations

- Comply with personal and environmental safety practices associated with clothing; eye protection; hand tools; power equipment; proper ventilation; and the handling, storage, and disposal of chemicals/materials in accordance with local, state, and federal safety and environmental regulations.

▶ TASK Inspect electric power steering assist system.

MLR A4B21

Time off_______________

Time on_______________

Student Instructions: Read through the entire procedure prior to starting. Prepare your work space and any tools or parts that may be needed to complete the task. When directed by your instructor, begin the procedure to complete the task and check the box as each step is finished. Track your time on this procedure for later comparison to the standard completion time (i.e., "flat rate" or customer pay time).

Total time_______________

Procedure:	Step Completed
1. Research service information for the proper procedure when using a scan tool to diagnose components of an electric power steering system.	
a. Describe the procedure for checking fault codes and proper operation of the electric power steering system.	☐

b. What type of tools are needed to inspect the steering system for proper function?	☐
2. Inspect the electric power steering system.	
a. Verify correct steering wheel operation. Turn the steering wheel from lock to lock. Does steering effort feel normal? Are there any abnormal noises?	☐
b. Are there any warning lights illuminated on the instrument cluster?	☐
c. Connect a scan tool. Follow the correct prompts to display the electric power steering diagnostic trouble codes (DTCs). Are there any codes stored in the history? Are there any codes stored as present?	☐
d. Once you have checked for DTCs, select the data display function. List the components, sensors, and all parameters listed under data display.	☐
e. Start the vehicle while turning the wheel back and forth, what readings are displayed for steering wheel position or angle? What readings are available for the torque sensor signal?	☐
f. Explain why vehicle speed is a parameter that must be monitored by the electric power steering ECU.	☐

g. Select functional tests. Describe what functional tests are available for the electric power steering system. Explain when a steering angle relearn procedure would need to be performed.	☐
3. Locate the electric power steering components.	
a. Describe the location of the following components: Electric Power Steering Motor, Rack and Pinion/Steering Gear, Electric Power Steering ECU, Steering Position Sensor.	☐
b. Perform a visual inspection of the electric power steering motor and rack and pinion. Describe the condition of each component. Inspect all wiring, connectors, and components for signs of damage.	☐
4. Based on your inspection, what are your recommendations for service?	

Non-Task-Specific Evaluations:	Step Completed
1. Tools and equipment were used as directed and returned in good working order.	☐
2. Complied with all general and task-specific safety standards, including proper use of any personal protective equipment.	☐
3. Completed the task in an appropriate time frame (Recommendation: 1.5 or 2 times flat rate).	☐
4. Left the work space clean and orderly.	☐
5. Cared for customer property and returned it undamaged.	☐

Student signature _______________________________ Date _______________________________

Comments:

Have your supervisor/instructor verify satisfactory completion of this procedure, any observations found, and any necessary action(s) recommended.

Tasksheet Scoring

Evaluation Items	Test		Retest	
	Pass	**Fail**	**Pass**	**Fail**
Task-Specific Evaluation	**(1 pt)**	**(0 pts)**	**(1 pt)**	**(0 pts)**
Research service information for the proper procedure when using a scan tool to diagnose components of an electric power steering system.				
Verify steering wheel operation and inspect for fault indicator lights prior to steering system inspection. Use a scan tool to inspect for diagnostic trouble codes, identify data display information, and describe functional tests.				
Locate the electric steering system components and perform visual inspection.				
Make recommendations for service based on your inspection.				
Non-Task-Specific Evaluation	**(0 pts)**	**(−1 pt)**	**(0 pts)**	**(−1 pt)**
Student successfully completed at least three of the non-task-specific steps.				
Student successfully completed all five of the non-task-specific steps.				
Total Score: <total # of points / 4 = %>				

Supervisor:

Supervisor/instructor signature _________________________ Date _________________

Comments:

Retest supervisor/instructor signature _________________________ Date _________________

Comments:

CDX Tasksheet Number: A4025

Student/Intern Information

Name _________________________________ Date ___________ Class _______________________

Vehicle, Customer, and Service Information

Vehicle used for this activity:

Year _______________ Make _________________________ Model _____________________

Odometer _________________________ VIN _____________________________

Task-Specific Safety Considerations
- Comply with personal and environmental safety practices associated with clothing; eye protection; hand tools; power equipment; proper ventilation; and the handling, storage, and disposal of chemicals/materials in accordance with local, state, and federal safety and environmental regulations.

Time off____________

Time on____________

Total time____________

▶ **TASK** Identify hybrid vehicle power steering system electrical circuits and safety precautions.

MLR
A4B22

Student Instructions: Read through the entire procedure prior to starting. Prepare your work space and any tools or parts that may be needed to complete the task. When directed by your instructor, begin the procedure to complete the task and check the box as each step is finished. Track your time on this procedure for later comparison to the standard completion time (i.e., "flat rate" or customer pay time).

Procedure:	Step Completed
1. Research the operation of hybrid vehicle power steering systems.	
a. Describe how the electric power steering system operates on your specific hybrid vehicle.	☐

b. How much voltage does the electric steering system operate with?	☐
c. What is the source of voltage for the steering motor?	☐
d. Explain the advantage of using an electric power steering over a conventional hydraulic system.	☐
e. List the components that make up the electric power steering system in your vehicle.	☐
2. Identify any safety precautions when working with hybrid vehicle power steering systems.	
a. Research any service precautions and/or TSBs to identify safety measures that must be taken when working on this system.	☐
b. What color is the wiring that comes from the power source and leads to the electric steering motor?	☐
c. Is there any special personal protective equipment that must be worn when servicing this system?	☐

Non-Task-Specific Evaluations:	Step Completed
1. Tools and equipment were used as directed and returned in good working order.	☐
2. Complied with all general and task-specific safety standards, including proper use of any personal protective equipment.	☐
3. Completed the task in an appropriate time frame (Recommendation: 1.5 or 2 times flat rate).	☐
4. Left the work space clean and orderly.	☐
5. Cared for customer property and returned it undamaged.	☐

Student signature _______________________________ Date _______________________________

Comments:

Have your supervisor/instructor verify satisfactory completion of this procedure, any observations found, and any necessary action(s) recommended.

Evaluation Instructions: The scoring box below is intended to act as a guide for both student and instructor. Each criterion listed will help students understand what is expected from them and evaluators articulate success at a particular task. The scoring is set up to allow a second attempt at each task (see the "Test" and "Retest" columns). Scoring is designed to reward students for correct completion of the task. Points are lost for failure to complete the employability requirements (see "Non-Task-Specific" criteria). When all criteria are evaluated, tally the points for a total at the bottom of each column.

Tasksheet Scoring

	Test		Retest	
Evaluation Items	**Pass**	**Fail**	**Pass**	**Fail**
Task-Specific Evaluation	**(1 pt)**	**(0 pts)**	**(1 pt)**	**(0 pts)**
Identify voltage requirements of the electric power steering system. Identify any service precautions.				
Describe the advantages of electric power steering versus hydraulic power steering.				
List the components that make up the electric power steering system.				
Identify any personal protective equipment needed when servicing high voltage steering systems.				
Non-Task-Specific Evaluation	**(0 pts)**	**(−1 pt)**	**(0 pts)**	**(−1 pt)**
Student successfully completed at least three of the non-task-specific steps.				
Student successfully completed all five of the non-task-specific steps.				
Total Score: <total # of points / 4 = %>				

Supervisor:

Supervisor/instructor signature ________________________________ Date ______________________

Comments:

Retest supervisor/instructor signature ________________________________ Date ______________________

Comments:

CDX Tasksheet Number: A4026

Student/Intern Information

Name _________________________________ Date ____________ Class _____________________

Vehicle, Customer, and Service Information

Vehicle used for this activity:

Year _______________ Make ___________________________ Model ___________________________

Odometer _________________________ VIN _____________________________________

Materials Required

- Blank work order
- Vehicle with available service history records
- Depending on the type of concern, special diagnostic/hand tools may be required. See your supervisor/instructor for instructions to identify what tools may be required.
- Service information database
- Personal protective equipment

Task-Specific Safety Considerations

- Comply with personal and environmental safety practices associated with clothing; eye protection; hand tools; power equipment; proper ventilation; and the handling, storage, and disposal of chemicals/materials in accordance with local, state, and federal safety and environmental regulations.

Time off_______________

Time on_______________

Total time_______________

▶ **TASK** Describe the function of suspension and steering control systems and components (i.e., active suspension and stability control).

MLR
A4B23

Student Instructions: Read through the entire procedure prior to starting. Prepare your work space and any tools or parts that may be needed to complete the task. When directed by your instructor, begin the procedure to complete the task and check the box as each step is finished. Track your time on this procedure for later comparison to the standard completion time (i.e., "flat rate" or customer pay time).

Procedure:	Step Completed
1. Describe the function of suspension control systems.	☐

2. Research the following types of suspension control systems and describe their function, purpose, and components.	
a. Magnetic dampers	☐
b. Hydraulic roll control	☐
c. Load leveling/ air suspension	☐
3. Describe the function of steering control systems.	
a. Research the following types of suspension control systems and describe their function, purpose, and components.	☐
b. Lane keep assist	☐
c. Vehicle stability control	☐
d. Active steering assist	☐
4. Optional: If your vehicle is equipped with any of the above systems, use a scan tool to read the data display, check for trouble codes, and locate any functional tests that may be available.	
a. Describe the system that is being used on your vehicle.	☐

<table>
<tr><td>b. Using the scan tool, check for any trouble codes. Are there any codes stored? What codes are stored?</td><td>☐</td></tr>
<tr><td>c. Are there any active/bidirectional controls available for your vehicles active steering or suspension system?</td><td>☐</td></tr>
</table>

Non-Task-Specific Evaluations:	Step Completed
1. Tools and equipment were used as directed and returned in good working order.	☐
2. Complied with all general and task-specific safety standards, including proper use of any personal protective equipment.	☐
3. Completed the task in an appropriate time frame (Recommendation: 1.5 or 2 times flat rate).	☐
4. Left the work space clean and orderly.	☐
5. Cared for customer property and returned it undamaged.	☐

Student signature ______________________________ Date ______________________________

Comments:

Have your supervisor/instructor verify satisfactory completion of this procedure, any observations found, and any necessary action(s) recommended.

Tasksheet Scoring

	Test		Retest	
Evaluation Items	**Pass**	**Fail**	**Pass**	**Fail**
Task-Specific Evaluation	**(1 pt)**	**(0 pts)**	**(1 pt)**	**(0 pts)**
Describe the purpose and function of active suspension control systems.				
Describe the purpose and function of active steering control systems. Research the components of each system.				
Use a scan tool to check the active steering and suspension systems for diagnostic trouble codes.				
Use a scan tool to check active steering and suspension systems for available functional tests and bidirectional controls.				
Non-Task-Specific Evaluation	**(0 pts)**	**(−1 pt)**	**(0 pts)**	**(−1 pt)**
Student successfully completed at least three of the non-task–specific steps.				
Student successfully completed all five of the non-task-specific steps.				
Total Score: <total # of points / 4 = %>				

Supervisor:

Supervisor/instructor signature _________________________________ Date _____________________

Comments:

Retest supervisor/instructor signature _________________________ Date _____________________

Comments:

4C. Wheel Alignment

Learning Objective/Task	CDX Tasksheet Number	ASE Education Foundation Reference Number; Priority Level
• Perform prealignment inspection; measure vehicle ride height.	A4027	A4C1, P-1
• Describe alignment angles (camber, caster, and toe)	A4028	A4C2, P-1

Materials Required

- Blank work order
- Vehicle with available service history records
- Depending on the type of concern, special diagnostic/hand tools may be required. See your supervisor/instructor for instructions to identify what tools may be required.
- Service information database
- Personal protective equipment

Safety Considerations

- When running any vehicles in the shop, make sure you use the shop's exhaust ventilation system to discharge all exhaust gas safely outside.
- Lifting equipment, such as vehicle jacks and stands, vehicle hoists, and engine hoists, are important tools that increase productivity and make the job easier. However, they can also cause severe injury or death if used improperly. Make sure you follow the manufacturer's operation procedures. Also, make sure you have your supervisor's/instructor's permission to use any particular type of lifting equipment.
- Comply with personal and environmental safety practices associated with clothing; eye protection; hand tools; power equipment; proper ventilation; and the handling, storage, and disposal of chemicals/materials in accordance with local, state, and federal safety and environmental regulations.
- If you need to start the vehicle, you should ensure that the parking brake is firmly applied; if necessary, use wheel chocks to prevent the vehicle from moving when the vehicle is started to verify the completion of these tasks.
- Extreme caution must be exercised when working around rotating components.

CDX Tasksheet Number: A4027

Student/Intern Information

Name ________________________________ Date ____________ Class ________________________

Vehicle, Customer, and Service Information

Vehicle used for this activity:

Year ______________ Make ______________________ Model ____________________

Odometer __________________________ VIN ____________________________

> **Materials Required**
> - Blank work order
> - Vehicle with available service history records
> - Depending on the type of concern, special diagnostic/hand tools may be required. See your supervisor/instructor for instructions to identify what tools may be required.
> - Service information database
> - Personal protective equipment
> - Tire pressure gauge
> - Measuring tape

Task-Specific Safety Considerations

- Lifting equipment, such as vehicle jacks and stands, vehicle hoists, and engine hoists, are important tools that increase productivity and make the job easier. However, they can also cause severe injury or death if used improperly. Make sure you follow the manufacturer's operation procedures. Also, make sure you have your supervisor's/instructor's permission to use any particular type of lifting equipment.
- Comply with personal and environmental safety practices associated with clothing; eye protection; hand tools; power equipment; proper ventilation; and the handling, storage, and disposal of chemicals/materials in accordance with local, state, and federal safety and environmental regulations.
- When lifting a vehicle using a drive-on style lift, ensure that either the vehicle is in Park, the parking brake is engaged, or the wheels are blocked with an approved tool.

▶ TASK Perform prealignment inspection; measure vehicle ride height.

MLR
A4C1

Time off____________

Time on____________

Student Instructions: Read through the entire procedure prior to starting. Prepare your work space and any tools or parts that may be needed to complete the task. When directed by your instructor, begin the procedure to complete the task and check the box as each step is finished. Track your time on this procedure for later comparison to the standard completion time (i.e., "flat rate" or customer pay time).

Total time____________

Procedure:	Step Completed
1. Research the alignment procedure and check for TSBs.	
a. Research and explain the correct order in which you should make adjustments to the vehicles steering/suspension components.	☐
b. Determine the specification for the vehicle ride height. Explain how you measure the vehicle ride height.	☐
c. Check for TSBs related to performing an alignment. List any TSBs that apply.	☐
2. Perform a prealignment inspection.	
a. Verify that the correct size tires are on the vehicle. Verify that all tires match each other. Where can you find tire size and pressure information on the vehicle? What size tires are supposed to be on your vehicle? Record each of the tires, size, tire brand, and tire model.	☐
b. Explain how incorrect or mismatched tires can affect the alignment, handling, and tire wear of your vehicle.	☐
c. Verify correct tire pressure and make adjustments as needed. What is the tire pressure specification for your vehicle?	☐
d. Record the tread depth of each of the tires on the vehicle.	☐

e. Are there any signs of irregular tire wear that may indicate an issue with tire pressure, improper alignment, or worn suspension components? What type of tire wear does the vehicle exhibit?	☐
f. Measure the vehicle ride height following proper procedure. Record your measurements. Compare them to your specifications. How do your measurements compare?	☐
3. Optional: As a technician, it is normal procedure to test drive vehicles before and after repairs are made. Following EXACT procedure, test drive the vehicle to verify the vehicle alignment. Does the vehicle exhibit a drift or pull condition?	

Non-Task-Specific Evaluations:	Step Completed
1. Tools and equipment were used as directed and returned in good working order.	☐
2. Complied with all general and task-specific safety standards, including proper use of any personal protective equipment.	☐
3. Completed the task in an appropriate time frame (Recommendation: 1.5 or 2 times flat rate).	☐
4. Left the work space clean and orderly.	☐
5. Cared for customer property and returned it undamaged.	☐

Student signature _______________________ Date _______________________

Comments:

Have your supervisor/instructor verify satisfactory completion of this procedure, any observations found, and any necessary action(s) recommended.

Evaluation Instructions: The scoring box below is intended to act as a guide for both student and instructor. Each criterion listed will help students understand what is expected from them and evaluators articulate success at a particular task. The scoring is set up to allow a second attempt at each task (see the "Test" and "Retest" columns). Scoring is designed to reward students for correct completion of the task. Points are lost for failure to complete the employability requirements (see "Non-Task-Specific" criteria). When all criteria are evaluated, tally the points for a total at the bottom of each column.

Tasksheet Scoring

	Test		Retest	
Evaluation Items	**Pass**	**Fail**	**Pass**	**Fail**
Task-Specific Evaluation	**(1 pt)**	**(0 pts)**	**(1 pt)**	**(0 pts)**
Research the alignment procedure, ride height specification and check for TSBs. Explain how mismatched tire size may affect alignment and vehicle handling. Explain other causes of vehicle drift or pull.				
Verify correct tire size and pressure.				
Inspect the tire tread depth.				
Inspect the vehicle ride height.				
Non-Task-Specific Evaluation	**(0 pts)**	**(−1 pt)**	**(0 pts)**	**(−1 pt)**
Student successfully completed at least three of the non-task-specific steps.				
Student successfully completed all five of the non-task-specific steps.				
Total Score: <total # of points / 4 = %>				

Supervisor:

Supervisor/instructor signature ____________________________ Date ____________________

Comments:

Retest supervisor/instructor signature ____________________________ Date ____________________

Comments:

CDX Tasksheet Number: A4028

Student/Intern Information

Name _______________________________ Date ____________ Class _______________________

Vehicle, Customer, and Service Information

Vehicle used for this activity:

Year ______________ Make _____________________________ Model _______________________

Odometer _________________________________ VIN _______________________________________

Materials Required

- Blank work order
- Vehicle with available service history records
- Depending on the type of concern, special diagnostic/hand tools may be required. See your supervisor/instructor for instructions to identify what tools may be required.
- Service information database
- Personal protective equipment

Task-Specific Safety Considerations

- Comply with personal and environmental safety practices associated with clothing; eye protection; hand tools; power equipment; proper ventilation; and the handling, storage, and disposal of chemicals/materials in accordance with local, state, and federal safety and environmental regulations.

▶ TASK Describe alignment angles (camber, caster, and toe).

MLR
A4C2

Time off_______________

Time on_______________

Total time_______________

Student Instructions: Read through the entire procedure prior to starting. Prepare your work space and any tools or parts that may be needed to complete the task. When directed by your instructor, begin the procedure to complete the task and check the box as each step is finished. Track your time on this procedure for later comparison to the standard completion time (i.e., "flat rate" or customer pay time).

Procedure:	Step Completed
1. Describe camber.	
a. Define camber and how it is measured.	☐

b. Will camber create tire wear? If so, what does this wear look like on a tire?	☐
c. If adjusted incorrectly, can camber cause a vehicle drift or pull condition?	☐
d. Describe the difference between positive and negative camber.	☐
e. Explain how negative camber may be beneficial in some instances.	☐
2. Describe caster.	
a. Define caster and how it is measured.	☐
b. Will caster create tire wear? If so, what does this wear look like on a tire?	☐
c. If adjusted incorrectly, can caster cause a vehicle drift or pull condition?	☐
d. Describe the difference between positive and negative caster.	☐
e. Explain how positive caster is beneficial to the way the vehicle drives. Is there a downside to having positive caster? How will a vehicle steer and handle with negative caster?	☐

3. Describe toe.	
a. Define toe and how it is measured.	☐
b. Will toe create tire wear? If so, what does this wear look like on a tire?	☐
c. If adjusted incorrectly, can toe cause a vehicle drift or pull condition?	☐
d. Describe the difference between positive and negative toe.	☐
e. Describe how making toe adjustments can alter the steering characteristics of the vehicle.	☐

Non-Task-Specific Evaluations:	Step Completed
1. Tools and equipment were used as directed and returned in good working order.	☐
2. Complied with all general and task-specific safety standards, including proper use of any personal protective equipment.	☐
3. Completed the task in an appropriate time frame (Recommendation: 1.5 or 2 times flat rate).	☐
4. Left the work space clean and orderly.	☐
5. Cared for customer property and returned it undamaged.	☐

Student signature _________________________________ Date _________________________________

Comments:

Have your supervisor/instructor verify satisfactory completion of this procedure, any observations found, and any necessary action(s) recommended.

Evaluation Instructions: The scoring box below is intended to act as a guide for both student and instructor. Each criterion listed will help students understand what is expected from them and evaluators articulate success at a particular task. The scoring is set up to allow a second attempt at each task (see the "Test" and "Retest" columns). Scoring is designed to reward students for correct completion of the task. Points are lost for failure to complete the employability requirements (see "Non-Task-Specific" criteria). When all criteria are evaluated, tally the points for a total at the bottom of each column.

Tasksheet Scoring

	Test		Retest	
Evaluation Items	**Pass**	**Fail**	**Pass**	**Fail**
Task-Specific Evaluation	**(1 pt)**	**(0 pts)**	**(1 pt)**	**(0 pts)**
Define camber, caster, and toe, and explain how they are measured.				
Explain tire wear and drift or pull conditions that may be created by incorrect adjustments of each angle.				
Describe the difference between positive and negative alignment angles.				
Describe performance and handling characteristics based on adjusting camber, caster, and toe.				
Non-Task-Specific Evaluation	**(0 pts)**	**(−1 pt)**	**(0 pts)**	**(−1 pt)**
Student successfully completed at least three of the non-task-specific steps.				
Student successfully completed all five of the non-task-specific steps.				
Total Score: <total # of points / 4 = %>				

Supervisor:

Supervisor/instructor signature _______________________ Date _______________

Comments:

Retest supervisor/instructor signature _______________________ Date _______________

Comments:

4D. Wheels and Tires

Learning Objective/Task	CDX Tasksheet Number	ASE Education Foundation Reference Number; Priority Level
• Inspect tire condition; identify tire wear patterns; check for correct tire size, application (load and speed ratings), and air pressure as listed on the tire information placard/label.	A4029	A4D1, P-1
• Rotate tires according to manufacturer's recommendations including vehicles equipped with tire pressure monitoring systems (TPMS).	A4030	A4D2, P-1
• Dismount, inspect, and remount tire on wheel; balance wheel and tire assembly.	A4031	A4D3, P-1
• Dismount, inspect, and remount tire on wheel equipped with tire pressure monitoring system sensor.	A4032	A4D4, P-1
• Inspect tire and wheel assembly for air loss; determine necessary action.	A4033	A4D5, P-1
• Repair tire following vehicle manufacturer approved procedure.	A4034	A4D6, P-1
• Identify indirect and direct tire pressure monitoring systems (TPMS); calibrate system; verify operation of instrument panel lamps.	A4035	A4D7, P-2
• Demonstrate knowledge of steps required to remove and replace sensors in a tire pressure monitoring system (TPMS) including relearn procedure.	A4036	A4D8, P-1

Materials Required

- Blank work order
- Vehicle with available service history records
- Depending on the type of concern, special diagnostic/hand tools may be required. See your supervisor/instructor for instructions to identify what tools may be required.
- Service information database
- Personal protective equipment

Safety Considerations

- When running any vehicles in the shop, make sure you use the shop's exhaust ventilation system to discharge all exhaust gas safely outside.
- Lifting equipment, such as vehicle jacks and stands, vehicle hoists, and engine hoists, are important tools that increase productivity and make the job easier. However, they can also cause severe injury or death if used improperly. Make sure you follow the manufacturer's operation procedures. Also, make sure you have your supervisor's/instructor's permission to use any particular type of lifting equipment.
- Comply with personal and environmental safety practices associated with clothing; eye protection; hand tools; power equipment; proper ventilation; and the handling, storage, and disposal of chemicals/materials in accordance with local, state, and federal safety and environmental regulations.
- If you need to start the vehicle, you should ensure that the parking brake is firmly applied; if necessary, use wheel chocks to prevent the vehicle from moving when the vehicle is started to verify the completion of these tasks.
- Extreme caution must be exercised when working around rotating components.

CDX Tasksheet Number: A4029

Student/Intern Information

Name _____________________________ Date ___________ Class _____________________

Vehicle, Customer, and Service Information

Vehicle used for this activity:

Year _____________ Make _____________________ Model _____________________

Odometer _____________________ VIN _____________________________

Materials Required

- Blank work order
- Vehicle with available service history records
- Depending on the type of concern, special diagnostic/hand tools may be required. See your supervisor/instructor for instructions to identify what tools may be required.
- Service information database
- Personal protective equipment
- Tread depth gauge
- Tire pressure gauge

Task-Specific Safety Considerations

- Lifting equipment, such as vehicle jacks and stands, vehicle hoists, and engine hoists, are important tools that increase productivity and make the job easier. However, they can also cause severe injury or death if used improperly. Make sure you follow the manufacturer's operation procedures. Also, make sure you have your supervisor's/instructor's permission to use any particular type of lifting equipment.
- Comply with personal and environmental safety practices associated with clothing; eye protection; hand tools; power equipment; proper ventilation; and the handling, storage, and disposal of chemicals/materials in accordance with local, state, and federal safety and environmental regulations.

▶ TASK Inspect tire condition; identify tire wear patterns; check for correct tire size, application (load and speed ratings), and air pressure as listed on the tire information placard/label.

**MLR
A4D1**

Time off______________

Time on______________

Total time____________

Student Instructions: Read through the entire procedure prior to starting. Prepare your work space and any tools or parts that may be needed to complete the task. When directed by your instructor, begin the procedure to complete the task and check the box as each step is finished. Track your time on this procedure for later comparison to the standard completion time (i.e., "flat rate" or customer pay time).

Procedure:	Step Completed
1. Locate the tire information label.	
a. Where is the tire information located in your specific vehicle?	☐
b. What size tires are supposed to be on the vehicle? Are the front tires different from the back tires?	☐
c. What speed rating is required for the tires on your vehicle?	☐
d. What load rating is required for the tires on your vehicle?	☐
e. What is the tire pressure specification for your vehicle?	☐
f. Is the vehicle equipped with a full-size spare or a temporary spare tire?	☐
2. Compare the tires on the vehicle to the specifications on the tire information label.	
a. Do the tires match the size, speed, and load specifications?	☐
b. Measure and adjust the tire pressure on your vehicle as needed. Record your measurements before making adjustments.	☐

3. Measure the tread depth of each tire.	
a. Record the tread depth of each tire. What is the minimum tread depth of a tire that is acceptable for road use?	☐
4. Inspect the tire condition for each of the following.	
a. Excessive cracking/dry rot	☐
b. Sidewall bulging	☐
c. Wheel damage	☐
d. Damaged sidewall/tread	☐
e. Punctured tread	☐
5. Inspect the tires for irregular wear. Inspect for the following conditions. Describe what causes the following conditions.	
a. Inside edge feathering	☐
b. Outside edge feathering	☐
c. Cupping	☐
d. Center wear	☐
e. Edge wear on both edges	☐
f. Flat spotting	☐
6. Based on your inspection, what is your recommendation for service?	

<table>
<tr><td>Non-Task-Specific Evaluations:</td><td>Step Completed</td></tr>
<tr><td>1. Tools and equipment were used as directed and returned in good working order.</td><td>☐</td></tr>
<tr><td>2. Complied with all general and task-specific safety standards, including proper use of any personal protective equipment.</td><td>☐</td></tr>
<tr><td>3. Completed the task in an appropriate time frame (Recommendation: 1.5 or 2 times flat rate).</td><td>☐</td></tr>
<tr><td>4. Left the work space clean and orderly.</td><td>☐</td></tr>
<tr><td>5. Cared for customer property and returned it undamaged.</td><td>☐</td></tr>
</table>

Student signature _________________________________ Date _________________________________

Comments:

Have your supervisor/instructor verify satisfactory completion of this procedure, any observations found, and any necessary action(s) recommended.

Evaluation Instructions: The scoring box below is intended to act as a guide for both student and instructor. Each criterion listed will help students understand what is expected from them and evaluators articulate success at a particular task. The scoring is set up to allow a second attempt at each task (see the "Test" and "Retest" columns). Scoring is designed to reward students for correct completion of the task. Points are lost for failure to complete the employability requirements (see "Non-Task-Specific" criteria). When all criteria are evaluated, tally the points for a total at the bottom of each column.

Tasksheet Scoring

Evaluation Items	Test		Retest	
	Pass	Fail	Pass	Fail
Task-Specific Evaluation	**(1 pt)**	**(0 pts)**	**(1 pt)**	**(0 pts)**
Locate the tire information label and record information.				
Inspect the tire pressure and tread depth. Adjust the tire pressure to meet specification.				
Inspect the tire for overall condition and irregular tire wear. Describe the cause of any irregular wear.				
Make recommendations for service based on your inspection.				
Non-Task-Specific Evaluation	**(0 pts)**	**(−1 pt)**	**(0 pts)**	**(−1 pt)**
Student successfully completed at least three of the non-task-specific steps.				
Student successfully completed all five of the non-task-specific steps.				
Total Score: <total # of points / 4 = %>				

Supervisor:

Supervisor/instructor signature ____________________ Date ____________________

Comments:

Retest supervisor/instructor signature ____________________ Date ____________________

Comments:

CDX Tasksheet Number: A4030

Student/Intern Information

Name _________________________________ Date ____________ Class _________________________

Vehicle, Customer, and Service Information

Vehicle used for this activity:

Year _______________ Make _________________________ Model _________________________

Odometer _________________________ VIN _________________________________

> **Materials Required**
> - Blank work order
> - Vehicle with available service history records
> - Depending on the type of concern, special diagnostic/hand tools may be required. See your supervisor/instructor for instructions to identify what tools may be required.
> - Service information database
> - Personal protective equipment
> - General hand/air tools
> - Impact sockets
> - Torque-limiting extension bar (torque sticks)
> - Torque wrench
> - TPMS tools

Task-Specific Safety Considerations

- Lifting equipment, such as vehicle jacks and stands, vehicle hoists, and engine hoists, are important tools that increase productivity and make the job easier. However, they can also cause severe injury or death if used improperly. Make sure you follow the manufacturer's operation procedures. Also, make sure you have your supervisor's/instructor's permission to use any particular type of lifting equipment.
- Comply with personal and environmental safety practices associated with clothing; eye protection; hand tools; power equipment; proper ventilation; and the handling, storage, and disposal of chemicals/materials in accordance with local, state, and federal safety and environmental regulations.

Time off_____________

Time on_____________

Total time_____________

▶ TASK Rotate tires according to manufacturer's recommendations including vehicles equipped with tire pressure monitoring systems (TPMS).

MLR
A4D2

Student Instructions: Read through the entire procedure prior to starting. Prepare your work space and any tools or parts that may be needed to complete the task. When directed by your instructor, begin the procedure to complete the task and check the box as each step is finished. Track your time on this procedure for later comparison to the standard completion time (i.e., "flat rate" or customer pay time).

Procedure:	Step Completed
1. Research the procedure for rotating tires on your specific vehicle. Research the TPMS reset procedure.	
a. List or print off the procedure for rotating tires on your vehicle. Describe the positioning of the tires before and after the rotation.	☐
b. What is the recommended interval for rotating the tires on your vehicle?	☐
c. What is the torque specification for the lug nuts?	☐
d. What is the tire pressure specification for your vehicle?	☐
e. List or print off the procedure for resetting the TPMS.	☐
f. Are there any special tools required for the TPMS reset procedure?	☐
g. Explain the sequence/pattern that is used when tightening lug nuts.	☐
2. Rotate the tires	
a. Lift the vehicle. Verify correct tire pressure before you rotate the tires. **Tip:** It is a good habit to check tire pressure first. This way, if you find that a tire is low, you can inspect the tire once you have it off. Don't wait to check the pressure until after you have rotated and tightened the wheels.	☐
b. Using the air impact and correct socket, loosen all lug nuts. What size socket did you use?	☐

c. Following procedure, rotate the tires in the correct pattern.	☐
d. *Start all lug nuts by hand.* Using the proper torque stick, use the impact to tighten the lug nuts in the correct sequence. Explain what may happen if lug nuts are not tightened in the correct sequence. **Tip:** To avoid breaking wheel studs, place the impact on the lowest torque setting when tightening the lug nuts.	☐
e. Lower the vehicle until the wheels are just touching the ground. At this point, use a torque wrench to verify that the lug nuts have been tightened to the correct specification.	☐
3. Reset the TPMS.	
a. Following procedure, reset the TPMS.	☐
b. Explain what may happen if the TPMS is not reset after a tire rotation service.	☐

Non-Task-Specific Evaluations:	Step Completed
1. Tools and equipment were used as directed and returned in good working order.	☐
2. Complied with all general and task-specific safety standards, including proper use of any personal protective equipment.	☐
3. Completed the task in an appropriate time frame (Recommendation: 1.5 or 2 times flat rate).	☐
4. Left the work space clean and orderly.	☐
5. Cared for customer property and returned it undamaged.	☐

Student signature _________________________________ Date _____________________________

Comments:

Have your supervisor/instructor verify satisfactory completion of this procedure, any observations found, and any necessary action(s) recommended.

Evaluation Instructions: The scoring box below is intended to act as a guide for both student and instructor. Each criterion listed will help students understand what is expected from them and evaluators articulate success at a particular task. The scoring is set up to allow a second attempt at each task (see the "Test" and "Retest" columns). Scoring is designed to reward students for correct completion of the task. Points are lost for failure to complete the employability requirements (see "Non-Task-Specific" criteria). When all criteria are evaluated, tally the points for a total at the bottom of each column.

Tasksheet Scoring

		Test		Retest	
Evaluation Items	**Pass**	**Fail**	**Pass**	**Fail**
Task-Specific Evaluation	**(1 pt)**	**(0 pts)**	**(1 pt)**	**(0 pts)**
Research the tire rotation and TPMS reset procedures. Identify the lug nut torque and tire pressure specifications.				
Remove the tires and rotate them in the correct order. Select the correct socket and use tools as directed.				
Install the tires. Start all lug nuts by hand. Use the correct torque stick and sequence when tightening lug nuts. Perform the final torqueing of the lug nuts.				
Perform the TPMS reset procedure.				
Non-Task-Specific Evaluation	**(0 pts)**	**(–1 pt)**	**(0 pts)**	**(–1 pt)**
Student successfully completed at least three of the non-task-specific steps.				
Student successfully completed all five of the non-task-specific steps.				
Total Score: <total # of points / 4 = %>				

Supervisor:

Supervisor/instructor signature _________________________________ Date _____________________

Comments:

Retest supervisor/instructor signature _________________________________ Date _____________________

Comments:

CDX Tasksheet Number: A4031

Student/Intern Information

Name _______________________________ Date _____________ Class _______________________

Vehicle, Customer, and Service Information

Vehicle used for this activity:

Year _______________ Make _________________________ Model _____________________

Odometer _________________________ VIN _________________________________

> ### Materials Required
>
> - Blank work order
> - Vehicle with available service history records
> - Depending on the type of concern, special diagnostic/hand tools may be required. See your supervisor/instructor for instructions to identify what tools may be required.
> - Service information database
> - Personal protective equipment
> - Wheel weight tool
> - Valve core tool
> - Wire brush
> - Balance weights

Task-Specific Safety Considerations

- Lifting equipment, such as vehicle jacks and stands, vehicle hoists, and engine hoists, are important tools that increase productivity and make the job easier. However, they can also cause severe injury or death if used improperly. Make sure you follow the manufacturer's operation procedures. Also, make sure you have your supervisor's/instructor's permission to use any particular type of lifting equipment.
- Comply with personal and environmental safety practices associated with clothing; eye protection; hand tools; power equipment; proper ventilation; and the handling, storage, and disposal of chemicals/materials in accordance with local, state, and federal safety and environmental regulations.
- Extreme caution must be exercised when working around rotating components.

▶ TASK Dismount, inspect, and remount tire on wheel; balance wheel and tire assembly.

MLR
A4D3

Student Instructions: Read through the entire procedure prior to starting. Prepare your work space and any tools or parts that may be needed to complete the task. When directed by your instructor, begin the procedure to complete the task and check the box as each step is finished. Track your time on this procedure for later comparison to the standard completion time (i.e., "flat rate" or customer pay time).

Procedure:	Step Completed
1. Dismount the tire.	
a. Remove the old weights using the wheel weight hammer/plier tool. Explain why it is necessary to remove the old weights prior to removing the tire from the wheel.	☐
b. Deflate the tire by removing the valve core with the valve core tool.	☐
c. Break down the bead on both sides of the tire. **Tip:** It may be helpful to apply lubricant around the tire bead as you break it down. Although this is not always necessary, it helps as you are removing the tire from the rim.	☐
d. Mount the wheel on the tire machine. Ensure that the wheel is held firmly in place. Lower the vertical arm and seat it against the outside flange of the wheel.	☐
e. Use the correct pry bar to remove each tire bead, starting with the top bead.	☐
2. Inspect tire	
a. Once the tire has been dismounted, perform a visual inspection of the tire bead and inside the tire. Ensure there are no punctures or evidence that the tire has been run low. Describe the condition of the tire. How might you know if a tire has been run low on pressure?	☐
b. Inspect the wheel flange for signs of excess rust, corrosion, or damage.	☐
3. Remount the tire.	
a. Before mounting the original or new tire, clean the bead sealing surface of the wheel. **Tip:** Wheels are prone to rust and corrosion build up, which will create air leaks. Use a wire brush or similar tool to thoroughly clean the wheel flange and bead of the tire.	☐
b. Remove the old rubber valve stem and replace it with a new one. Take special care not to gouge the wheel when removing or installing a valve stem.	☐
c. If necessary, apply bead sealer to the wheel flange to ensure a good seal.	☐
d. Apply a small amount of tire lube to the beads. Install the tire onto the wheel.	☐
e. Once the tire is on the wheel, inflate the tire to specification. **Caution:** As the bead seats, a loud noise may be heard. Keep your fingers away from the tire bead as it seats. Some tires must be placed in a safety cage when they are being inflated. Refer to your instructor for further information.	☐

4. Balance the wheel/tire assembly.	
a. What type of wheel weights are needed for your wheel?	☐
b. Mount the wheel to the balancer using the correct adapters. Ensure that the wheel is tight before moving on. Double check to make sure all weights have been removed.	☐
c. Follow the prompts on the screen to calibrate the machine. What measurements must be taken before balancing a wheel and tire assembly?	☐
d. Allow the machine to spin the wheel and take the initial measurements. How much weight is needed to balance the wheel/tire? Are there any other warning indicators at this time?	☐
e. Apply the needed weights in the proper position. Allow the machine to spin the wheel/tire assembly once more to verify correct placement of the weights. If adjustments are needed a second time, it is good practice to remove all weights and start over rather than continually adding weights.	☐

Non-Task-Specific Evaluations:	Step Completed
1. Tools and equipment were used as directed and returned in good working order.	☐
2. Complied with all general and task-specific safety standards, including proper use of any personal protective equipment.	☐
3. Completed the task in an appropriate time frame (Recommendation: 1.5 or 2 times flat rate).	☐
4. Left the work space clean and orderly.	☐
5. Cared for customer property and returned it undamaged.	☐

Student signature ________________________________ Date ________________________________

Comments:

Have your supervisor/instructor verify satisfactory completion of this procedure, any observations found, and any necessary action(s) recommended.

Tasksheet Scoring

		Test		Retest	
Evaluation Items	**Pass**	**Fail**	**Pass**	**Fail**
Task-Specific Evaluation	**(1 pt)**	**(0 pts)**	**(1 pt)**	**(0 pts)**
Dismount the tire.				
Inspect the tire and wheel.				
Clean the wheel flange, apply bead sealer, and install the tire.				
Calibrate the wheel balancer and then balance the wheel/tire.				
Non-Task-Specific Evaluation	**(0 pts)**	**(−1 pt)**	**(0 pts)**	**(−1 pt)**
Student successfully completed at least three of the non-task-specific steps.				
Student successfully completed all five of the non-task-specific steps.				
Total Score: <total # of points / 4 = %>				

Supervisor:

Supervisor/instructor signature _________________________ Date ________________

Comments:

Retest supervisor/instructor signature _________________________ Date ________________

Comments:

CDX Tasksheet Number: A4032

Student/Intern Information

Name _________________________________ Date ____________ Class _________________________

Vehicle, Customer, and Service Information

Vehicle used for this activity:

Year _______________ Make _______________________ Model _______________________

Odometer _______________________ VIN _______________________________

Materials Required

- Blank work order
- Vehicle with available service history records
- Depending on the type of concern, special diagnostic/hand tools may be required. See your supervisor/instructor for instructions to identify what tools may be required.
- Service information database
- Personal protective equipment
- Wheel weight tool
- Valve core tool
- Wire brush
- Inch pound torque wrench
- Balance weights

Task-Specific Safety Considerations

- Comply with personal and environmental safety practices associated with clothing; eye protection; hand tools; power equipment; proper ventilation; and the handling, storage, and disposal of chemicals/materials in accordance with local, state, and federal safety and environmental regulations.

▶ TASK Dismount, inspect, and remount tire on wheel equipped with tire pressure monitoring system sensor.

MLR
A4D4

Time off____________

Time on____________

Total time____________

Student Instructions: Read through the entire procedure prior to starting. Prepare your work space and any tools or parts that may be needed to complete the task. When directed by your instructor, begin the procedure to complete the task and check the box as each step is finished. Track your time on this procedure for later comparison to the standard completion time (i.e., "flat rate" or customer pay time).

Procedure:	Step Completed
1. Determine if your vehicle is equipped with a tire pressure sensor.	
a. Refer to service information and describe where the tire pressure sensor is located on your vehicle. Is the sensor part of the valve stem? Is the sensor a band-style mounted around the drop center of the wheel?	☐
b. According to the service information, what is the recommended procedure for removing and replacing a tire on a wheel equipped with a pressure sensor? Are there any special service precautions?	☐
2. Dismount the tire.	
a. Remove the old weights. Use the wheel weight hammer/plier tool to remove the old weights. Explain why it is necessary to remove the old weights prior to removing the tire from the wheel.	☐
b. Deflate the tire by removing the valve core with the valve core tool.	☐
c. Break down the bead on both sides of the tire. **Tip:** It may be helpful to apply lubricant around the tire bead as you break it down. Although this is not always necessary, it helps as you are removing the tire from the rim. **Caution:** Tire pressure sensors may be easily damaged at this point. Refer to the service information for proper procedure. You may be required to loosen the sensor retaining nut and drop the sensor down into the tire prior to breaking down the bead.	☐
d. Mount the wheel on the tire machine. Ensure that the wheel is held firmly in place. Lower the vertical arm and seat it against the outside flange of the wheel.	☐
e. Use the correct pry bar to remove each tire bead, starting with the top bead. Remove the tire pressure sensor if it was loosened in the previous steps.	☐
3. Inspect the tire and tire pressure sensor.	
a. Once the tire has been dismounted, perform a visual inspection of the tire bead and inside the tire. Ensure there are no punctures or evidence that the tire has been run low. Describe the condition of the tire. How might you know if a tire has been run low on pressure?	☐

b. Inspect the wheel flange for signs of excess rust, corrosion, or damage.	☐
c. Inspect the pressure sensor for signs of damage. Inspect the valve stem seal for excess corrosion and replace with a new seal if needed.	☐
4. Remount the tire.	
a. Before mounting the original or new tire, clean the bead sealing surface of the wheel. **Tip:** Wheels are prone to rust and corrosion build up which will create air leaks. Use a wire brush or similar tool to thoroughly clean the wheel flange and bead of the tire.	☐
b. If necessary, apply bead sealer to the wheel flange to ensure a good seal.	☐
c. Apply a small amount of tire lube to the beads. Install the tire onto the wheel.	☐
d. Push down on the top bead and install the tire pressure sensor if it was removed in previous steps. Torque the sensor nut to specification. What specification did you use?	☐
e. Once the tire is on the wheel, inflate the tire to specification. **Caution:** As the bead seats, a loud noise may be heard. Keep your fingers away from the tire bead as it seats. Some tires must be placed in a safety cage when they are being inflated. Refer to your instructor for further information.	☐
5. Balance the wheel/tire assembly.	
a. What type of wheel weights are needed for your wheel?	☐
b. Mount the wheel to the balancer using the correct adapters. Ensure that the wheel is tight before moving on. Double check to make sure all weights have been removed.	☐
c. Follow the prompts on the screen to calibrate the machine. What measurements must be taken before balancing a wheel and tire assembly?	☐
d. Allow the machine to spin the wheel and take the initial measurements. How much weight is needed to balance the wheel/tire? Are there any other warning indicators at this time?	☐

| e. Apply the needed weights in the proper position. Allow the machine to spin the wheel/tire assembly once more to verify correct placement of the weights. If adjustments are needed a second time, it is good practice to remove all weights and start over rather than continually adding weights. | ☐ |

Non-Task-Specific Evaluations:	**Step Completed**
1. Tools and equipment were used as directed and returned in good working order.	☐
2. Complied with all general and task-specific safety standards, including proper use of any personal protective equipment.	☐
3. Completed the task in an appropriate time frame (Recommendation: 1.5 or 2 times flat rate).	☐
4. Left the work space clean and orderly.	☐
5. Cared for customer property and returned it undamaged.	☐

Student signature _________________________ Date _________________________

Comments:

Have your supervisor/instructor verify satisfactory completion of this procedure, any observations found, and any necessary action(s) recommended.

Evaluation Instructions: The scoring box below is intended to act as a guide for both student and instructor. Each criterion listed will help students understand what is expected from them and evaluators articulate success at a particular task. The scoring is set up to allow a second attempt at each task (see the "Test" and "Retest" columns). Scoring is designed to reward students for correct completion of the task. Points are lost for failure to complete the employability requirements (see "Non-Task-Specific" criteria). When all criteria are evaluated, tally the points for a total at the bottom of each column.

Tasksheet Scoring

	Test		Retest	
Evaluation Items	**Pass**	**Fail**	**Pass**	**Fail**
Task-Specific Evaluation	**(1 pt)**	**(0 pts)**	**(1 pt)**	**(0 pts)**
Dismount the tire. Take special care not to damage the tire pressure sensor. Inspect the tire, wheel, and sensor.				
Clean the wheel flange, apply bead sealer, and install the tire.				
Calibrate the wheel balancer.				
Balance the wheel/tire.				
Non-Task-Specific Evaluation	**(0 pts)**	**(−1 pt)**	**(0 pts)**	**(−1 pt)**
Student successfully completed at least three of the non-task-specific steps.				
Student successfully completed all five of the non-task-specific steps.				
Total Score: <total # of points / 4 = %>				

CDX Tasksheet Number: A4033

Student/Intern Information

Name ___________________________ Date __________ Class _______________________

Vehicle, Customer, and Service Information

Vehicle used for this activity:

Year _____________ Make ___________________________ Model ___________________________

Odometer ___________________________ VIN ___________________________

> **Materials Required**
> - Blank work order
> - Vehicle with available service history records
> - Depending on the type of concern, special diagnostic/hand tools may be required. See your supervisor/instructor for instructions to identify what tools may be required.
> - Service information database
> - Personal protective equipment
> - Tire pressure gauge
> - General hand/air tools
> - Spray bottle
> - Dunk tank

Task-Specific Safety Considerations

- Comply with personal and environmental safety practices associated with clothing; eye protection; hand tools; power equipment; proper ventilation; and the handling, storage, and disposal of chemicals/materials in accordance with local, state, and federal safety and environmental regulations.

Time off________

Time on________

Total time________

▶ TASK Inspect tire and wheel assembly for air loss; determine necessary action.

MLR
A4D5

Student Instructions: Read through the entire procedure prior to starting. Prepare your work space and any tools or parts that may be needed to complete the task. When directed by your instructor, begin the procedure to complete the task and check the box as each step is finished. Track your time on this procedure for later comparison to the standard completion time (i.e., "flat rate" or customer pay time).

Procedure:	Step Completed
1. Perform a preliminary inspection.	
a. If the vehicle is equipped with a TPMS, are there any warning indicators illuminated? Does the gauge cluster offer any information on tire pressure for each individual tire?	☐
b. Lift the vehicle. Check the tire pressure on all four wheels and compare them to specifications. Are any tires low on pressure?	☐
c. Use a spray bottle filled with soapy water and spray the tire that shows air pressure loss. Spray around the beads, both front and back. Spray around the entire tread surface. Unscrew the valve stem cap and spray the valve core and at the base of the valve stem. A large formation of bubbles will accumulate around a leaking portion of the tire. Describe your results. What area is leaking?	☐
2. Perform further inspection.	
a. If further inspection is required, remove the wheel assembly using the correct socket and air impact.	☐
b. Use a dunk tank filled with water and dip the wheel/tire assembly in the water. Slowly rotate the tire and look for signs of air bubbles escaping from the assembly. **Tip:** It is helpful to ensure that the tire is fully inflated when performing this checking. Carefully inspect the wheel as well as the tire for leaks.	☐
3. Based on your inspection, what is your recommendation for service?	☐
a. Explain what parts of a tire are repairable and what type of damage or leaks would require that a tire be replaced.	☐
b. Explain what type of repair is needed to fix the pressure loss in your wheel/tire assembly.	☐

Non-Task-Specific Evaluations:	Step Completed
1. Tools and equipment were used as directed and returned in good working order.	☐
2. Complied with all general and task-specific safety standards, including proper use of any personal protective equipment.	☐
3. Completed the task in an appropriate time frame (Recommendation: 1.5 or 2 times flat rate).	☐
4. Left the work space clean and orderly.	☐
5. Cared for customer property and returned it undamaged.	☐

Student signature _______________________________ Date _______________________________

Comments:

Have your supervisor/instructor verify satisfactory completion of this procedure, any observations found,

and any necessary action(s) recommended.

Evaluation Instructions: The scoring box below is intended to act as a guide for both student and instructor. Each criterion listed will help students understand what is expected from them and evaluators articulate success at a particular task. The scoring is set up to allow a second attempt at each task (see the "Test" and "Retest" columns). Scoring is designed to reward students for correct completion of the task. Points are lost for failure to complete the employability requirements (see "Non-Task-Specific" criteria). When all criteria are evaluated, tally the points for a total at the bottom of each column.

Tasksheet Scoring

	Test		Retest	
Evaluation Items	**Pass**	**Fail**	**Pass**	**Fail**
Task-Specific Evaluation	**(1 pt)**	**(0 pts)**	**(1 pt)**	**(0 pts)**
Perform a preliminary tire pressure inspection. Identify if the vehicle has a TPMS and determine if any warning lights are illuminated.				
Use soapy water or a dunk tank to determine cause of pressure loss in a tire.				
Explain what parts of a tire can be repaired and what constitutes tire replacement.				
Make recommendations for service based on your inspection.				
Non-Task-Specific Evaluation	**(0 pts)**	**(−1 pt)**	**(0 pts)**	**(−1 pt)**
Student successfully completed at least three of the non-task-specific steps.				
Student successfully completed all five of the non-task-specific steps.				
Total Score: <total # of points / 4 = %>				

Supervisor:

Supervisor/instructor signature _____________________________ Date _________________

Comments:

Retest supervisor/instructor signature _____________________________ Date _________________

Comments:

CDX Tasksheet Number: A4034

Student/Intern Information

Name _________________________________ Date ____________ Class _____________________

Vehicle, Customer, and Service Information

Vehicle used for this activity:

Year _______________ Make _______________________ Model _____________________

Odometer _____________________________ VIN ___________________________________

Materials Required

- Blank work order
- Vehicle with available service history records
- Depending on the type of concern, special diagnostic/hand tools may be required. See your supervisor/instructor for instructions to identify what tools may be required.
- Service information database
- Personal protective equipment
- General hand/air tools
- Tire machine and equipment
- Wheel balancer
- Patch and tire repair tools

Task-Specific Safety Considerations

- Lifting equipment, such as vehicle jacks and stands, vehicle hoists, and engine hoists, are important tools that increase productivity and make the job easier. However, they can also cause severe injury or death if used improperly. Make sure you follow the manufacturer's operation procedures. Also, make sure you have your supervisor's/instructor's permission to use any particular type of lifting equipment.
- Comply with personal and environmental safety practices associated with clothing; eye protection; hand tools; power equipment; proper ventilation; and the handling, storage, and disposal of chemicals/materials in accordance with local, state, and federal safety and environmental regulations.

▶ TASK Repair tire following vehicle manufacturer approved procedure.

MLR
A4D6

Time off________________

Time on________________

Total time________________

Student Instructions: Read through the entire procedure prior to starting. Prepare your work space and any tools or parts that may be needed to complete the task. When directed by your instructor, begin the procedure to complete the task and check the box as each step is finished. Track your time on this procedure for later comparison to the standard completion time (i.e., "flat rate" or customer pay time).

Procedure:	Step Completed
1. Describe the tire repair procedure.	
a. What parts of a tire are able to be repaired?	☐
b. Damage to what areas of a tire would require that tire to be replaced?	☐
c. Explain an acceptable method to repair a tire according to your service information. Give an example of a tire repair that is not recommended.	☐
2. Repair the tire following the manufacturer's approved procedure.	
a. Once you have identified the tire that is low on pressure and the source of its leak, determine if the tire is repairable.	☐
b. What method did you use to determine the source of the pressure leak?	☐
c. Remove the tire from the vehicle. Refer to the instructions on Tasksheets A4031/4032 and dismount the tire that needs repaired.	☐
d. Once the tire is removed, further inspect it for damage. If there are signs that the tire has been driven on excessively with low pressure, it is advisable to replace the tire.	☐
e. If the tire is being repaired because of a puncture, remove the foreign object with a pair of pliers.	☐
f. Buff the area where the patch will go. **Tip:** Only buff an area large enough for the tire patch to fit. Be careful not to remove excess rubber and expose the metal reinforcement area underneath.	☐
g. Ream the hole out using the correct size reamer to match your patch plug.	☐
h. If necessary, clean the buffed area with an approved tire cleaner solvent. Apply a small amount of vulcanizing cement to the buffed area.	☐

i. Insert the patch plug into the hole and use the stitcher to seat the patch against the tire.	☐
j. Cut off the protruding plug stem flush with the tire tread on the outside of the tire.	☐
k. Refer to instructions from Tasksheets A4031/4032 and remount and balance the tire.	☐
l. Verify that the tire is leak-free. Mount the tire on the vehicle using the correct torque specifications and sequence. What torque specifications did you use?	☐

Non-Task-Specific Evaluations:	Step Completed
1. Tools and equipment were used as directed and returned in good working order.	☐
2. Complied with all general and task-specific safety standards, including proper use of any personal protective equipment.	☐
3. Completed the task in an appropriate time frame (Recommendation: 1.5 or 2 times flat rate).	☐
4. Left the work space clean and orderly.	☐
5. Cared for customer property and returned it undamaged.	☐

Student signature _________________________________ Date _________________________________

Comments:

Have your supervisor/instructor verify satisfactory completion of this procedure, any observations found, and any necessary action(s) recommended.

Evaluation Instructions: The scoring box below is intended to act as a guide for both student and instructor. Each criterion listed will help students understand what is expected from them and evaluators articulate success at a particular task. The scoring is set up to allow a second attempt at each task (see the "Test" and "Retest" columns). Scoring is designed to reward students for correct completion of the task. Points are lost for failure to complete the employability requirements (see "Non-Task-Specific" criteria). When all criteria are evaluated, tally the points for a total at the bottom of each column.

Tasksheet Scoring

	Test		Retest	
Evaluation Items	**Pass**	**Fail**	**Pass**	**Fail**
Task-Specific Evaluation	**(1 pt)**	**(0 pts)**	**(1 pt)**	**(0 pts)**
Explain an acceptable method to repair a tire according to your service information.				
Determine the source of pressure loss. Determine if tire is repairable.				
Dismount the tire following proper procedure and use approved method for repairing a punctured tire.				
Remount the tire, verify the tire is now leak-free. Balance the wheel/tire assembly. Mount the wheel on the vehicle using the correct torque specifications and sequence.				
Non-Task-Specific Evaluation	**(0 pts)**	**(−1 pt)**	**(0 pts)**	**(−1 pt)**
Student successfully completed at least three of the non-task-specific steps.				
Student successfully completed all five of the non-task-specific steps.				
Total Score: <total # of points / 4 = %>				

Supervisor:

Supervisor/instructor signature _______________________________ Date _________________

Comments:

Retest supervisor/instructor signature _______________________________ Date _________________

Comments:

CDX Tasksheet Number: A4035

Student/Intern Information

Name _________________________________ Date ____________ Class _____________________

Vehicle, Customer, and Service Information

Vehicle used for this activity:

Year ______________ Make _____________________________ Model _____________________________

Odometer _________________________________ VIN _________________________________

Materials Required

- Blank work order
- Vehicle with available service history records
- Depending on the type of concern, special diagnostic/hand tools may be required. See your supervisor/instructor for instructions to identify what tools may be required.
- Service information database
- Personal protective equipment
- TPMS tools/scan tool

Task-Specific Safety Considerations

- Comply with personal and environmental safety practices associated with clothing; eye protection; hand tools; power equipment; proper ventilation; and the handling, storage, and disposal of chemicals/materials in accordance with local, state, and federal safety and environmental regulations.

▶ **TASK** Identify indirect and direct tire pressure monitoring systems (TPMS); calibrate system; verify operation of instrument panel lamps.

**MLR
A4D7**

Time off________________

Time on________________

Total time________________

Student Instructions: Read through the entire procedure prior to starting. Prepare your work space and any tools or parts that may be needed to complete the task. When directed by your instructor, begin the procedure to complete the task and check the box as each step is finished. Track your time on this procedure for later comparison to the standard completion time (i.e., "flat rate" or customer pay time).

Procedure:	Step Completed
1. Research the procedure for calibrating the TPMS. Identify differences in direct and indirect type TPMS.	
a. List or print off the procedure for calibrating the TPMS in your specific vehicle.	☐

b. Are there any special tools or precautions to complete this task?	☐
c. Is your vehicle equipped with a direct or indirect TPMS? How can you tell?	☐
d. Explain the difference between a direct and indirect TPMS. Describe how a low tire pressure condition is recognized in both systems.	☐
e. Explain the conditions that need to be met in order to set a TPMS warning light on the dash.	☐
2. Verify the operation of the instrument panel lamps.	
a. Cycle the key to the ON position. Verify that the TPMS light comes on and turns off in 3–5 seconds.	☐
b. Does the TPMS light remain on if the vehicle is started?	☐
c. Explain what a continuously flashing TPMS light means. What does it mean if the light remains on steady?	☐
d. If no light comes on, try to meet the conditions required to set a light. Relieve some tire pressure and test drive vehicle if required. Caution: If a test drive is required, follow the EXACT procedure. Were you able to set a TPMS light?	☐
3. Calibrate the TPMS.	
a. Following procedure and using the correct TPMS tools, record the serial numbers, tire pressure, and tire temperature from all tires that have a tire pressure sensor.	☐
b. Following procedure, inflate all tires to specification and perform a TPMS calibration.	☐

Non-Task-Specific Evaluations:	Step Completed
1. Tools and equipment were used as directed and returned in good working order.	☐
2. Complied with all general and task-specific safety standards, including proper use of any personal protective equipment.	☐
3. Completed the task in an appropriate time frame (Recommendation: 1.5 or 2 times flat rate).	☐
4. Left the work space clean and orderly.	☐
5. Cared for customer property and returned it undamaged.	☐

Student signature _______________________________ Date _______________________________

Comments:

Have your supervisor/instructor verify satisfactory completion of this procedure, any observations found, and any necessary action(s) recommended.

Evaluation Instructions: The scoring box below is intended to act as a guide for both student and instructor. Each criterion listed will help students understand what is expected from them and evaluators articulate success at a particular task. The scoring is set up to allow a second attempt at each task (see the "Test" and "Retest" columns). Scoring is designed to reward students for correct completion of the task. Points are lost for failure to complete the employability requirements (see "Non-Task-Specific" criteria). When all criteria are evaluated, tally the points for a total at the bottom of each column.

Tasksheet Scoring

Evaluation Items	Test		Retest	
	Pass	Fail	Pass	Fail
Task-Specific Evaluation	**(1 pt)**	**(0 pts)**	**(1 pt)**	**(0 pts)**
Research the procedure for calibrating the TPMS. Verify correct operation of the instrument panel warning lights.				
Identify the differences between direct and indirect TPMS.				
Identify the TPMS system equipped on your vehicle.				
Perform a TPMS calibration.				
Non-Task-Specific Evaluation	**(0 pts)**	**(−1 pt)**	**(0 pts)**	**(−1 pt)**
Student successfully completed at least three of the non-task-specific steps.				
Student successfully completed all five of the non-task-specific steps.				
Total Score: <total # of points / 4 = %>				

Supervisor:

Supervisor/instructor signature _____________________________ Date ___________________

Comments:

Retest supervisor/instructor signature _____________________________ Date ___________________

Comments:

CDX Tasksheet Number: A4036

Student/Intern Information

Name _________________________________ Date ___________ Class _________________________

Vehicle, Customer, and Service Information

Vehicle used for this activity:

Year _______________ Make _______________________ Model _________________________

Odometer _________________________ VIN _________________________________

Materials Required

- Blank work order
- Vehicle with available service history records
- Depending on the type of concern, special diagnostic/hand tools may be required. See your supervisor/instructor for instructions to identify what tools may be required.
- Service information database
- Personal protective equipment
- General hand/air tools
- Tire machine and equipment
- Wheel balancer
- TPMS tools/ scan tool

Task-Specific Safety Considerations

- Lifting equipment, such as vehicle jacks and stands, vehicle hoists, and engine hoists, are important tools that increase productivity and make the job easier. However, they can also cause severe injury or death if used improperly. Make sure you follow the manufacturer's operation procedures. Also, make sure you have your supervisor's/instructor's permission to use any particular type of lifting equipment.
- Comply with personal and environmental safety practices associated with clothing; eye protection; hand tools; power equipment; proper ventilation; and the handling, storage, and disposal of chemicals/materials in accordance with local, state, and federal safety and environmental regulations.

Time off________________

Time on________________

Total time________________

▶ TASK Demonstrate knowledge of steps required to remove and replace sensors in a tire pressure monitoring system (TPMS) including relearn procedure.

MLR
A4D8

Student Instructions: Read through the entire procedure prior to starting. Prepare your work space and any tools or parts that may be needed to complete the task. When directed by your instructor, begin the procedure to complete the task and check the box as each step is finished. Track your time on this procedure for later comparison to the standard completion time (i.e., "flat rate" or customer pay time).

Procedure:	Step Completed
1. Research the TPMS sensor removal and calibration/relearn procedures.	
a. List or print off the procedure for replacing a tire pressure sensor with a new sensor and preforming a relearn procedure.	☐
b. What special tools are needed for sensor removal, installation, and calibration?	☐
c. Is there a torque specification for the sensor? If so, what is the specification?	☐
2. Remove the wheel/tire from the vehicle and dismount the tire. Replace the faulty TPMS sensor.	
a. Determine what wheel has a faulty TPMS sensor. Remove the wheel and prepare to dismount the tire.	☐
b. Refer to instructions from Tasksheets A4031/4032 and remove the faulty tire pressure sensor.	☐
c. Record the serial number from the faulty sensor as well as the new replacement sensor, as this information may be required for the relearn procedure.	☐
d. Install the new sensor and torque to specification. Remount and balance the tire.	☐
e. Install the wheel on the vehicle and torque to specification using the correct sequence. What torque specification did you use?	☐
3. Perform the relearn procedure.	
a. Using proper procedure, perform the relearn procedure for the TPMS and new sensor.	☐
b. Verify that there are no TPMS warning lights illuminated. Verify that all tire pressures are set to specification.	☐

Non-Task-Specific Evaluations:	Step Completed
1. Tools and equipment were used as directed and returned in good working order.	☐
2. Complied with all general and task-specific safety standards, including proper use of any personal protective equipment.	☐
3. Completed the task in an appropriate time frame (Recommendation: 1.5 or 2 times flat rate).	☐
4. Left the work space clean and orderly.	☐
5. Cared for customer property and returned it undamaged.	☐

Student signature ___________________________ Date ___________________________

Comments:

Have your supervisor/instructor verify satisfactory completion of this procedure, any observations found, and any necessary action(s) recommended.

Evaluation Instructions: The scoring box below is intended to act as a guide for both student and instructor. Each criterion listed will help students understand what is expected from them and evaluators articulate success at a particular task. The scoring is set up to allow a second attempt at each task (see the "Test" and "Retest" columns). Scoring is designed to reward students for correct completion of the task. Points are lost for failure to complete the employability requirements (see "Non-Task-Specific" criteria). When all criteria are evaluated, tally the points for a total at the bottom of each column.

Tasksheet Scoring

		Test		Retest	
Evaluation Items	Pass	Fail	Pass	Fail
Task-Specific Evaluation	**(1 pt)**	**(0 pts)**	**(1 pt)**	**(0 pts)**
Research the TPMS relearn procedure.				
Dismount the tire and remove the faulty TPMS sensor. Record the serial number from the faulty sensor.				
Install the new sensor and remount the tire. Balance the wheel/tire assembly.				
Perform the TPMS relearn procedure.				
Non-Task-Specific Evaluation	**(0 pts)**	**(−1 pt)**	**(0 pts)**	**(−1 pt)**
Student successfully completed at least three of the non-task-specific steps.				
Student successfully completed all five of the non-task-specific steps.				
Total Score: <total # of points / 4 = %>				

Supervisor:

Supervisor/instructor signature ________________________________ Date ________________

Comments:

Retest supervisor/instructor signature ________________________________ Date ________________

Comments:

5A. Brakes General

Learning Objective/Task	CDX Tasksheet Number	ASE Education Foundation Reference Number; Priority Level
• Research vehicle service information, including fluid type, vehicle service history, service precautions, and technical service bulletins.	A5001	A5A1, P-1
• Describe procedure for performing a road test to check brake system operation, including an antilock brake system (ABS).	A5002	A5A2, P-1
• Install wheel and torque lug nuts.	A5003	A5A3, P-1
• Identify brake system components and configuration.	A5004	A5A4, P-1

Materials Required

- Blank work order
- Vehicle with available service history records
- Depending on the type of concern, special diagnostic/hand tools may be required. See your supervisor/instructor for instructions to identify what tools may be required.
- Service information database
- Personal protective equipment

Safety Considerations

- When running any vehicles in the shop, make sure you use the shop's exhaust ventilation system to discharge all exhaust gas safely outside.
- Lifting equipment, such as vehicle jacks and stands, vehicle hoists, and engine hoists, are important tools that increase productivity and make the job easier. However, they can also cause severe injury or death if used improperly. Make sure you follow the manufacturer's operation procedures. Also, make sure you have your supervisor's/instructor's permission to use any particular type of lifting equipment.
- Comply with personal and environmental safety practices associated with clothing; eye protection; hand tools; power equipment; proper ventilation; and the handling, storage, and disposal of chemicals/materials in accordance with local, state, and federal safety and environmental regulations.
- If you need to start the vehicle, you should ensure that the parking brake is firmly applied; if necessary, use wheel chocks to prevent the vehicle from moving when the vehicle is started to verify the completion of these tasks.
- Extreme caution must be exercised when working around rotating components.

CDX Tasksheet Number: A5001

Student/Intern Information

Name _________________________________ Date ____________ Class _________________________

Vehicle, Customer, and Service Information

Vehicle used for this activity:

Year _______________ Make _____________________________ Model _______________________________

Odometer _________________________________ VIN _______________________________________

Materials Required

- Blank work order
- Vehicle with available service history records
- Depending on the type of concern, special diagnostic/hand tools may be required. See your supervisor/instructor for instructions to identify what tools may be required.
- Service information database
- Personal protective equipment

Task-Specific Safety Considerations

- Comply with personal and environmental safety practices associated with clothing; eye protection; hand tools; power equipment; proper ventilation; and the handling, storage, and disposal of chemicals/materials in accordance with local, state, and federal safety and environmental regulations.

Time off____________

Time on____________

Total time____________

▶ **TASK** Research vehicle service information, including fluid type, vehicle service history, service precautions, and technical service bulletins.

MLR
A5A1

Student Instructions: Read through the entire procedure prior to starting. Prepare your workspace and any tools or parts that may be needed to complete the task. When directed by your instructor, begin the procedure to complete the task and check the box as each step is finished. Track your time on this procedure for later comparison to the standard completion time (i.e., "flat rate" or customer pay time).

Procedure:	Step Completed
1. Research the vehicle service information.	
a. List the brake fluid type used in your vehicle.	☐

b. Are there any precautions that should be followed when working with brake fluid?	☐
c. What is the recommended maintenance interval for servicing the brake fluid in your vehicle?	☐
2. Research the vehicle service history.	
a. Determine if routine maintenance has been completed.	☐
b. Determine if there is any outstanding/completed recall work.	☐
c. Determine if there are any major repairs in the service history.	☐
d. Identify the service precautions related to any maintenance or recall related task.	☐
3. Determine any TSBs issued for your specific vehicle.	
a. Identify one brake-related TSB and list the TSB number, vehicle concern, and repair procedure.	☐

Non-Task-Specific Evaluations:	**Step Completed**
1. Tools and equipment were used as directed and returned in good working order.	☐
2. Complied with all general and task-specific safety standards, including proper use of any personal protective equipment.	☐
3. Completed the task in an appropriate time frame (Recommendation: 1.5 or 2 times flat rate).	☐
4. Left the work space clean and orderly.	☐
5. Cared for customer property and returned it undamaged.	☐

Student signature _______________________________ Date _________________________________

Comments:

Have your supervisor/instructor verify satisfactory completion of this procedure, any observations found, and any necessary action(s) recommended.

Evaluation Instructions: The scoring box below is intended to act as a guide for both student and instructor. Each criterion listed will help students understand what is expected from them and evaluators articulate success at a particular task. The scoring is set up to allow a second attempt at each task (see the "Test" and "Retest" columns). Scoring is designed to reward students for correct completion of the task. Points are lost for failure to complete the employability requirements (see "Non-Task-Specific" criteria). When all criteria are evaluated, tally the points for a total at the bottom of each column.

Tasksheet Scoring

Evaluation Items	Test		Retest	
	Pass	**Fail**	**Pass**	**Fail**
Task-Specific Evaluation	**(1 pt)**	**(0 pts)**	**(1 pt)**	**(0 pts)**
Research the brake fluid type and identify any service precautions.				
Research the vehicle service history.				
Identify any completed/outstanding recalls.				
Identify any brake related TSBs.				
Non-Task-Specific Evaluation	**(0 pts)**	**(−1 pt)**	**(0 pts)**	**(−1 pt)**
Student successfully completed at least three of the non-task-specific steps.				
Student successfully completed all five of the non-task-specific steps.				
Total Score: <total # of points / 4 = %>				

Supervisor:

Supervisor/instructor signature ________________________________ Date ____________________

Comments:

Retest supervisor/instructor signature ________________________________ Date ____________________

Comments:

CDX Tasksheet Number: A5002

Student/Intern Information

Name _________________________________ Date ____________ Class _____________________

Vehicle, Customer, and Service Information

Vehicle used for this activity:

Year ______________ Make ______________________________ Model ______________________

Odometer _________________________________ VIN _________________________________

Materials Required

- Blank work order
- Vehicle with available service history records
- Depending on the type of concern, special diagnostic/hand tools may be required. See your supervisor/instructor for instructions to identify what tools may be required.
- Service information database
- Personal protective equipment

Task-Specific Safety Considerations

- Comply with personal and environmental safety practices associated with clothing; eye protection; hand tools; power equipment; proper ventilation; and the handling, storage, and disposal of chemicals/materials in accordance with local, state, and federal safety and environmental regulations.
- **Caution:** If a vehicle requires a test drive to verify correct operation of particular system, you must follow EXACT procedures and all rules that apply. Refer to your instructor for further details.

Time off_____________

Time on_____________

Total time_____________

▶ **TASK** Describe procedure for performing a road test to check brake system operation, including an antilock brake system (ABS).

MLR A5A2

Student Instructions: Read through the entire procedure prior to starting. Prepare your work space and any tools or parts that may be needed to complete the task. When directed by your instructor, begin the procedure to complete the task and check the box as each step is finished. Track your time on this procedure for later comparison to the standard completion time (i.e., "flat rate" or customer pay time).

Procedure:	Step Completed
1. Describe what causes the following brake system concerns.	
a. Brake pull	☐

<table>
<tr><td>b. Brake fade</td><td>☐</td></tr>
<tr><td>c. Brake noise</td><td>☐</td></tr>
<tr><td>d. Increased pedal effort</td><td>☐</td></tr>
<tr><td>e. Low, spongy brake pedal</td><td>☐</td></tr>
<tr><td>f. Vibration during high speed braking</td><td>☐</td></tr>
<tr><td>2. Explain how the ABS system performs a self-test, when the test is performed, and how the driver is notified if there is a fault discovered during the self-test.</td><td></td></tr>
<tr><td>3. Explain what different braking scenarios you may perform to check the braking ability under a variety of conditions.</td><td></td></tr>
<tr><td>4. Research the conditions that must be met for the ABS to activate for your vehicle.</td><td></td></tr>
</table>

<table>
<tr><td>Non-Task-Specific Evaluations:</td><td>Step Completed</td></tr>
<tr><td>1. Tools and equipment were used as directed and returned in good working order.</td><td>☐</td></tr>
<tr><td>2. Complied with all general and task-specific safety standards, including proper use of any personal protective equipment.</td><td>☐</td></tr>
<tr><td>3. Completed the task in an appropriate time frame (Recommendation: 1.5 or 2 times flat rate).</td><td>☐</td></tr>
<tr><td>4. Left the work space clean and orderly.</td><td>☐</td></tr>
<tr><td>5. Cared for customer property and returned it undamaged.</td><td>☐</td></tr>
</table>

Student signature _______________________________ Date _______________________________

Comments:

Have your supervisor/instructor verify satisfactory completion of this procedure, any observations found, and any necessary action(s) recommended.

Tasksheet Scoring

	Test		Retest	
Evaluation Items	**Pass**	**Fail**	**Pass**	**Fail**
Task-Specific Evaluation	**(1 pt)**	**(0 pts)**	**(1 pt)**	**(0 pts)**
Describe causes for various brake system concerns.				
Describe the function of the ABS self-test and any braking tests that may be performed to check braking performance under a variety of conditions.				
Describe how the driver is notified of ABS malfunctions.				
Describe the criteria for ABS activation.				
Non-Task-Specific Evaluation	**(0 pts)**	**(−1 pt)**	**(0 pts)**	**(−1 pt)**
Student successfully completed at least three of the non-task-specific steps.				
Student successfully completed all five of the non-task-specific steps.				
Total Score: <total # of points / 4 = %>				

Supervisor:

Supervisor/instructor signature _________________________________ Date _____________________

Comments:

Retest supervisor/instructor signature _________________________________ Date _____________________

Comments:

CDX Tasksheet Number: A5003

Student/Intern Information

Name _________________________________ Date ___________ Class _____________________

Vehicle, Customer, and Service Information

Vehicle used for this activity:

Year ______________ Make _______________________ Model ___________________

Odometer _________________________ VIN _______________________________

Materials Required

- Blank work order
- Vehicle with available service history records
- Depending on the type of concern, special diagnostic/hand tools may be required. See your supervisor/instructor for instructions to identify what tools may be required.
- Service information database
- Personal protective equipment
- Vehicle lifting equipment
- General hand/air tools
- Impact sockets
- Torque limiting extension bar (torque stick)
- Torque wrench

Task-Specific Safety Considerations

- Lifting equipment, such as vehicle jacks and stands, vehicle hoists, and engine hoists, are important tools that increase productivity and make the job easier. However, they can also cause severe injury or death if used improperly. Make sure you follow the manufacturer's operation procedures. Also, make sure you have your supervisor's/instructor's permission to use any particular type of lifting equipment.
- Comply with personal and environmental safety practices associated with clothing; eye protection; hand tools; power equipment; proper ventilation; and the handling, storage, and disposal of chemicals/materials in accordance with local, state, and federal safety and environmental regulations.

▶ **TASK** Install wheel and torque lug nuts.

MLR
A5A3

Time off_____________

Time on_____________

Total time_____________

Student Instructions: Read through the entire procedure prior to starting. Prepare your work space and any tools or parts that may be needed to complete the task. When directed by your instructor, begin the procedure to complete the task and check the box as each step is finished. Track your time on this procedure for later comparison to the standard completion time (i.e., "flat rate" or customer pay time).

Procedure:	Step Completed
1. Research the procedure for installing a wheel and for torqueing lug nuts.	
a. Use service information to look up the torque specification for the lug nuts on your vehicle.	☐
b. Describe the pattern or sequence that you would use to tighten lug nuts.	☐
c. What kind of tools are needed for installing a wheel properly?	☐
d. What size and type of socket will you use to remove and install the lug nuts?	☐
2. Remove the wheel.	
a. Lift the vehicle to a comfortable working height.	☐
b. Select the correct size and type socket. Remove the wheel using a 1/2" air impact.	☐
c. Before installing the wheel, inspect the wheel studs for signs of damage. List any action required.	☐
3. Install the wheel.	
a. Mount the wheel to the hub by starting all lug nuts by hand.	☐
b. Hold the wheel in the centered position and tighten the lug nuts in the correct sequence. Use a torque stick to provide the initial torque for the lug nuts. Use the lowest torque setting on the air impact to avoid damaging of wheel studs.	☐
c. Lower the vehicle until it just contacts the ground. Use a torque wrench to provide the final torque of the lug nuts. Lower the vehicle the rest of the way.	☐

Non-Task-Specific Evaluations:	**Step Completed**
1. Tools and equipment were used as directed and returned in good working order.	☐
2. Complied with all general and task-specific safety standards, including proper use of any personal protective equipment.	☐
3. Completed the task in an appropriate time frame (Recommendation: 1.5 or 2 times flat rate).	☐
4. Left the work space clean and orderly.	☐
5. Cared for customer property and returned it undamaged.	☐

Student signature _______________________________ Date _______________________________

Comments:

Have your supervisor/instructor verify satisfactory completion of this procedure, any observations found, and any necessary action(s) recommended.

Evaluation Instructions: The scoring box below is intended to act as a guide for both student and instructor. Each criterion listed will help students understand what is expected from them and evaluators articulate success at a particular task. The scoring is set up to allow a second attempt at each task (see the "Test" and "Retest" columns). Scoring is designed to reward students for correct completion of the task. Points are lost for failure to complete the employability requirements (see "Non-Task-Specific" criteria). When all criteria are evaluated, tally the points for a total at the bottom of each column.

Taksheet Scoring

	Test		Retest	
Evaluation Items	**Pass**	**Fail**	**Pass**	**Fail**
Task-Specific Evaluation	**(1 pt)**	**(0 pts)**	**(1 pt)**	**(0 pts)**
Research the lug nut torque specification and tightening sequence.				
Select the correct socket size and type to remove and install wheels.				
Lift the vehicle and remove the wheel.				
Install the wheel using the air impact and torque stick and perform final torqueing of the lug nuts using a torque wrench.				
Non-Task-Specific Evaluation	**(0 pts)**	**(−1 pt)**	**(0 pts)**	**(−1 pt)**
Student successfully completed at least three of the non-task-specific steps.				
Student successfully completed all five of the non-task-specific steps.				
Total Score: <total # of points / 4 = %>				

Supervisor:

Supervisor/instructor signature _______________________________ Date _______________________

Comments:

Retest supervisor/instructor signature _______________________________ Date _______________________

Comments:

CDX Tasksheet Number: A5004

Student/Intern Information

Name _________________________________ Date ___________ Class _____________________

Vehicle, Customer, and Service Information

Vehicle used for this activity:

Year _______________ Make _____________________ Model _____________________

Odometer _____________________ VIN _____________________________________

Task-Specific Safety Considerations

- Lifting equipment such as vehicle jacks and stands, vehicle hoists, and engine hoists are important tools that increase productivity and make the job easier. However, they can also cause severe injury or death if used improperly. Make sure you follow the manufacturer's operation procedures. Also, make sure you have your supervisor's/instructor's permission to use any particular type of lifting equipment.
- Comply with personal and environmental safety practices associated with clothing; eye protection; hand tools; power equipment; proper ventilation; and the handling, storage, and disposal of chemicals/materials in accordance with local, state, and federal safety and environmental regulations.

Time off_____________

Time on_____________

Total time_____________

▶ **TASK** Identify brake system components and configuration.

MLR
A5A4

Student Instructions: Read through the entire procedure prior to starting. Prepare your work space and any tools or parts that may be needed to complete the task. When directed by your instructor, begin the procedure to complete the task and check the box as each step is finished. Track your time on this procedure for later comparison to the standard completion time (i.e., "flat rate" or customer pay time).

Procedure:	Step Completed
1. Identify the brake system configuration on your vehicle. Answer the following questions:	
a. Is the vehicle equipped with disc brakes?	☐
b. Is the vehicle equipped with drum brakes?	☐
c. Is the vehicle equipped with ABS?	☐
d. Is the vehicle equipped with hydraulic or vacuum power brake assist?	☐
e. Is the parking brake part of the brake caliper or does it rely on brake shoes inside a drum or rotor hat?	☐
2. Locate the following items on your vehicle:	
a. Brake rotor	☐
b. Brake caliper	☐
c. Brake hose and line	☐
d. Master cylinder	☐
e. Brake booster	☐
f. Brake drum	☐
g. Metering and proportioning valve. Explain how each of these components work.	☐
h. Parking brake lever, cable, and parking brake	☐
i. ABS module	☐

Non-Task-Specific Evaluations:	Step Completed
1. Tools and equipment were used as directed and returned in good working order.	☐
2. Complied with all general and task-specific safety standards, including proper use of any personal protective equipment.	☐
3. Completed the task in an appropriate time frame (Recommendation: 1.5 or 2 times flat rate).	☐
4. Left the work space clean and orderly.	☐
5. Cared for customer property and returned it undamaged.	☐

Student signature _______________________________ Date _______________________________

Comments:

Have your supervisor/instructor verify satisfactory completion of this procedure, any observations found,

and any necessary action(s) recommended.

Evaluation Instructions: The scoring box below is intended to act as a guide for both student and instructor. Each criterion listed will help students understand what is expected from them and evaluators articulate success at a particular task. The scoring is set up to allow a second attempt at each task (see the "Test" and "Retest" columns). Scoring is designed to reward students for correct completion of the task. Points are lost for failure to complete the employability requirements (see "Non-Task-Specific" criteria). When all criteria are evaluated, tally the points for a total at the bottom of each column.

Tasksheet Scoring

	Test		Retest	
Evaluation Items	**Pass**	**Fail**	**Pass**	**Fail**
Task-Specific Evaluation	**(1 pt)**	**(0 pts)**	**(1 pt)**	**(0 pts)**
Determine if vehicle is equipped with disc or drum brakes.				
Determine if the vehicle is equipped with ABS.				
Locate the brake system components on the vehicle.				
Explain the difference between a metering and proportioning valve.				
Non-Task-Specific Evaluation	**(0 pts)**	**(−1 pt)**	**(0 pts)**	**(−1 pt)**
Student successfully completed at least three of the non-task-specific steps.				
Student successfully completed all five of the non-task-specific steps.				
Total Score: <total # of points / 4 = %>				

Supervisor:

Supervisor/instructor signature ____________________________ Date __________________

Comments:

Retest supervisor/instructor signature ________________________ Date _______________

Comments:

5B. Hydraulic System

Learning Objective/Task	CDX Tasksheet Number	ASE Education Foundation Reference Number; Priority Level
• Describe proper brake pedal height, travel, and feel.	A5005	A5B1, P-1
• Check master cylinder for external leaks and proper operation.	A5006	A5B2, P-1
• Inspect brake lines, flexible hoses, and fittings for leaks, dents, kinks, rust, cracks, bulging, wear, and loose fittings/supports.	A5007	A5B3, P-1
• Select, handle, store, and fill brake fluids to proper level; use proper fluid type per manufacturer specification.	A5008	A5B4, P-1
• Identify components of hydraulic brake warning light system.	A5009	A5B5, P-3
• Bleed and/or flush brake system.	A5010	A5B6, P-1
• Test brake fluid for contamination.	A5011	A5B7, P-1

Materials Required

- Blank work order
- Vehicle with available service history records
- Depending on the type of concern, special diagnostic/hand tools may be required. See your supervisor/instructor for instructions to identify what tools may be required.
- Service information database
- Personal protective equipment

Safety Considerations

- When running any vehicles in the shop, make sure you use the shop's exhaust ventilation system to discharge all exhaust gas safely outside.
- Lifting equipment, such as vehicle jacks and stands, vehicle hoists, and engine hoists, are important tools that increase productivity and make the job easier. However, they can also cause severe injury or death if used improperly. Make sure you follow the manufacturer's operation procedures. Also, make sure you have your supervisor's/instructor's permission to use any particular type of lifting equipment.
- Comply with personal and environmental safety practices associated with clothing; eye protection; hand tools; power equipment; proper ventilation; and the handling, storage, and disposal of chemicals/materials in accordance with local, state, and federal safety and environmental regulations.
- If you need to start the vehicle, you should ensure that the parking brake is firmly applied; if necessary, use wheel chocks to prevent the vehicle from moving when the vehicle is started to verify the completion of these tasks.
- Extreme caution must be exercised when working around rotating components.
- **Caution:** Most types of brake fluid are harmful to painted surfaces. Be sure to prevent brake fluid from coming into contact with a vehicle's paint. Use fender covers to minimize this risk and be sure to wipe up any spilled brake fluid immediately with a wet rag.

- **Caution:** Brake dust may contain asbestos, which has been determined to cause cancer when inhaled or ingested. Treat all brake dust as if it contains asbestos and use OSHA-approved asbestos removal equipment. Do not allow brake dust to become airborne by using anything that would disturb the dust. Also, wear protective gloves during this procedure and dispose of or clean them in an approved manner.

CDX Tasksheet Number: A5005

Student/Intern Information

Name _________________________ Date _________ Class _________________________

Vehicle, Customer, and Service Information

Vehicle used for this activity:

Year _____________ Make _____________________ Model _________________________

Odometer _____________________ VIN _________________________________

Task-Specific Safety Considerations
- Comply with personal and environmental safety practices associated with clothing; eye protection; hand tools; power equipment; proper ventilation; and the handling, storage, and disposal of chemicals/materials in accordance with local, state, and federal safety and environmental regulations.
- When running any vehicles in the shop, make sure you use the shop's exhaust ventilation system to discharge all exhaust gas safely outside.

Time off_____________

Time on_____________

Total time_____________

▶ **TASK** Describe proper brake pedal height, travel, and feel.

MLR
A5B1

Student Instructions: Read through the entire procedure prior to starting. Prepare your work space and any tools or parts that may be needed to complete the task. When directed by your instructor, begin the procedure to complete the task and check the box as each step is finished. Track your time on this procedure for later comparison to the standard completion time (i.e., "flat rate" or customer pay time).

Procedure:	Step Completed
1. Describe the brake pedal feel.	

a. Start the vehicle and let it run for 10 seconds. Turn the vehicle off, and depress the brake pedal 3–5 times. Explain any changes that are felt in the brake pedal. Explain what should happen when you perform this test.	☐
b. Apply the brake and start the vehicle. Describe what happens to the pedal when the engine is started. Pump the pedal and describe how the pedal feels and how far it is traveling. Is it sinking to the floor gradually? Does it sink just a little, yet still feel firm? Explain how the pedal should during this test.	☐
c. What does it mean if a brake pedal feels soft and gradually sinks to the floor? How would you fix this concern?	☐
2. Research the specifications for brake pedal height, freeplay, and adjustment.	
a. Using service information, list the specifications for brake pedal height and freeplay.	☐
b. List or print off the procedure for making adjustments to brake pedal freeplay.	☐
3. Measure the brake pedal height and freeplay.	
a. Following procedure, measure the brake pedal height. List the measurement you made and compare it to specification.	☐
b. Following procedure, measure the brake pedal freeplay. List the measurement you made and compare it to specification.	☐

| 4. Based on your inspection, what is your recommendation for service? | |

Non-Task-Specific Evaluations:	**Step Completed**
1. Tools and equipment were used as directed and returned in good working order.	☐
2. Complied with all general and task-specific safety standards, including proper use of any personal protective equipment.	☐
3. Completed the task in an appropriate time frame (Recommendation: 1.5 or 2 times flat rate).	☐
4. Left the work space clean and orderly.	☐
5. Cared for customer property and returned it undamaged.	☐

Student signature _________________________ Date _________________________

Comments:

Have your supervisor/instructor verify satisfactory completion of this procedure, any observations found, and any necessary action(s) recommended.

Evaluation Instructions: The scoring box below is intended to act as a guide for both student and instructor. Each criterion listed will help students understand what is expected from them and evaluators articulate success at a particular task. The scoring is set up to allow a second attempt at each task (see the "Test" and "Retest" columns). Scoring is designed to reward students for correct completion of the task. Points are lost for failure to complete the employability requirements (see "Non-Task-Specific" criteria). When all criteria are evaluated, tally the points for a total at the bottom of each column.

Tasksheet Scoring

Evaluation Items	Test		Retest	
	Pass	Fail	Pass	Fail
Task-Specific Evaluation	**(1 pt)**	**(0 pts)**	**(1 pt)**	**(0 pts)**
Evaluate the brake pedal feel and travel in your vehicle and describe causes for brake pedal feel concerns.				
Research the brake pedal height and freeplay specifications.				
Measure the brake pedal height and freeplay.				
Make recommendations for service based on your inspection.				
Non-Task-Specific Evaluation	**(0 pts)**	**(−1 pt)**	**(0 pts)**	**(−1 pt)**
Student successfully completed at least three of the non-task-specific steps.				
Student successfully completed all five of the non-task-specific steps.				
Total Score: <total # of points / 4 = %>				

Supervisor:

Supervisor/instructor signature _______________________________ Date _____________________

Comments:

Retest supervisor/instructor signature _______________________ Date _____________________

Comments:

CDX Tasksheet Number: A5006

Student/Intern Information

Name _________________________________ Date ____________ Class _____________________

Vehicle, Customer, and Service Information

Vehicle used for this activity:

Year _______________ Make ___________________________ Model _____________________________

Odometer _______________________________ VIN _______________________________________

Materials Required

- Blank work order
- Vehicle with available service history records
- Depending on the type of concern, special diagnostic/hand tools may be required. See your supervisor/instructor for instructions to identify what tools may be required.
- Service information database
- Personal protective equipment

Task-Specific Safety Considerations

- Comply with personal and environmental safety practices associated with clothing; eye protection; hand tools; power equipment; proper ventilation; and the handling, storage, and disposal of chemicals/materials in accordance with local, state, and federal safety and environmental regulations.
- When running any vehicles in the shop, make sure you use the shop's exhaust ventilation system to discharge all exhaust gas safely outside.
- **Caution:** Most types of brake fluid are harmful to painted surfaces. Be sure to prevent brake fluid from coming into contact with a vehicle's paint. Use fender covers to minimize this risk and be sure to wipe up any spilled brake fluid immediately with a wet rag.

▶ **TASK** Check master cylinder for external leaks and proper operation.

MLR
A5B2

Time off____________

Time on____________

Total time____________

Student Instructions: Read through the entire procedure prior to starting. Prepare your work space and any tools or parts that may be needed to complete the task. When directed by your instructor, begin the procedure to complete the task and check the box as each step is finished. Track your time on this procedure for later comparison to the standard completion time (i.e., "flat rate" or customer pay time).

Procedure:	Step Completed
1. Inspect the master cylinder for external leaks.	
a. Begin by placing a fender cover on your vehicle near your work area. Verify correct fluid level in the master cylinder. Is the level okay?	☐
b. Inspect the master cylinder for leaks around the housing and brake lines.	☐
c. Inspect the brake booster for signs of brake fluid contamination. In some instances, brake fluid may even be seen leaking through the firewall near the brake pedal pushrod.	☐
d. Explain the results of your inspection.	☐
2. Inspect the master cylinder for proper operation.	
a. Start the vehicle and pump the brake pedal 3–4 times. Depress the brake pedal for 30 seconds. The pedal should not drop. It should remain firm in the same position. Does the brake pedal feel normal?	☐
b. Explain what may be causing the brake pedal to drop in a vehicle while holding it for 30 seconds.	☐
c. Explain the results based on your inspection.	☐

Non-Task-Specific Evaluations:	Step Completed
1. Tools and equipment were used as directed and returned in good working order.	☐
2. Complied with all general and task-specific safety standards, including proper use of any personal protective equipment.	☐
3. Completed the task in an appropriate time frame (Recommendation: 1.5 or 2 times flat rate).	☐
4. Left the work space clean and orderly.	☐
5. Cared for customer property and returned it undamaged.	☐

Student signature _______________________________ Date _________________________________

Comments:

Have your supervisor/instructor verify satisfactory completion of this procedure, any observations found, and any necessary action(s) recommended.

Evaluation Instructions: The scoring box below is intended to act as a guide for both student and instructor. Each criterion listed will help students understand what is expected from them and evaluators articulate success at a particular task. The scoring is set up to allow a second attempt at each task (see the "Test" and "Retest" columns). Scoring is designed to reward students for correct completion of the task. Points are lost for failure to complete the employability requirements (see "Non-Task-Specific" criteria). When all criteria are evaluated, tally the points for a total at the bottom of each column.

Tasksheet Scoring

	Test		Retest	
Evaluation Items	**Pass**	**Fail**	**Pass**	**Fail**
Task-Specific Evaluation	**(1 pt)**	**(O pts)**	**(1 pt)**	**(O pts)**
Verify the correct brake fluid level prior to inspection.				
Inspect the master cylinder for signs of leaking fluid and for correct operation.				
Explain the causes of improper brake pedal feel.				
Make recommendations for service based on your inspection.				
Non-Task-Specific Evaluation	**(O pts)**	**(−1 pt)**	**(O pts)**	**(−1 pt)**
Student successfully completed at least three of the non-task-specific steps.				
Student successfully completed all five of the non-task-specific steps.				
Total Score: <total # of points / 4 = %>				

Supervisor:

Supervisor/instructor signature _________________________ Date _____________

Comments:

Retest supervisor/instructor signature _________________________ Date _____________

Comments:

CDX Tasksheet Number: A5007

Student/Intern Information

Name _________________________________ Date ___________ Class _________________________

Vehicle, Customer, and Service Information

Vehicle used for this activity:

Year ______________ Make _____________________________ Model _____________________________

Odometer _________________________________ VIN _____________________________

Task-Specific Safety Considerations

- Lifting equipment, such as vehicle jacks and stands, vehicle hoists, and engine hoists, are important tools that increase productivity and make the job easier. However, they can also cause severe injury or death if used improperly. Make sure you follow the manufacturer's operation procedures. Also, make sure you have your supervisor's/instructor's permission to use any particular type of lifting equipment.
- Comply with personal and environmental safety practices associated with clothing; eye protection; hand tools; power equipment; proper ventilation; and the handling, storage, and disposal of chemicals/materials in accordance with local, state, and federal safety and environmental regulations.

Time off_______________

Time on_______________

Total time_______________

▶ **TASK** Inspect brake lines, flexible hoses, and fittings for leaks, dents, kinks, rust, cracks, bulging, wear, and loose fittings/supports.

MLR
A5B3

Student Instructions: Read through the entire procedure prior to starting. Prepare your work space and any tools or parts that may be needed to complete the task. When directed by your instructor, begin the procedure to complete the task and check the box as each step is finished. Track your time on this procedure for later comparison to the standard completion time (i.e., "flat rate" or customer pay time).

Procedure:	Step Completed
1. Inspect the brake lines, hoses, and fittings.	
a. Begin by lifting the hood of the vehicle. Inspect the fluid level in the master cylinder. How is the level? Locate the brake lines that attach to the master cylinder.	☐
b. Trace the brake lines down to each front wheel. Ensure that the brake lines are not loose from their retainers along the firewall. Inspect for leaks, dents, kinks, or rust along these lines.	☐
c. At each front wheel, locate the fitting where the hard metal line meets the flexible rubber hose. Inspect this junction for leaks and be sure that the bracket holding this junction is securely fixed to the vehicle. **Tip:** You may have to remove the wheels to thoroughly inspect the condition of the brake hoses. Be sure to tighten the lug nuts according to proper procedure and use the correct torque specifications.	☐
d. Inspect the rubber brake hose for correct routing, kinks, or bulges. Have a helper apply the brake pedal. Watch the rubber hose for signs of bulging.	☐
e. Inspect the point where the flexible hose attaches to the brake caliper for signs of leaking fluid.	☐
f. Continue to trace the metal brake lines to the back of the vehicle. Along the way, ensure there are no leaks at the metering/proportioning valve and all lines are not loose from their retainers. Inspect for leaks, kinks, dents, or rust. Inspect for signs of improper repair.	☐
g. Explain what improper brake line repair may look like. Explain what type of fittings should not be used when making brake line repairs and why.	☐
h. Inspect the junction at the rear of the vehicle and ensure that it is securely fixed to the vehicle. Inspect the lines that branch out to each wheel. If a flexible hose is being used, perform the same test to inspect for signs of bulging.	☐
2. Based on your inspection, what is your recommendation for service?	☐
3. Describe the faults associated with brake line failure	
a. Describe what customers may feel while braking if their vehicle has a flexible rubber brake hose that is bulging. How will the brake pedal feel? How will the vehicle react when the brakes are applied?	☐

<table>
<tr><td>b. Describe what customers may feel while braking if their vehicle has a brake line/hose that is leaking. How will the brake pedal feel? How will the vehicle react when the brakes are applied?</td><td>☐</td></tr>
</table>

Non-Task-Specific Evaluations:	**Step Completed**
1. Tools and equipment were used as directed and returned in good working order.	☐
2. Complied with all general and task-specific safety standards, including proper use of any personal protective equipment.	☐
3. Completed the task in an appropriate time frame (Recommendation: 1.5 or 2 times flat rate).	☐
4. Left the work space clean and orderly.	☐
5. Cared for customer property and returned it undamaged.	☐

Student signature _________________________ Date _____________________________

Comments:

Have your supervisor/instructor verify satisfactory completion of this procedure, any observations found, and any necessary action(s) recommended.

Evaluation Instructions: The scoring box below is intended to act as a guide for both student and instructor. Each criterion listed will help students understand what is expected from them and evaluators articulate success at a particular task. The scoring is set up to allow a second attempt at each task (see the "Test" and "Retest" columns). Scoring is designed to reward students for correct completion of the task. Points are lost for failure to complete the employability requirements (see "Non-Task-Specific" criteria). When all criteria are evaluated, tally the points for a total at the bottom of each column.

Tasksheet Scoring

	Test		Retest	
Evaluation Items	**Pass**	**Fail**	**Pass**	**Fail**
Task-Specific Evaluation	**(1 pt)**	**(0 pts)**	**(1 pt)**	**(0 pts)**
Inspect the hard lines for kinks, bends, rust, and leaks.				
Inspect the brake hoses for improper routing, kinks, bulging, or leaks. Describe improper brake line repair.				
Make recommendations for service based on your inspection.				
Describe the faults associated with brake line failure.				
Non-Task-Specific Evaluation	**(0 pts)**	**(−1 pt)**	**(0 pts)**	**(−1 pt)**
Student successfully completed at least three of the non-task-specific steps.				
Student successfully completed all five of the non-task-specific steps.				
Total Score: <total # of points / 4 = %>				

Supervisor:

Supervisor/instructor signature _______________________________ Date _______________

Comments:

Retest supervisor/instructor signature _______________________ Date _______________

Comments:

CDX Tasksheet Number: A5008

Student/Intern Information

Name _________________________________ Date ___________ Class _____________________

Vehicle, Customer, and Service Information

Vehicle used for this activity:

Year _______________ Make ______________________ Model _____________________

Odometer ____________________ VIN _________________________

Materials Required

- Blank work order
- Vehicle with available service history records
- Depending on the type of concern, special diagnostic/hand tools may be required. See your supervisor/instructor for instructions to identify what tools may be required.
- Service information database
- Personal protective equipment
- Fender cover
- Brake fluid

Task-Specific Safety Considerations

- Comply with personal and environmental safety practices associated with clothing; eye protection; hand tools; power equipment; proper ventilation; and the handling, storage, and disposal of chemicals/materials in accordance with local, state, and federal safety and environmental regulations.
- **Caution:** Most types of brake fluid are harmful to painted surfaces. Be sure to prevent brake fluid from coming into contact with a vehicle's paint. Use fender covers to minimize this risk and be sure to wipe up any spilled brake fluid immediately with a wet rag.

Time off_____________

Time on_____________

Total time_____________

▶ TASK Select, handle, store, and fill brake fluids to proper level; use proper fluid type per manufacturer specification.

MLR
A5B4

Student Instructions: Read through the entire procedure prior to starting. Prepare your work space and any tools or parts that may be needed to complete the task. When directed by your instructor, begin the procedure to complete the task and check the box as each step is finished. Track your time on this procedure for later comparison to the standard completion time (i.e., "flat rate" or customer pay time).

Procedure:	Step Completed
1. Research the brake fluid type required for your vehicle.	
a. Using your service information, list the type of brake fluid required for your vehicle.	☐
b. Explain the different types of brake fluid that are available for use. Explain what makes them different from each other. Are any of the fluids compatible with each other?	☐
c. Describe how you should keep brake fluid stored. Are there any special precautions to handling or storing brake fluid?	☐
d. Define hygroscopic. Explain what would happen to brake fluid if it became contaminated with water. How will brake feel be affected?	☐
2. Inspect the brake fluid level in your vehicle.	
a. Lift the hood on your vehicle. Before inspecting the fluid level, place a fender cover over your vehicle. Explain what will happen if brake fluid gets on the paint of the vehicle.	☐
b. Inspect the brake fluid level and list any action required.	☐
c. Select the correct brake fluid and top off the master cylinder.	☐

Non-Task-Specific Evaluations:	Step Completed
1. Tools and equipment were used as directed and returned in good working order.	☐
2. Complied with all general and task-specific safety standards, including proper use of any personal protective equipment.	☐
3. Completed the task in an appropriate time frame (Recommendation: 1.5 or 2 times flat rate).	☐
4. Left the work space clean and orderly.	☐
5. Cared for customer property and returned it undamaged.	☐

Student signature _____________________ Date _____________________

Comments:

Have your supervisor/instructor verify satisfactory completion of this procedure, any observations found, and any necessary action(s) recommended.

Tasksheet Scoring

	Test		Retest	
Evaluation Items	Pass	Fail	Pass	Fail
Task-Specific Evaluation	**(1 pt)**	**(0 pts)**	**(1 pt)**	**(0 pts)**
Research the brake fluid type required in your vehicle.				
Explain the precautions that should be followed when using brake fluid.				
Display an understanding of the hygroscopic characteristics of brake fluids. Explain the symptoms related to contaminated brake fluid.				
Inspect the brake fluid level and top off using the correct fluid type.				
Non-Task-Specific Evaluation	**(0 pts)**	**(−1 pt)**	**(0 pts)**	**(−1 pt)**
Student successfully completed at least three of the non-task-specific steps.				
Student successfully completed all five of the non-task-specific steps.				
Total Score: <total # of points / 4 = %>				

Supervisor:

Supervisor/instructor signature _________________________ Date _______________

Comments:

Retest supervisor/instructor signature _________________________ Date _______________

Comments:

CDX Tasksheet Number: A5009

Student/Intern Information

Name _________________________________ Date ___________ Class _________________________

Vehicle, Customer, and Service Information

Vehicle used for this activity:

Year _______________ Make _______________________ Model _____________________

Odometer _______________________ VIN _______________________________

> **Materials Required**
> - Blank work order
> - Vehicle with available service history records
> - Depending on the type of concern, special diagnostic/hand tools may be required. See your supervisor/instructor for instructions to identify what tools may be required.
> - Service information database
> - Personal protective equipment
> - Brake pedal depressor
> - General hand tools

Task-Specific Safety Considerations
- Comply with personal and environmental safety practices associated with clothing; eye protection; hand tools; power equipment; proper ventilation; and the handling, storage, and disposal of chemicals/materials in accordance with local, state, and federal safety and environmental regulations.

▶ TASK Identify components of hydraulic brake warning light system.

MLR
A5B5

Time off_______________

Time on_______________

Total time_______________

Student Instructions: Read through the entire procedure prior to starting. Prepare your work space and any tools or parts that may be needed to complete the task. When directed by your instructor, begin the procedure to complete the task and check the box as each step is finished. Track your time on this procedure for later comparison to the standard completion time (i.e., "flat rate." or customer pay time).

Procedure:	Step Completed
1. Identify the brake warning indicators on the gauge cluster.	
a. Cycle the ignition to the ON position. List the brake related warning indicators that are illuminated. Did the lights go out after 3–5 seconds? List any lights that remain on.	☐

b. Draw the symbols that were illuminated during step 1a.	☐
c. Is the vehicle equipped with ABS? How do you know?	☐
d. Lightly engage the parking brake. Did a warning indicator come on? Describe the light. What color is it? Is it flashing? Is there an audible chime that comes on with the light?	☐
e. Start the vehicle. Are there any lights illuminated indicating that the brake pedal must be pressed to shift the vehicle from park? If so, describe what the light looks like.	☐
2. Inspect the brake fluid level sensor.	
a. Describe the location of the brake fluid level sensor.	☐
b. Unplug the sensor. Cycle the ignition to the ON position. Are there any warning lights illuminated on the dash? Plug the sensor in and verify that there are no warning lights illuminated.	☐
3. Inspect the brake lights.	
a. Use a helper or a brake pedal depressor to depress the brakes. Inspect the brake lights on the rear of the vehicle. How many lights are illuminated? Is there a third brake light?	☐

b. Describe the location of the brake switch. Unplug the switch. Cycle the ignition to the ON position. Are there any warning lights illuminated on the dash? Depress the brake pedal. Are the brake lights illuminated on the rear of the vehicle? Plug the switch in and verify that the brake lights are working properly.	☐
c. Unplug a brake light bulb from its socket. Turn the vehicle on and depress the brake pedal. Are there any warning lights on the dash indicating that a bulb is not functioning correctly? If so, describe what the light looks like. Once this test is complete, plug the bulb back into its socket and verify that it is working correctly.	☐
4. Research different types of brake warning indicators. Draw the symbol for the following indicators and describe the conditions that would cause the light to illuminate.	
a. Parking brake light	☐
b. Brake light bulb failure	☐
c. Press brake to shift	☐
d. ABS failure	☐
e. Hydraulics failure	☐

f. Brake pad warning	☐
5. Using your service information, explain how the brake system pressure is monitored. What components on your vehicle measure brake fluid pressure?	☐

Non-Task-Specific Evaluations:	**Step Completed**
1. Tools and equipment were used as directed and returned in good working order.	☐
2. Complied with all general and task-specific safety standards, including proper use of any personal protective equipment.	☐
3. Completed the task in an appropriate time frame (Recommendation: 1.5 or 2 times flat rate).	☐
4. Left the work space clean and orderly.	☐
5. Cared for customer property and returned it undamaged.	☐

Student signature _________________________________ Date _________________________________

Comments:

Have your supervisor/instructor verify satisfactory completion of this procedure, any observations found, and any necessary action(s) recommended.

Evaluation Instructions: The scoring box below is intended to act as a guide for both student and instructor. Each criterion listed will help students understand what is expected from them and evaluators articulate success at a particular task. The scoring is set up to allow a second attempt at each task (see the "Test" and "Retest" columns). Scoring is designed to reward students for correct completion of the task. Points are lost for failure to complete the employability requirements (see "Non-Task-Specific" criteria). When all criteria are evaluated, tally the points for a total at the bottom of each column.

Tasksheet Scoring

	Test		Retest	
Evaluation Items	Pass	Fail	Pass	Fail
Task-Specific Evaluation	**(1 pt)**	**(0 pts)**	**(1 pt)**	**(0 pts)**
Identify the brake warning indicators on the gauge cluster.				
Inspect the brake fluid level sensor.				
Research the different types of brake warning indicators. Draw the symbols and describe the conditions that would cause the light to illuminate.				
Explain how brake system pressure is monitored.				
Non-Task-Specific Evaluation	**(0 pts)**	**(−1 pt)**	**(0 pts)**	**(−1 pt)**
Student successfully completed at least three of the non-task-specific steps.				
Student successfully completed all five of the non-task-specific steps.				
Total Score: <total # of points / 4 = %>				

Supervisor:

Supervisor/instructor signature ________________________________ Date ________________

Comments:

Retest supervisor/instructor signature ________________________________ Date ________________

Comments:

CDX Tasksheet Number: A5010

Student/Intern Information

Name _________________________________ Date ____________ Class _____________________

Vehicle, Customer, and Service Information

Vehicle used for this activity:

Year _______________ Make ______________________________ Model _______________________

Odometer _________________________________ VIN ___________________________________

> **Materials Required**
> - Blank work order
> - Vehicle with available service history records
> - Depending on the type of concern, special diagnostic/hand tools may be required. See your supervisor/instructor for instructions to identify what tools may be required.
> - Service information database
> - Personal protective equipment
> - General hand tools
> - Scan tool (if applicable)
> - Brake flush equipment
> - Brake fluid
> - Fender cover
> - Fluid catch pan

Task-Specific Safety Considerations
- Lifting equipment, such as vehicle jacks and stands, vehicle hoists, and engine hoists, are important tools that increase productivity and make the job easier. However, they can also cause severe injury or death if used improperly. Make sure you follow the manufacturer's operation procedures. Also, make sure you have your supervisor's/instructor's permission to use any particular type of lifting equipment.
- Comply with personal and environmental safety practices associated with clothing; eye protection; hand tools; power equipment; proper ventilation; and the handling, storage, and disposal of chemicals/materials in accordance with local, state, and federal safety and environmental regulations.
- If you need to start the vehicle, you should ensure that the parking brake is firmly applied; if necessary, use wheel chocks to prevent the vehicle from moving when the vehicle is started to verify the completion of these tasks.
- Extreme caution must be exercised when working around rotating components.
- **Caution:** Most types of brake fluid are harmful to painted surfaces. Be sure to prevent brake fluid from coming into contact with a vehicle's paint. Use fender covers to minimize this risk and be sure to wipe up any spilled brake fluid immediately with a wet rag.

▶ TASK Bleed and/or flush brake system.

MLR
A5B6

Student Instructions: Read through the entire procedure prior to starting. Prepare your work space and any tools or parts that may be needed to complete the task. When directed by your instructor, begin the procedure to complete the task and check the box as each step is finished. Track your time on this procedure for later comparison to the standard completion time (i.e., "flat rate" or customer pay time).

Procedure:	Step Completed
1. Research the procedure for bleeding brakes.	
a. Explain when it would be necessary for a brake system to be bled.	☐
b. Explain when it would be necessary to flush the brake system. Is there a maintenance service interval for replacing the brake fluid on your vehicle?	☐
c. List or print off the procedure for bleeding brakes on your vehicle.	☐
d. Are the any special precautions or tools that are needed when bleeding brakes equipped with ABS?	☐
e. What type of brake fluid is required for your vehicle?	☐
f. Explain the sequence in which you would bleed brakes. Which wheel would you start with and how would you continue with the procedure?	☐
2. Bleed the brake system.	
a. Following proper procedure, bleed the brakes on your vehicle.	☐

b. Begin by topping off the master cylinder. Be certain that the fluid level is correct at all times during the bleed procedure. Always use a fender cover when working under the hood of a vehicle.	☐
c. Begin bleeding at the wheel furthest from the master cylinder. Special care must be taken when loosening bleeder screw. Screws are prone to rust and corrosion and can be easily broken. If no fluid is coming from the bleed screw, there is a chance that it has been clogged with dirt or rust build up. You may remove a bleed screw and attempt to clean it out using compressed air or a similar method of loosening the debris. Always wear eye protection when using solvents or compressed air.	☐
d. Follow the procedure to bleed the brakes. Use a catch pan to collect fluid from the bleed procedure. Once a solid stream of brake fluid is seen coming from the bleed screw, tighten the bleed screw and move to the next wheel. Repeat the procedure until you have bled the entire brake system. Ensure that brake fluid level is correct once the procedure is complete.	☐
3. Flush the brake system.	
a. Research and describe how the brake flush equipment is used.	☐
b. Following proper procedure, use the flush equipment to remove old, dirty, or contaminated fluid from the brake system.	☐
c. Once the flush procedure is complete, remove the equipment and secure any bleed screws that were loosened. Be sure to verify correct fluid level in the master cylinder.	☐

Non-Task-Specific Evaluations:	**Step Completed**
1. Tools and equipment were used as directed and returned in good working order.	☐
2. Complied with all general and task-specific safety standards, including proper use of any personal protective equipment.	☐
3. Completed the task in an appropriate time frame (Recommendation: 1.5 or 2 times flat rate).	☐
4. Left the work space clean and orderly.	☐
5. Cared for customer property and returned it undamaged.	☐

Student signature _______________________________ Date _______________________________

Comments:

Have your supervisor/instructor verify satisfactory completion of this procedure, any observations found, and any necessary action(s) recommended.

Tasksheet Scoring

Evaluation Items	Test		Retest	
	Pass	**Fail**	**Pass**	**Fail**
Task-Specific Evaluation	**(1 pt)**	**(0 pts)**	**(1 pt)**	**(0 pts)**
Research the procedure for bleeding brakes and using flush equipment. Describe maintenance service intervals.				
Identify special precautions or tools needed to complete the task.				
Bleed the brake system.				
Flush the brake system.				
Non-Task-Specific Evaluation	**(0 pts)**	**(−1 pt)**	**(0 pts)**	**(−1 pt)**
Student successfully completed at least three of the non-task-specific steps.				
Student successfully completed all five of the non-task-specific steps.				
Total Score: <total # of points / 4 = %>				

Supervisor:

Supervisor/instructor signature ________________________________ Date ________________

Comments:

Retest supervisor/instructor signature ________________________________ Date ________________

Comments:

CDX Tasksheet Number: A5O11

Student/Intern Information

Name _________________________ Date _________ Class _________________________

Vehicle, Customer, and Service Information

Vehicle used for this activity:

Year _____________ Make _________________ Model _________________

Odometer _________________ VIN _________________________

<table>
<tr><td>

Materials Required

- Blank work order
- Vehicle with available service history records
- Depending on the type of concern, special diagnostic/hand tools may be required. See your supervisor/instructor for instructions to identify what tools may be required.
- Service information database
- Personal protective equipment
- Electronic brake fluid tester or litmus style test strips
- Brake fluid
- Fender cover

</td></tr>
</table>

Task-Specific Safety Considerations

- Comply with personal and environmental safety practices associated with clothing; eye protection; hand tools; power equipment; proper ventilation; and the handling, storage, and disposal of chemicals/materials in accordance with local, state, and federal safety and environmental regulations.
- **Caution:** Most types of brake fluid are harmful to painted surfaces. Be sure to prevent brake fluid from coming into contact with a vehicle's paint. Use fender covers to minimize this risk and be sure to wipe up any spilled brake fluid immediately with a wet rag.

▶ **TASK** Test brake fluid for contamination.

MLR
A5B7

Time off_____________

Time on_____________

Total time_____________

Student Instructions: Read through the entire procedure prior to starting. Prepare your workspace and any tools or parts that may be needed to complete the task. When directed by your instructor, begin the procedure to complete the task and check the box as each step is finished. Track your time on this procedure for later comparison to the standard completion time (i.e., "flat rate" or customer pay time).

Procedure:	Step Completed
1. Describe the types of brake fluid contamination.	
a. Explain what hygroscopic means. How does this apply to brake fluid?	☐
b. How can excessive moisture in a vehicle's brake fluid affect the braking system?	☐
c. Determine the maintenance service interval for the brake fluid in your vehicle.	☐
2. Test the brake fluid for contamination.	
a. Lift the hood of the vehicle. Place a fender cover over the fender and surrounding work area.	☐
b. Before sampling the fluid in your vehicle, take a small sample of new brake fluid. Describe your readings. These readings can be used to compare the fluid in the vehicle.	☐
c. Clean the area around the master cylinder reservoir cap with a shop rag. Remove the cap.	☐
d. Take a sample of the brake fluid in your vehicle. Describe your results.	☐
e. What color is the fluid in the vehicle? How does this compare to new fluid?	☐
f. Based on your inspection, what is your recommendation for service?	☐

Non-Task-Specific Evaluations:	**Step Completed**
1. Tools and equipment were used as directed and returned in good working order.	☐
2. Complied with all general and task-specific safety standards, including proper use of any personal protective equipment.	☐
3. Completed the task in an appropriate time frame (Recommendation: 1.5 or 2 times flat rate).	☐
4. Left the work space clean and orderly.	☐
5. Cared for customer property and returned it undamaged.	☐

Student signature ________________________ Date ________________________

Comments:

Have your supervisor/instructor verify satisfactory completion of this procedure, any observations found, and any necessary action(s) recommended.

Tasksheet Scoring

	Test		Retest	
Evaluation Items	**Pass**	**Fail**	**Pass**	**Fail**
Task-Specific Evaluation	**(1 pt)**	**(0 pts)**	**(1 pt)**	**(0 pts)**
Define hygroscopic and describe the effects of moisture in a brake system.				
Determine the maintenance service interval for the brake fluid in your vehicle.				
Sample known good brake fluid, then test brake fluid for contamination.				
Make recommendations for service based on your inspection.				
Non-Task-Specific Evaluation	**(0 pts)**	**(−1 pt)**	**(0 pts)**	**(−1 pt)**
Student successfully completed at least three of the non-task-specific steps.				
Student successfully completed all five of the non-task-specific steps.				
Total Score: <total # of points / 4 = %>				

Supervisor:

Supervisor/instructor signature _________________________ Date _________________

Comments:

Retest supervisor/instructor signature _________________________ Date _________________

Comments:

5C. Drum Brakes

Learning Objective/Task	CDX Tasksheet Number	ASE Education Foundation Reference Number; Priority Level
• Remove, clean, and inspect brake drum; measure brake drum diameter; determine serviceability.	A5012	A5C1, P-1
• Refinish brake drum and measure final drum diameter; compare with specification.	A5013	A5C2, P-1
• Remove, clean, inspect, and/or replace brake shoes, springs, pins, clips, levers, adjusters/self-adjusters, other related brake hardware, and backing support plates; lubricate and reassemble.	A5014	A5C3, P-1
• Inspect wheel cylinders for leaks and proper operation; remove and replace as needed.	A5015	A5C4, P-2
• Preadjust brake shoes and parking brake; install brake drums or drum/hub assemblies and wheel bearings; make final checks and adjustments.	A5016	A5C5, P-1

Materials Required

- Blank work order
- Vehicle with available service history records
- Depending on the type of concern, special diagnostic/hand tools may be required. See your supervisor/instructor for instructions to identify what tools may be required.
- Service information database
- Personal protective equipment

Safety Considerations

- When running any vehicles in the shop, make sure you use the shop's exhaust ventilation system to discharge all exhaust gas safely outside.
- Lifting equipment, such as vehicle jacks and stands, vehicle hoists, and engine hoists, are important tools that increase productivity and make the job easier. However, they can also cause severe injury or death if used improperly. Make sure you follow the manufacturer's operation procedures. Also, make sure you have your supervisor's/instructor's permission to use any particular type of lifting equipment.
- Comply with personal and environmental safety practices associated with clothing; eye protection; hand tools; power equipment; proper ventilation; and the handling, storage, and disposal of chemicals/materials in accordance with local, state, and federal safety and environmental regulations.
- If you need to start the vehicle, you should ensure that the parking brake is firmly applied; if necessary, use wheel chocks to prevent the vehicle from moving when the vehicle is started to verify the completion of these tasks.
- Extreme caution must be exercised when working around rotating components.

CDX Tasksheet Number: A5012

Student/Intern Information

Name _________________________________ Date ___________ Class _________________________

Vehicle, Customer, and Service Information

Vehicle used for this activity:

Year _______________ Make _____________________________ Model _____________________

Odometer _________________________________ VIN _________________________________

Materials Required

- Blank work order
- Vehicle with available service history records
- Depending on the type of concern, special diagnostic/hand tools may be required. See your supervisor/instructor for instructions to identify what tools may be required.
- Service information database
- Personal protective equipment
- General hand/air tools
- Brake wash equipment
- Brake drum micrometer

Task-Specific Safety Considerations

- Lifting equipment, such as vehicle jacks and stands, vehicle hoists, and engine hoists, are important tools that increase productivity and make the job easier. However, they can also cause severe injury or death if used improperly. Make sure you follow the manufacturer's operation procedures. Also, make sure you have your supervisor's/instructor's permission to use any particular type of lifting equipment.
- Comply with personal and environmental safety practices associated with clothing; eye protection; hand tools; power equipment; proper ventilation; and the handling, storage, and disposal of chemicals/materials in accordance with local, state, and federal safety and environmental regulations.
- **Caution:** Brake dust may contain asbestos, which has been determined to cause cancer when inhaled or ingested. Treat all brake dust as if it contains asbestos and use OSHA-approved asbestos removal equipment. Do not allow brake dust to become airborne by using anything that would disturb the dust. Also, wear protective gloves during this procedure and dispose of or clean them in an approved manner.

Time off_______________

Time on_______________

Total time_______________

▶ TASK Remove, clean, and inspect brake drum; measure brake drum diameter; determine serviceability.

MLR
A5C1

Student Instructions: Read through the entire procedure prior to starting. Prepare your work space and any tools or parts that may be needed to complete the task. When directed by your instructor, begin the procedure to complete the task and check the box as each step is finished. Track your time on this procedure for later comparison to the standard completion time (i.e., "flat rate" or customer pay time).

Procedure:	Step Completed
1. Remove, clean, and inspect the brake drum.	
a. Lift vehicle and remove wheels.	☐
b. Remove the brake drum. **Tip:** At times brake drums could be difficult to remove. Use a hammer to tap around drum to loosen. Also it is helpful in situations to key down brake shoe adjusters.	☐
c. Use approved brake wash equipment to wash the shoes and the drum.	☐
d. Inspect the brake drum for signs of heat damage, oil contamination, and excessive rust.	☐
2. Measure the brake drum's diameter.	
a. Using a brake drum micrometer, measure the inner diameter in four places. Use largest measurement to determine serviceability. Note your findings here.	☐
b. Use vehicle service information to determine specification for serviceability. What is the maximum drum diameter specification?	☐
3. Based on your measurement in Step 2, determine whether the drum can be machined and reused.	

Non-Task-Specific Evaluations:	Step Completed
1. Tools and equipment were used as directed and returned in good working order.	☐
2. Complied with all general and task-specific safety standards, including proper use of any personal protective equipment.	☐
3. Completed the task in an appropriate time frame (Recommendation: 1.5 or 2 times flat rate).	☐
4. Left the work space clean and orderly.	☐
5. Cared for customer property and returned it undamaged.	☐

Student signature _________________________________ Date _________________________________

Comments:

Have your supervisor/instructor verify satisfactory completion of this procedure, any observations found,

and any necessary action(s) recommended.

Evaluation Instructions: The scoring box below is intended to act as a guide for both student and instructor. Each criterion listed will help students understand what is expected from them and evaluators articulate success at a particular task. The scoring is set up to allow a second attempt at each task (see the "Test" and "Retest" columns). Scoring is designed to reward students for correct completion of the task. Points are lost for failure to complete the employability requirements (see "Non-Task-Specific" criteria). When all criteria are evaluated, tally the points for a total at the bottom of each column.

Tasksheet Scoring

	Test		Retest	
Evaluation Items	**Pass**	**Fail**	**Pass**	**Fail**
Task-Specific Evaluation	**(1 pt)**	**(0 pts)**	**(1 pt)**	**(0 pts)**
Lift the vehicle and remove the wheels, then the brake drum.				
Use brake wash equipment to clean the brake shoes and drums.				
Inspect the brake drum condition and measure its diameter.				
Make recommendations for service based on your inspection.				
Non-Task-Specific Evaluation	**(0 pts)**	**(−1 pt)**	**(0 pts)**	**(−1 pt)**
Student successfully completed at least three of the non-task-specific steps.				
Student successfully completed all five of the non-task-specific steps.				
Total Score: <total # of points / 4 = %>				

Supervisor:

Supervisor/instructor signature ________________________________ Date ________________________

Comments:

Retest supervisor/instructor signature ________________________________ Date ________________________

Comments:

CDX Tasksheet Number: A5013

Student/Intern Information

Name _________________________________ Date _____________ Class _____________________

Vehicle, Customer, and Service Information

Vehicle used for this activity:

Year _______________ Make _______________________ Model _____________________

Odometer _______________________ VIN _______________________________

Materials Required

- Blank work order
- Vehicle with available service history records
- Depending on the type of concern, special diagnostic/hand tools may be required. See your supervisor/instructor for instructions to identify what tools may be required.
- Service information database
- Personal protective equipment
- General hand/air tools
- Bench brake lathe

Task-Specific Safety Considerations

- Comply with personal and environmental safety practices associated with clothing; eye protection; hand tools; power equipment; proper ventilation; and the handling, storage, and disposal of chemicals/materials in accordance with local, state, and federal safety and environmental regulations.

▶ **TASK** Refinish brake drum and measure final drum diameter; compare with specification.

**MLR
A5C2**

Time off_______________

Time on_______________

Total time_______________

Student Instructions: Read through the entire procedure prior to starting. Prepare your workspace and any tools or parts that may be needed to complete the task. When directed by your instructor, begin the procedure to complete the task and check the box as each step is finished. Track your time on this procedure for later comparison to the standard completion time (i.e., "flat rate" or customer pay time).

Procedure:	Step Completed
1. Research and list the specifications for maximum brake drum diameter.	☐

2. Measure brake drum diameter.	
a. Lift the vehicle and remove the rear wheels.	☐
b. Remove the brake drums.	☐
c. Use a brake drum micrometer to measure the drum diameter in four different spots. Record your measurements. Will you be able to refinish the drum and still meet diameter specifications? Note the measurements here.	☐
3. Refinish the brake drum.	
a. Remove any dust and oil from the brake drum prior to refinishing.	☐
b. Mount the brake drum to the lathe. If needed, attach the correct cutting attachment.	☐
c. Install a vibration damper band around the outside of the drum. Explain what may happen if this band is not installed.	☐
d. Turn the lathe on. Move the cutting bit inward until the bit just begins to touch the friction surface area of the drum. Zero the dial indicator.	☐
e. Perform a "scratch test" to ensure that the surface being refinished is mostly flat.	☐
f. Move the cutting bit in 0.010-0.020" and start the fast/rough cut. Continue to do rough cuts until the drum surface has been cut evenly and there are no low/high spots or discoloration present.	☐
g. Perform a slow finish cut. Use sand paper once the cut has been finished to remove any remaining irregularities in the drum surface.	☐
h. Clean the drum with approved solvents. Measure the diameter once more to ensure that specifications for diameter have not been exceeded. Reassemble all brake components.	☐

Non-Task-Specific Evaluations:	Step Completed
1. Tools and equipment were used as directed and returned in good working order.	☐
2. Complied with all general and task-specific safety standards, including proper use of any personal protective equipment.	☐
3. Completed the task in an appropriate time frame (Recommendation: 1.5 or 2 times flat rate).	☐
4. Left the work space clean and orderly.	☐
5. Cared for customer property and returned it undamaged.	☐

Student signature _________________________________ Date _______________________________

Comments:

Have your supervisor/instructor verify satisfactory completion of this procedure, any observations found, and any necessary action(s) recommended.

Evaluation Instructions: The scoring box below is intended to act as a guide for both student and instructor. Each criterion listed will help students understand what is expected from them and evaluators articulate success at a particular task. The scoring is set up to allow a second attempt at each task (see the "Test" and "Retest" columns). Scoring is designed to reward students for correct completion of the task. Points are lost for failure to complete the employability requirements (see "Non-Task-Specific" criteria). When all criteria are evaluated, tally the points for a total at the bottom of each column.

Tasksheet Scoring

	Test		Retest	
Evaluation Items	**Pass**	**Fail**	**Pass**	**Fail**
Task-Specific Evaluation	**(1 pt)**	**(0 pts)**	**(1 pt)**	**(0 pts)**
Research the brake drum diameter specifications. Remove the brake drums, measure their diameter, and compare it to specifications.				
Mount the drum to the lathe and install the vibration damper band.				
Perform a rough cut until the drum surface is even.				
Perform a finish cut. Remeasure the drum diameter and compare it to specifications.				
Non-Task-Specific Evaluation	**(0 pts)**	**(−1 pt)**	**(0 pts)**	**(−1 pt)**
Student successfully completed at least three of the non-task-specific steps.				
Student successfully completed all five of the non-task-specific steps.				
Total Score: <total # of points / 4 = %>				

Supervisor:

Supervisor/instructor signature _______________________________ Date _______________________

Comments:

Retest supervisor/instructor signature _________________________ Date ____________________

Comments:

CDX Tasksheet Number: A5014

Student/Intern Information

Name _________________________________ Date _____________ Class _______________________

Vehicle, Customer, and Service Information

Vehicle used for this activity:

Year _______________ Make _________________________________ Model _______________________

Odometer _________________________________ VIN _______________________________________

Materials Required

- Blank work order
- Vehicle with available service history records
- Depending on the type of concern, special diagnostic/hand tools may be required. See your supervisor/instructor for instructions to identify what tools may be required.
- Service information database
- Personal protective equipment
- Brake spring tools

Task-Specific Safety Considerations

- Comply with personal and environmental safety practices associated with clothing; eye protection; hand tools; power equipment; proper ventilation; and the handling, storage, and disposal of chemicals/materials in accordance with local, state, and federal safety and environmental regulations.

▶ TASK Remove, clean, inspect, and/or replace brake shoes, springs, pins, clips, levers, adjusters/self-adjusters, other related brake hardware, and backing support plates; lubricate and reassemble.

MLR
A5C3

Time off______________

Time on______________

Total time______________

Student Instructions: Read through the entire procedure prior to starting. Prepare your workspace and any tools or parts that may be needed to complete the task. When directed by your instructor, begin the procedure to complete the task and check the box as each step is finished. Track your time on this procedure for later comparison to the standard completion time (i.e., "flat rate" or customer pay time).

Procedure:	Step Completed
1. Research the procedure for replacing the brake shoes.	
a. From your research, determine what type of brake shoes are on the vehicle. Are they leading/trailing or duo-servo?	☐
b. Explain the difference between leading/trailing and duo-servo drum brake systems.	☐
2. Remove and replace the brake shoes.	
a. Lift the vehicle and remove the wheels.	☐
b. Remove the brake drum and use brake wash equipment to wash the drum, shoes, and all associated hardware.	☐
c. Use brake spring tools to remove the shoe retainers and all other necessary components.	☐
d. Inspect all components for rust, wear, and damage. Replace components as needed.	☐
e. Apply brake grease to the contact patches on the backing plate.	☐
f. Install the shoes, related springs, and hardware.	☐
g. Adjust the brake shoes until light drag/resistance is felt as you rotate the brake drum.	☐

Non-Task-Specific Evaluations:	Step Completed
1. Tools and equipment were used as directed and returned in good working order.	☐
2. Complied with all general and task-specific safety standards, including proper use of any personal protective equipment.	☐
3. Completed the task in an appropriate time frame (Recommendation: 1.5 or 2 times flat rate).	☐
4. Left the work space clean and orderly.	☐
5. Cared for customer property and returned it undamaged.	☐

Student signature _________________________________ Date _________________________________

Comments:

Have your supervisor/instructor verify satisfactory completion of this procedure, any observations found, and any necessary action(s) recommended.

Tasksheet Scoring

	Test		Retest	
Evaluation Items	**Pass**	**Fail**	**Pass**	**Fail**
Task-Specific Evaluation	**(1 pt)**	**(0 pts)**	**(1 pt)**	**(0 pts)**
Research the brake shoe replacement procedure.				
Determine the brake shoe configuration and operation.				
Explain different brake shoe configurations.				
Remove and replace the brake shoes. Adjust the brake shoes as necessary.				
Non-Task-Specific Evaluation	**(0 pts)**	**(−1 pt)**	**(0 pts)**	**(−1 pt)**
Student successfully completed at least three of the non-task-specific steps.				
Student successfully completed all five of the non-task-specific steps.				
Total Score: <total # of points / 4 = %>				

Supervisor:

Supervisor/instructor signature _________________________ Date _______________

Comments:

Retest supervisor/instructor signature _________________________ Date _______________

Comments:

CDX Tasksheet Number: A5015

Student/Intern Information

Name _________________________________ Date _____________ Class _____________________

Vehicle, Customer, and Service Information

Vehicle used for this activity:

Year _______________ Make _____________________________ Model _____________________

Odometer _________________________________ VIN _____________________________

Task-Specific Safety Considerations

- Comply with personal and environmental safety practices associated with clothing; eye protection; hand tools; power equipment; proper ventilation; and the handling, storage, and disposal of chemicals/materials in accordance with local, state, and federal safety and environmental regulations.

▶ **TASK** Inspect wheel cylinders for leaks and proper operation; remove and replace as needed.

MLR
A5C4

Time off______________

Time on______________

Total time____________

Student Instructions: Read through the entire procedure prior to starting. Prepare your workspace and any tools or parts that may be needed to complete the task. When directed by your instructor, begin the procedure to complete the task and check the box as each step is finished. Track your time on this procedure for later comparison to the standard completion time (i.e., "flat rate" or customer pay time).

Procedure:	Step Completed
1. Inspect the wheel cylinders for leaks and proper operation.	
a. Lift the vehicle and remove the wheels.	☐
b. Remove the brake drum and perform a visual inspection. Are there any signs of fluid leaking from the wheel cylinders or fluid contamination on the brake shoes?	☐
c. Determine any necessary action; clean the wheel cylinders with brake parts washer and remove/replace the leaking wheel cylinder.	☐
2. Research the procedure for removing/replacing the wheel cylinder. Remove and replace the wheel cylinder.	
a. With the drum removed, remove the brake line from the wheel cylinder. If possible, use a soft grip pair of pliers to clamp the rubber brake hose portion to prevent excessive brake fluid from escaping the line.	☐
b. Remove the fasteners holding the wheel cylinder in place and remove the wheel cylinder. **Tip:** On some vehicles you may be required to remove the brake shoes in order to remove the wheel cylinder.	☐
c. Compare the new wheel cylinder with the one that has been removed. Install the wheel cylinder and torque the fasteners to specification.	☐
d. Install the brake line into the wheel cylinder and remove the soft-grip pliers.	☐
e. Adjust the brake shoes as needed and install the brake drum.	☐
f. Any time you replace hydraulic components or open lines in brake systems, you will need to bleed air from the system. Bleed the brakes following the recommended procedure.	☐
g. Install the wheel and torque the lug nuts to specification.	☐

Non-Task-Specific Evaluations:	Step Completed
1. Tools and equipment were used as directed and returned in good working order.	☐
2. Complied with all general and task-specific safety standards, including proper use of any personal protective equipment.	☐
3. Completed the task in an appropriate time frame (Recommendation: 1.5 or 2 times flat rate).	☐
4. Left the work space clean and orderly.	☐
5. Cared for customer property and returned it undamaged.	☐

Student signature ________________________________ Date ________________________________

Comments:

Have your supervisor/instructor verify satisfactory completion of this procedure, any observations found, and any necessary action(s) recommended.

Evaluation Instructions: The scoring box below is intended to act as a guide for both student and instructor. Each criterion listed will help students understand what is expected from them and evaluators articulate success at a particular task. The scoring is set up to allow a second attempt at each task (see the "Test" and "Retest" columns). Scoring is designed to reward students for correct completion of the task. Points are lost for failure to complete the employability requirements (see "Non-Task-Specific" criteria). When all criteria are evaluated, tally the points for a total at the bottom of each column.

Tasksheet Scoring

Evaluation Items	Test		Retest	
	Pass	Fail	Pass	Fail
Task-Specific Evaluation	**(1 pt)**	**(0 pts)**	**(1 pt)**	**(0 pts)**
Lift the vehicle and remove the wheel and brake drum.				
Perform a visual inspection of the brake shoes and wheel cylinders.				
Remove and replace the wheel cylinder.				
Bleed the brake system.				
Non-Task-Specific Evaluation	**(0 pts)**	**(−1 pt)**	**(0 pts)**	**(−1 pt)**
Student successfully completed at least three of the non-task-specific steps.				
Student successfully completed all five of the non-task-specific steps.				
Total Score: <total # of points / 4 = %>				

Supervisor:

Supervisor/instructor signature _________________________________ Date _____________________

Comments:

Retest supervisor/instructor signature _________________________ Date _____________________

Comments:

CDX Tasksheet Number: A5016

Student/Intern Information

Name _________________________________ Date ____________ Class _____________________

Vehicle, Customer, and Service Information

Vehicle used for this activity:

Year _______________ Make _______________________ Model ____________________

Odometer _______________________ VIN _______________________________

Materials Required

- Blank work order
- Vehicle with available service history records
- Depending on the type of concern, special diagnostic/hand tools may be required. See your supervisor/instructor for instructions to identify what tools may be required.
- Service information database
- Personal protective equipment
- General hand/air tools

Task-Specific Safety Considerations

- Comply with personal and environmental safety practices associated with clothing; eye protection; hand tools; power equipment; proper ventilation; and the handling, storage, and disposal of chemicals/materials in accordance with local, state, and federal safety and environmental regulations.

▶ **TASK** Preadjust brake shoes and parking brake; install brake drums or drum/hub assemblies and wheel bearings; make final checks and adjustments.

MLR
A5C5

Time off________________

Time on________________

Total time________________

Student Instructions: Read through the entire procedure prior to starting. Prepare your workspace and any tools or parts that may be needed to complete the task. When directed by your instructor, begin the procedure to complete the task and check the box as each step is finished. Track your time on this procedure for later comparison to the standard completion time (i.e., "flat rate" or customer pay time).

Procedure:	Step Completed
1. Research the procedure for adjusting the brake shoes. Include the preload specifications for drum/hub assemblies that use wheel bearings.	☐
2. Pre-adjust the brake shoes.	
a. When reassembling drum brakes, place the drum on the vehicle and adjust the brake shoes from the access hole on the backing plate. Adjust the shoes until they begin to touch the drum and light drag can be felt. Back the shoes off the drum two or three clicks.	☐
b. If a drum/hub assembly is being used, ensure that the bearings, races, and spindle are cleaned and greased.	☐
c. Install the brake drum and torque the preload nut to specification to seat the bearings. Follow procedures to ensure that the bearing is seated correctly.	☐

Non-Task-Specific Evaluations:	Step Completed
1. Tools and equipment were used as directed and returned in good working order.	☐
2. Complied with all general and task-specific safety standards, including proper use of any personal protective equipment.	☐
3. Completed the task in an appropriate time frame (Recommendation: 1.5 or 2 times flat rate).	☐
4. Left the work space clean and orderly.	☐
5. Cared for customer property and returned it undamaged.	☐

Student signature _______________________________ Date _______________________________

Comments:

Have your supervisor/instructor verify satisfactory completion of this procedure, any observations found, and any necessary action(s) recommended.

Evaluation Instructions: The scoring box below is intended to act as a guide for both student and instructor. Each criterion listed will help students understand what is expected from them and evaluators articulate success at a particular task. The scoring is set up to allow a second attempt at each task (see the "Test" and "Retest" columns). Scoring is designed to reward students for correct completion of the task. Points are lost for failure to complete the employability requirements (see "Non-Task-Specific" criteria). When all criteria are evaluated, tally the points for a total at the bottom of each column.

Tasksheet Scoring

		Test		Retest	
Evaluation Items	**Pass**	**Fail**	**Pass**	**Fail**
Task-Specific Evaluation	**(1 pt)**	**(O pts)**	**(1 pt)**	**(O pts)**
Research the bearing preload procedure and specifications.				
Pre-adjust the brake shoes.				
Install the brake drums and verify that the shoes have been adjusted properly.				
Follow the procedure to preload bearings.				
Non-Task-Specific Evaluation	**(O pts)**	**(−1 pt)**	**(O pts)**	**(−1 pt)**
Student successfully completed at least three of the non-task-specific steps.				
Student successfully completed all five of the non-task-specific steps.				
Total Score: <total # of points / 4 = %>				

Supervisor:

Supervisor/instructor signature _________________________________ Date _____________________

Comments:

Retest supervisor/instructor signature _____________________________ Date _____________________

Comments:

5D. Disc Brakes

Learning Objective/Task	CDX Tasksheet Number	ASE Education Foundation Reference Number; Priority Level
• Remove and clean caliper assembly; inspect for leaks and damage/wear; determine necessary action.	A5017	A5D1, P-1
• Inspect caliper mounting and slides/pins for proper operation, wear, and damage; determine necessary action.	A5018	A5D2, P-1
• Remove, inspect, and/or replace brake pads and retaining hardware; determine necessary action.	A5019	A5D3, P-1
• Lubricate and reinstall caliper, brake pads, and related hardware; seat brake pads and inspect for leaks.	A5020	A5D4, P-1
• Clean and inspect rotor and mounting surface, measure rotor thickness, thickness variation, and lateral runout; determine necessary action.	A5021	A5D5, P-1
• Remove and reinstall/replace rotor.	A5022	A5D6, P-1
• Refinish rotor on vehicle; measure final rotor thickness and compare with specification.	A5023	A5D7, P-1
• Refinish rotor off vehicle; measure final rotor thickness and compare with specification.	A5024	A5D8, P-1
• Retract and readjust caliper piston on an integral parking brake system.	A5025	A5D9, P-2
• Check brake pad wear indicator; determine necessary action.	A5026	A5D10, P-1
• Describe importance of operating vehicle to burnish/break-in replacement brake pads according to manufacturer's recommendation.	A5027	A5D11, P-1

Materials Required

- Blank work order
- Vehicle with available service history records
- Depending on the type of concern, special diagnostic/hand tools may be required. See your supervisor/instructor for instructions to identify what tools may be required.
- Service information database
- Personal protective equipment

Safety Considerations

- When running any vehicles in the shop, make sure you use the shop's exhaust ventilation system to discharge all exhaust gas safely outside.
- Lifting equipment such as vehicle jacks and stands, vehicle hoists, and engine hoists are important tools that increase productivity and make the job easier. However, they can also cause severe injury or death if used improperly. Make sure you follow the manufacturer's operation procedures. Also, make sure you have your supervisor's/instructor's permission to use any particular type of lifting equipment.

- Comply with personal and environmental safety practices associated with clothing; eye protection; hand tools; power equipment; proper ventilation; and the handling, storage, and disposal of chemicals/materials in accordance with local, state, and federal safety and environmental regulations.
- If you need to start the vehicle, you should ensure that the parking brake is firmly applied; if necessary, use wheel chocks to prevent the vehicle from moving when the vehicle is started to verify the completion of these tasks.
- Extreme caution must be exercised when working around rotating components.
- **Caution:** Most types of brake fluid are harmful to painted surfaces. Be sure to prevent brake fluid from coming into contact with a vehicle's paint. Use fender covers to minimize this risk and be sure to wipe up any spilled brake fluid immediately with a wet rag.
- **Caution:** Brake dust may contain asbestos, which has been determined to cause cancer when inhaled or ingested. Treat all brake dust as if it contains asbestos and use Occupational Safety and Health Administration (OSHA)-approved asbestos removal equipment. Do not allow brake dust to become airborne by using anything that would disturb the dust. Also, wear protective gloves during this procedure and dispose of or clean them in an approved manner.

CDX Tasksheet Number: A5017

Vehicle, Customer, and Service Information

Vehicle used for this activity:

Year ______________ Make ________________________ Model ____________________________

Odometer ________________________________ VIN ____________________________________

Materials Required

- Blank work order
- Vehicle with available service history records
- Depending on the type of concern, special diagnostic/hand tools may be required. See your supervisor/instructor for instructions to identify what tools may be required.
- Service Information Database
- Personal Protective Equipment
- General hand/air tools
- Impact sockets
- Fluid catch pan
- Torque wrench
- Brake parts wash
- Wire brush and file
- Brake hose pliers
- Caliper piston compressor tool

Task-Specific Safety Considerations

- Lifting equipment such as vehicle jacks and stands, vehicle hoists, and engine hoists are important tools that increase productivity and make the job easier. However, they can also cause severe injury or death if used improperly. Make sure you follow the manufacturer's operation procedures. Also, make sure you have your supervisor's/instructor's permission to use any particular type of lifting equipment.
- Comply with personal and environmental safety practices associated with clothing; eye protection; hand tools; power equipment; proper ventilation; and the handling, storage, and disposal of chemicals/materials in accordance with local, state, and federal safety and environmental regulations.
- **Caution:** Brake dust may contain asbestos, which has been determined to cause cancer when inhaled or ingested. Treat all brake dust as if it contains asbestos and use OSHA-approved asbestos removal equipment. Do not allow brake dust to become airborne by using anything that would disturb the dust. Also, wear protective gloves during this procedure and dispose of or clean them in an approved manner.

 Remove and clean caliper assembly; inspect for leaks and damage/wear; determine necessary action.

MLR
A5D1

Time off_______________

Time on_______________

Total time_______________

Student Instructions: Read through the entire procedure prior to starting. Prepare your workspace and any tools or parts that may be needed to complete the task. When directed by your instructor, begin the procedure to complete the task and check the box as each step is finished. Track your time on this procedure for later comparison to the standard completion time (i.e., "flat rate" or customer pay time).

Procedure:	Step Completed
1. Research the procedure for removing and replacing the brake caliper on your vehicle.	
a. List or print off the procedure for removing and replacing the brake caliper on your vehicle.	☐
b. List the torque specification for the caliper mounting bracket and lug nuts.	☐
2. Remove the brake caliper.	
a. Lift the vehicle. Remove the wheel to access the brake caliper in need of service.	☐
b. Following proper procedure, remove the brake caliper.	☐
c. Using the brake caliper piston tool, attempt to collapse the piston inside the caliper. The piston should collapse all the way into the caliper with minimal force.	☐
d. Explain what may cause a piston to not collapse into the caliper.	☐
e. Remove the caliper bracket bolts, bracket, and brake pads.	☐
f. Remove the brake pads from the caliper bracket. The pads should come out with minimal force. Excess rust may build up behind the brake pads causing difficulty when attempting to remove them. Describe the condition of the brake pads and caliper mount.	☐

g. If removing the brake caliper completely, use a pair of brake hose pliers to pinch off the flexible brake hose. Loosen the bolt holding the line to the caliper. Remove the brake hose and two washers. **Tip:** Never reuse washers, as they may create a leak.	☐
3. Clean and inspect the brake caliper and bracket.	
a. Remove the metal shims from the caliper bracket. Use a wire brush, wire wheel, or sand blast cabinet and remove any rust or grease build up on the shims.	☐
b. Remove built-up rust or grease from the caliper bracket using a wire brush, file, wire wheel, or sand blast cabinet. **Tip:** If using a sand blast cabinet, ensure that no blast media enters the slide pin recesses. Keep the slide pins and boots intact while the cleaning procedure is performed. Use a parts washer solvent and compressed air to remove any leftover media or grease.	☐
c. Inspect the slide pins for proper function. Pins should move freely in their bores. Remove the pins and clean them using approved solvents. Apply new brake grease to the pins and reinstall.	☐
d. Inspect the slide pin boots for signs of cracking or tears. List any action required.	☐
e. Inspect the brake caliper assembly. Inspect the rubber boot covering the piston for signs of leaks or damage. List any action required.	☐
4. Based on your cleaning and inspection, what are your recommendations for service?	☐

Non-Task-Specific Evaluations:	Step Completed
1. Tools and equipment were used as directed and returned in good working order.	☐
2. Complied with all general and task-specific safety standards, including proper use of any personal protective equipment.	☐
3. Completed the task in an appropriate time frame (Recommendation: 1.5 or 2 times flat rate).	☐
4. Left the work space clean and orderly.	☐
5. Cared for customer property and returned it undamaged.	☐

Student signature _______________________________ Date _______________________

Comments:

Have your supervisor/instructor verify satisfactory completion of this procedure, any observations found,

and any necessary action(s) recommended.

Tasksheet Scoring

	Test		Retest	
Evaluation Items	**Pass**	**Fail**	**Pass**	**Fail**
Task-Specific Evaluation	**(1 pt)**	**(0 pts)**	**(1 pt)**	**(0 pts)**
Research the procedure for removing a brake caliper. Research torque specifications.				
Remove the brake caliper and the caliper bracket.				
Clean and inspect all brake caliper components.				
Make recommendations for service based on your inspection.				
Non-Task-Specific Evaluation	**(0 pts)**	**(−1 pt)**	**(0 pts)**	**(−1 pt)**
Student successfully completed at least three of the non-task-specific steps.				
Student successfully completed all five of the non-task-specific steps.				
Total Score: <total # of points / 4 = %>				

Supervisor:

Supervisor/instructor signature _________________________ Date _______________

Comments:

Retest supervisor/instructor signature _________________________ Date _______________

Comments:

CDX Tasksheet Number: A5018

Student/Intern Information

Name _______________________________ Date ___________ Class _______________________

Vehicle, Customer, and Service Information

Vehicle used for this activity:

Year _______________ Make _____________________________ Model _______________________

Odometer _________________________________ VIN _______________________________

Materials Required

- Blank work order
- Vehicle with available service history records
- Depending on the type of concern, special diagnostic/hand tools may be required. See your supervisor/instructor for instructions to identify what tools may be required.
- Service information database
- Personal protective equipment
- General hand/air tools
- Caliper piston tool

Task-Specific Safety Considerations

- Lifting equipment such as vehicle jacks and stands, vehicle hoists, and engine hoists are important tools that increase productivity and make the job easier. However, they can also cause severe injury or death if used improperly. Make sure you follow the manufacturer's operation procedures. Also, make sure you have your supervisor's/instructor's permission to use any particular type of lifting equipment.
- Comply with personal and environmental safety practices associated with clothing; eye protection; hand tools; power equipment; proper ventilation; and the handling, storage, and disposal of chemicals/materials in accordance with local, state, and federal safety and environmental regulations.

▶ TASK Inspect caliper mounting and slides/pins for proper operation, wear, and damage; determine necessary action.

MLR
A5D2

Time off_______________

Time on_______________

Student Instructions: Read through the entire procedure prior to starting. Prepare your workspace and any tools or parts that may be needed to complete the task. When directed by your instructor, begin the procedure to complete the task and check the box as each step is finished. Track your time on this procedure for later comparison to the standard completion time (i.e., "flat rate" or customer pay time).

Total time_______________

Procedure:	Step Completed
1. Inspect the caliper mounting.	
a. Research and describe what type of caliper is used on your vehicle. Is it a fixed or floating type? How many pistons are equipped on each caliper?	☐
b. Lift the vehicle and remove the wheels.	☐
c. Perform a visual inspection of the brake calipers and caliper mounts. Inspect for signs of damage, torn slide pin boots, and loose/damaged fasteners.	☐
d. Inspect caliper for seized slide pins. Depending on design, the caliper should move freely, back and forth, on its slide pins. Components held in place by the caliper slide pins should move freely.	☐
e. Remove the brake caliper for further inspection. **Tip:** Never allow a brake caliper to hang from its brake hose. Hang the caliper with a strap/hook.	☐
f. Compress the caliper pistons and ensure that there are no seized pistons.	☐
g. Inspect all slide pins. Ensure that the pins are not damaged, are free of rust, and are moving freely in the pin bores.	☐
2. Based on your inspection, what is your recommendation for service?	
3. Reassemble the caliper components.	
a. Clean the slide pins using a wire wheel, shop rags, or solvent. Apply the proper brake grease to the pins.	☐
b. Install the brake caliper and torque all fasteners to specification. What torque specification did you use to tighten the caliper mounting bolts?	☐
c. Install the wheel and torque the lug nuts to specification.	☐
d. Lower the vehicle. ALWAYS pump the brake pedal after brake service, prior to starting the vehicle. Ensure that you have good brake pressure before attempting to move a vehicle. If pistons were collapsed during brake service, it is normal to have to pump the brake pedal two or three times to regain normal pedal feel.	☐

Non-Task-Specific Evaluations:	Step Completed
1. Tools and equipment were used as directed and returned in good working order.	☐
2. Complied with all general and task-specific safety standards, including proper use of any personal protective equipment.	☐
3. Completed the task in an appropriate time frame (Recommendation: 1.5 or 2 times flat rate).	☐
4. Left the work space clean and orderly.	☐
5. Cared for customer property and returned it undamaged.	☐

Student signature _______________________________ Date _______________________________

Comments:

Have your supervisor/instructor verify satisfactory completion of this procedure, any observations found,

and any necessary action(s) recommended.

Evaluation Instructions: The scoring box below is intended to act as a guide for both student and instructor. Each criterion listed will help students understand what is expected from them and evaluators articulate success at a particular task. The scoring is set up to allow a second attempt at each task (see the "Test" and "Retest" columns). Scoring is designed to reward students for correct completion of the task. Points are lost for failure to complete the employability requirements (see "Non-Task-Specific" criteria). When all criteria are evaluated, tally the points for a total at the bottom of each column.

Tasksheet Scoring

	Test		Retest	
Evaluation Items	**Pass**	**Fail**	**Pass**	**Fail**
Task-Specific Evaluation	**(1 pt)**	**(0 pts)**	**(1 pt)**	**(0 pts)**
Research brake caliper designs and evaluate the caliper design used on your vehicle. Perform a visual inspection of the caliper mounts and slide pins.				
Remove the brake caliper and inspect the condition of the slide pins.				
Clean and reassemble the brake caliper components.				
Make recommendations for service based on your inspection.				
Non-Task-Specific Evaluation	**(0 pts)**	**(−1 pt)**	**(0 pts)**	**(−1 pt)**
Student successfully completed at least three of the non-task-specific steps.				
Student successfully completed all five of the non-task-specific steps.				
Total Score: <total # of points / 4 = %>				

Supervisor:

Supervisor/instructor signature ________________________________ Date ________________________

Comments:

Retest supervisor/instructor signature ________________________________ Date ________________________

Comments:

CDX Tasksheet Number: A5019

Student/Intern Information

Name _________________________________ Date ___________ Class _____________________

Vehicle, Customer, and Service Information

Vehicle used for this activity:

Year _______________ Make _______________________ Model _____________________

Odometer _______________________ VIN _________________________________

Materials Required

- Blank work order
- Vehicle with available service history records
- Depending on the type of concern, special diagnostic/hand tools may be required. See your supervisor/instructor for instructions to identify what tools may be required.
- Service information database
- Personal protective equipment
- Brake pad gauge or ruler
- General hand/air tools

Task-Specific Safety Considerations

- Lifting equipment such as vehicle jacks and stands, vehicle hoists, and engine hoists are important tools that increase productivity and make the job easier. However, they can also cause severe injury or death if used improperly. Make sure you follow the manufacturer's operation procedures. Also, make sure you have your supervisor's/instructor's permission to use any particular type of lifting equipment.
- Comply with personal and environmental safety practices associated with clothing; eye protection; hand tools; power equipment; proper ventilation; and the handling, storage, and disposal of chemicals/materials in accordance with local, state, and federal safety and environmental regulations.

▶ TASK Remove, inspect, and/or replace brake pads and retaining hardware; determine necessary action.

MLR
A5D3

Time off________________

Time on________________

Total time________________

Student Instructions: Read through the entire procedure prior to starting. Prepare your workspace and any tools or parts that may be needed to complete the task. When directed by your instructor, begin the procedure to complete the task and check the box as each step is finished. Track your time on this procedure for later comparison to the standard completion time (i.e., "flat rate" or customer pay time).

Procedure:	Step Completed
1. Research the procedure for inspecting and removing brake pads.	
a. List or print off the procedure for replacing the brake pads on your vehicle.	☐
b. List all torque specifications associated with this task.	☐
c. Is there a specification for minimum brake pad thickness?	☐
2. Inspect the brake pads.	
a. Lift the vehicle. Remove the wheels.	☐
b. Perform a visual inspection of the brake pads. Use a brake pad gauge and measure the thickness of the brake pads. Describe the condition of the pads. Are any brake pads worn unevenly?	☐
c. Describe what would cause brake pads to wear unevenly.	☐
d. Inspect the brake caliper and caliper mounts to verify correct operation. Inspect for excess rust build up. Caliper slide pins should be clean and allow the brake caliper and components to move freely. Ensure that the caliper piston retracts completely.	☐
3. Remove and replace the brake pads.	
a. Following procedure, remove the brake pads.	☐
b. Remove the retaining hardware and either clean or replace with new hardware. It is critical that the brake pads and hardware are clean and move freely for correct brake operation.	☐
c. Apply the proper brake grease to the brake hardware. Clean slide pins and apply grease.	☐
d. Install the brake pads. **Tip:** Brake pads must be installed in the correct orientation. Some pads are equipped with wear indicators that must be installed in the correct direction.	☐

e. Install the caliper. The caliper piston must be retracted before it can be installed. Inspect the caliper piston boot/seal for signs of cracking, tears, or fluid leaks. **Tip:** Some caliper pistons require a special tool to twist and retract the piston into the bore.	☐
f. Torque all fasteners to specification. Install the wheel(s).	☐
g. Lower the vehicle. ALWAYS pump the brake pedal after brake service, prior to starting the vehicle. Ensure that you have good brake pressure before attempting to move a vehicle. If pistons were collapsed during brake service, it is normal to have to pump the brake pedal two or three times to regain normal pedal feel.	☐

Non-Task-Specific Evaluations:	**Step Completed**
1. Tools and equipment were used as directed and returned in good working order.	☐
2. Complied with all general and task-specific safety standards, including proper use of any personal protective equipment.	☐
3. Completed the task in an appropriate time frame (Recommendation: 1.5 or 2 times flat rate).	☐
4. Left the work space clean and orderly.	☐
5. Cared for customer property and returned it undamaged.	☐

Student signature ________________________ Date ________________________

Comments:

Have your supervisor/instructor verify satisfactory completion of this procedure, any observations found, and any necessary action(s) recommended.

Evaluation Instructions: The scoring box below is intended to act as a guide for both student and instructor. Each criterion listed will help students understand what is expected from them and evaluators articulate success at a particular task. The scoring is set up to allow a second attempt at each task (see the "Test" and "Retest" columns). Scoring is designed to reward students for correct completion of the task. Points are lost for failure to complete the employability requirements (see "Non-Task-Specific" criteria). When all criteria are evaluated, tally the points for a total at the bottom of each column.

Tasksheet Scoring

	Test		Retest	
Evaluation Items	**Pass**	**Fail**	**Pass**	**Fail**
Task-Specific Evaluation	**(1 pt)**	**(0 pts)**	**(1 pt)**	**(0 pts)**
Research the procedure for inspecting and removing brake pads.				
Perform a visual inspection of all brake pads. Provide a measurement for brake pad thickness. Determine causes for abnormal brake pad wear.				
Remove and inspect the brake pads.				
Replace the brake pads.				
Non-Task-Specific Evaluation	**(0 pts)**	**(−1 pt)**	**(0 pts)**	**(−1 pt)**
Student successfully completed at least three of the non-task-specific steps.				
Student successfully completed all five of the non-task-specific steps.				
Total Score: <total # of points / 4 = %>				

Supervisor:

Supervisor/instructor signature _________________________ Date _________________

Comments:

Retest supervisor/instructor signature _________________________ Date _________________

Comments:

Student/Intern Information

Name _________________________________ Date _____________ Class _____________________________

Vehicle, Customer, and Service Information

Vehicle used for this activity:

Year _______________ Make _______________________________ Model _______________________________

Odometer _________________________________ VIN ____

Materials Required

- Blank work order
- Vehicle with available service history records
- Depending on the type of concern, special diagnostic/hand tools may be required. See your supervisor/instructor for instructions to identify what tools may be required.
- Service information database
- Personal protective equipment
- Brake wash equipment
- General hand/air tools

Task-Specific Safety Considerations

- Lifting equipment such as vehicle jacks and stands, vehicle hoists, and engine hoists are important tools that increase productivity and make the job easier. However, they can also cause severe injury or death if used improperly. Make sure you follow the manufacturer's operation procedures. Also, make sure you have your supervisor's/instructor's permission to use any particular type of lifting equipment.
- Comply with personal and environmental safety practices associated with clothing; eye protection; hand tools; power equipment; proper ventilation; and the handling, storage, and disposal of chemicals/materials in accordance with local, state, and federal safety and environmental regulations.
- **Caution:** Brake dust may contain asbestos, which has been determined to cause cancer when inhaled or ingested. Treat all brake dust as if it contains asbestos and use Occupational Safety and Health Administration (OSHA)-approved asbestos removal equipment. Do not allow brake dust to become airborne by using anything that would disturb the dust. Also, wear protective gloves during this procedure and dispose of or clean them in an approved manner.

▶ **TASK** Lubricate and reinstall caliper, brake pads, and related hardware;
seat brake pads and inspect for leaks.

MLR
A5D4

Student Instructions: Read through the entire procedure prior to starting. Prepare your workspace and any tools or parts that may be needed to complete the task. When directed by your instructor, begin the procedure to complete the task and check the box as each step is finished. Track your time on this procedure for later comparison to the standard completion time (i.e., "flat rate" or customer pay time).

Procedure:	Step Completed
1. Assemble the brake pads, calipers, and related hardware.	
a. When preparing to assemble disc brakes on a vehicle, it is important to inspect and clean all components for proper function. Clean all brake components using the approved methods.	☐
b. Apply brake grease to the caliper slide pins and a small amount to the pad hardware located inside the caliper bracket.	☐
c. Install the brake pads into the caliper bracket. Pads should fit easily and require minimal force to move back and forth. Some brackets require the use of a shim to ensure that the pads do not move excessively. **Caution:** Some brake pads must be installed in the correct orientation. Refer to your service information for further instruction.	☐
d. Install the brake caliper. This step may require the piston to be retracted. Depending on caliper design, pads may have to be inserted into the caliper rather than the caliper being placed over the pads. Refer to service information for instructions.	☐
e. Install the wheel and torque the lug nuts to specification. What specification did you use? **Tip:** Use a torque limiting extension bar and an air impact for the initial torque.	☐
f. Lower the vehicle and use a torque wrench to tighten lug nuts to their final specification.	☐
g. ALWAYS pump the brake pedal after brake service, prior to starting the vehicle. Ensure that you have good brake pressure before attempting to move a vehicle. If pistons were collapsed during brake service, it is normal to have to pump the brake pedal two or three times to regain normal pedal feel.	☐

Non-Task-Specific Evaluations:	Step Completed
1. Tools and equipment were used as directed and returned in good working order.	☐
2. Complied with all general and task-specific safety standards, including proper use of any personal protective equipment.	☐
3. Completed the task in an appropriate time frame (Recommendation: 1.5 or 2 times flat rate).	☐
4. Left the work space clean and orderly.	☐
5. Cared for customer property and returned it undamaged.	☐

Student signature _________________________________ Date _________________________________

Comments:

Have your supervisor/instructor verify satisfactory completion of this procedure, any observations found, and any necessary action(s) recommended.

Tasksheet Scoring

	Test		Retest	
Evaluation Items	**Pass**	**Fail**	**Pass**	**Fail**
Task-Specific Evaluation	**(1 pt)**	**(0 pts)**	**(1 pt)**	**(0 pts)**
Clean and lubricate all necessary brake components.				
Install the brake hardware, brake pads and brake caliper in the correct orientation. Torque the fasteners to specification.				
Tighten the wheel using the correct methods and torque specifications.				
Seat the brake pads after service.				
Non-Task-Specific Evaluation	**(0 pts)**	**(−1 pt)**	**(0 pts)**	**(−1 pt)**
Student successfully completed at least three of the non-task-specific steps.				
Student successfully completed all five of the non-task-specific steps.				
Total Score: <total # of points / 4 = %>				

Supervisor:

Supervisor/instructor signature _________________________ Date _________________

Comments:

Retest supervisor/instructor signature _________________________ Date _________________

Comments:

CDX Tasksheet Number: A5021

Student/Intern Information

Name _________________________ Date __________ Class _________________

Vehicle, Customer, and Service Information

Vehicle used for this activity:

Year ______________ Make _________________________ Model _________________

Odometer _________________________ VIN _________________________

Materials Required

- Blank work order
- Vehicle with available service history records
- Depending on the type of concern, special diagnostic/hand tools may be required. See your supervisor/instructor for instructions to identify what tools may be required.
- Service information database
- Personal protective equipment
- Micrometer set
- Dial indicator
- General hand/air tools

Task-Specific Safety Considerations

- Lifting equipment such as vehicle jacks and stands, vehicle hoists, and engine hoists are important tools that increase productivity and make the job easier. However, they can also cause severe injury or death if used improperly. Make sure you follow the manufacturer's operation procedures. Also, make sure you have your supervisor's/instructor's permission to use any particular type of lifting equipment.
- Comply with personal and environmental safety practices associated with clothing; eye protection; hand tools; power equipment; proper ventilation; and the handling, storage, and disposal of chemicals/materials in accordance with local, state, and federal safety and environmental regulations.

▶ TASK Clean and inspect rotor and mounting surface, measure rotor thickness, thickness variation, and lateral runout; determine necessary action. **MLR** **A5D5**

Time off___________

Time on___________

Total time___________

Student Instructions: Read through the entire procedure prior to starting. Prepare your workspace and any tools or parts that may be needed to complete the task. When directed by your instructor, begin the procedure to complete the task and check the box as each step is finished. Track your time on this procedure for later comparison to the standard completion time (i.e., "flat rate" or customer pay time).

Procedure:	Step Completed
1. Research specifications for rotor measurements.	
a. List the specifications for minimum rotor thickness, thickness variation, and lateral runout for the front rotors on your vehicle.	☐
b. Describe how you would check a rotor for thickness variation and explain what causes this problem.	☐
c. Describe how you would check lateral runout on a rotor and explain what causes this problem.	☐
2. Measure rotor thickness and thickness variation.	
a. Lift the vehicle and remove the front wheels. Use a micrometer and measure the rotor thickness. Take the measurement near the middle of the rotor. What is the measurement? How does it compare to the minimum thickness specification?	☐
b. Continue to make measurements of rotor thickness around the entire rotor. Be sure to check in at least five or six spots, measuring near the middle of the rotor. List your results.	☐
c. Explain what symptoms a customer may have if the rotors on their vehicle have excessive thickness variation. What is the maximum allowable variation for your vehicle?	☐
3. Measure the rotor's lateral runout.	
a. With the wheels removed, mount the dial indicator so a measurement can be taken. Be sure to zero the dial when preparing to take a measurement.	☐

	Step Completed
b. Slowly rotate the rotor and watch as the dial indicator begins to work. Record your maximum runout and compare it to your specification.	☐
c. Explain what symptoms a customer may have if the rotors on their vehicle have excessive lateral runout. What is the maximum allowable runout for your vehicle?	☐
4. Based on the measurements you made, what are your recommendations for service?	☐

Non-Task-Specific Evaluations:	Step Completed
1. Tools and equipment were used as directed and returned in good working order.	☐
2. Complied with all general and task-specific safety standards, including proper use of any personal protective equipment.	☐
3. Completed the task in an appropriate time frame (Recommendation: 1.5 or 2 times flat rate).	☐
4. Left the work space clean and orderly.	☐
5. Cared for customer property and returned it undamaged.	☐

Student signature _______________________________ Date _______________________________

Comments:

Have your supervisor/instructor verify satisfactory completion of this procedure, any observations found, and any necessary action(s) recommended.

Tasksheet Scoring

	Test		Retest	
Evaluation Items	Pass	Fail	Pass	Fail
Task-Specific Evaluation	**(1 pt)**	**(0 pts)**	**(1 pt)**	**(0 pts)**
Research specifications for rotor measurements.				
Measure the rotor's thickness and thickness variation. Measure the rotor's lateral runout.				
Use precision measuring equipment.				
Make recommendations for service based on your inspection.				
Non-Task-Specific Evaluation	**(0 pts)**	**(−1 pt)**	**(0 pts)**	**(−1 pt)**
Student successfully completed at least three of the non-task-specific steps.				
Student successfully completed all five of the non-task-specific steps.				
Total Score: <total # of points / 4 = %>				

Supervisor:

Supervisor/instructor signature ________________________________ Date ________________

Comments:

Retest supervisor/instructor signature ________________________________ Date ________________

Comments:

CDX Tasksheet Number: A5022

Student/Intern Information

Name _________________________________ Date _____________ Class _____________________

Vehicle, Customer, and Service Information

Vehicle used for this activity:

Year _______________ Make _______________________ Model _____________________

Odometer _______________________ VIN _________________________________

Task-Specific Safety Considerations

- Lifting equipment such as vehicle jacks and stands, vehicle hoists, and engine hoists are important tools that increase productivity and make the job easier. However, they can also cause severe injury or death if used improperly. Make sure you follow the manufacturer's operation procedures. Also, make sure you have your supervisor's/instructor's permission to use any particular type of lifting equipment.
- Comply with personal and environmental safety practices associated with clothing; eye protection; hand tools; power equipment; proper ventilation; and the handling, storage, and disposal of chemicals/materials in accordance with local, state, and federal safety and environmental regulations.

▶ **TASK** Remove and reinstall/replace rotor.

MLR
A5D6

Time off_______________

Time on_______________

Total time_______________

Student Instructions: Read through the entire procedure prior to starting. Prepare your workspace and any tools or parts that may be needed to complete the task. When directed by your instructor, begin the procedure to complete the task and check the box as each step is finished. Track your time on this procedure for later comparison to the standard completion time (i.e., "flat rate" or customer pay time).

Procedure:	Step Completed
1. Research the rotor replacement procedure.	
a. What type of disc rotor is used on your vehicle? Is the rotor solid or vented? Does the rotor contain grease seals and a wheel bearing?	☐
b. List or print off the procedure for replacing the front rotors on your vehicle.	☐
c. List any torque specifications associated with this task.	☐
d. If wheel bearings will be serviced at the same time as rotor replacement, list the procedure for cleaning, inspecting, and torqueing the bearings.	☐
2. Remove the front brake rotors.	
a. Lift the vehicle and remove the front wheels.	☐
b. Remove the brake caliper and caliper bracket. Always support the caliper once removed. Do not allow the caliper to hang by the brake hose.	☐
c. Remove the rotor. **Tip:** Some rotors are held in place by a set screw or a lock washer around a wheel stud.	☐
3. Install the front brake rotors.	
a. Before installing a rotor, the hub surface must be clean and free of rust. Use a wire brush or a sanding disc to clean the hub/rotor contact area. If servicing a wheel bearing at this time, clean and inspect the grease seal, wheel bearing, and spindle.	☐
b. Following procedure, install the front rotors. Be sure to replace the set screw or follow instructions to properly preload the wheel bearings.	☐
c. Install the brake calipers and torque mounting bolts to specification.	☐

d. Install the wheel and torque the lug nuts to specification. What specification did you use? **Tip:** Use a torque limiting extension bar and an air impact for the initial torque.	☐
e. Lower the vehicle and use a torque wrench to tighten lug nuts to their final specification.	☐
f. ALWAYS pump the brake pedal after brake service, prior to starting the vehicle. Ensure that you have good brake pressure before attempting to move a vehicle. If pistons were collapsed during brake service, it is normal to have to pump the brake pedal two or three times to regain normal pedal feel.	☐

Non-Task-Specific Evaluations:	**Step Completed**
1. Tools and equipment were used as directed and returned in good working order.	☐
2. Complied with all general and task-specific safety standards, including proper use of any personal protective equipment.	☐
3. Completed the task in an appropriate time frame (Recommendation: 1.5 or 2 times flat rate).	☐
4. Left the work space clean and orderly.	☐
5. Cared for customer property and returned it undamaged.	☐

Student signature _________________________ Date _________________________

Comments:

Have your supervisor/instructor verify satisfactory completion of this procedure, any observations found, and any necessary action(s) recommended.

Evaluation Instructions: The scoring box below is intended to act as a guide for both student and instructor. Each criterion listed will help students understand what is expected from them and evaluators articulate success at a particular task. The scoring is set up to allow a second attempt at each task (see the "Test" and "Retest" columns). Scoring is designed to reward students for correct completion of the task. Points are lost for failure to complete the employability requirements (see "Non-Task-Specific" criteria). When all criteria are evaluated, tally the points for a total at the bottom of each column.

Tasksheet Scoring

	Test		Retest	
Evaluation Items	**Pass**	**Fail**	**Pass**	**Fail**
Task-Specific Evaluation	**(1 pt)**	**(0 pts)**	**(1 pt)**	**(0 pts)**
Research the rotor replacement procedure and list torque specifications associated with this task.				
Describe the type of rotor used on your vehicle.				
Remove the rotors. Clean the hub surface area. Clean and inspect the wheel bearings.				
Install the rotors.				
Non-Task-Specific Evaluation	**(0 pts)**	**(−1 pt)**	**(0 pts)**	**(−1 pt)**
Student successfully completed at least three of the non-task-specific steps.				
Student successfully completed all five of the non-task-specific steps.				
Total Score: <total # of points / 4 = %>				

Supervisor:

Supervisor/instructor signature _________________________________ Date _______________________

Comments:

Retest supervisor/instructor signature _________________________________ Date _______________________

Comments:

CDX Tasksheet Number: A5023

Student/Intern Information

Name _________________________________ Date ___________ Class _____________________________

Vehicle, Customer, and Service Information

Vehicle used for this activity:

Year _______________ Make _______________________________ Model _____________________________

Odometer _________________________________ VIN _____________________________________

Materials Required

- Blank work order
- Vehicle with available service history records
- Depending on the type of concern, special diagnostic/hand tools may be required. See your supervisor/instructor for instructions to identify what tools may be required.
- Service information database
- Personal protective equipment
- General hand/air tools
- On-car brake lathe
- Micrometer set

Task-Specific Safety Considerations

- Lifting equipment such as vehicle jacks and stands, vehicle hoists, and engine hoists are important tools that increase productivity and make the job easier. However, they can also cause severe injury or death if used improperly. Make sure you follow the manufacturer's operation procedures. Also, make sure you have your supervisor's/instructor's permission to use any particular type of lifting equipment.
- Comply with personal and environmental safety practices associated with clothing; eye protection; hand tools; power equipment; proper ventilation; and the handling, storage, and disposal of chemicals/materials in accordance with local, state, and federal safety and environmental regulations.
- Extreme caution must be exercised when working around rotating components.

▶ **TASK** Refinish rotor on vehicle; measure final rotor thickness and compare with specification.

MLR
A5D7

Time off_______________

Time on_______________

Total time_______________

Student Instructions: Read through the entire procedure prior to starting. Prepare your workspace and any tools or parts that may be needed to complete the task. When directed by your instructor, begin the procedure to complete the task and check the box as each step is finished. Track your time on this procedure for later comparison to the standard completion time (i.e., "flat rate" or customer pay time).

Procedure:	Step Completed
1. Research rotor specifications.	
a. List the minimum rotor thickness that is acceptable for your vehicles front rotors.	☐
2. Prepare to refinish the rotors.	
a. Lift the vehicle until the tires just begin to leave the ground. Reach inside the vehicle and place the transmission in neutral. **Note:** This step applies when resurfacing rotors on a drive axle of the vehicle.	☐
b. Remove the wheels. Use a micrometer and measure the rotor thickness and compare it to your specification. Is the rotor thick enough to machine?	☐
c. Remove the brake caliper. Do not allow the brake caliper to hang by the brake hose.	☐
d. Prepare the on-car lathe for making the initial cut. Select the correct adapter to fit your hub bolt pattern and fit the lathe to the rotor.	☐
e. Perform the compensation procedure for the lathe that you are using. Explain why this compensation procedure is necessary.	☐
f. Perform a "scratch test" by adjusting the cutting tips to the point where they just begin to contact the rotor surface. Continue to lightly cut the rotor as you move the cutting arms in toward the center of the rotor.	☐
g. Prepare to make the initial cut. How much material should be removed on the first cut?	☐
3. Refinish the rotor.	
a. Adjust the cutting arms to remove 0.002" of material at a time. If applicable, start by making a fast cut until the rotor surface has been machined true and follow this by making a slow pass to ensure a smooth finish. Most on-car lathes use cutting bits that are for high speed cutting.	☐

b. Once the rotor is refinished, disconnect the lathe. Measure the rotor thickness once more and compare it to your minimum thickness specification. Is the rotor usable?	☐
c. Reassemble the brakes. Torque all fasteners to specification.	☐
d. Install the wheel and torque the lug nuts to specification. What specification did you use? **Tip:** Use a torque limiting extension bar and an air impact for the initial torque.	☐
e. Lower the vehicle and use a torque wrench to tighten lug nuts to their final specification. **Caution:** You may have placed the vehicle in neutral prior to brake service.	☐
f. ALWAYS pump the brake pedal after brake service, prior to starting the vehicle. Ensure that you have good brake pressure before attempting to move a vehicle. If pistons were collapsed during brake service, it is normal to have to pump the brake pedal two or three times to regain normal pedal feel.	☐

Non-Task-Specific Evaluations:	Step Completed
1. Tools and equipment were used as directed and returned in good working order.	☐
2. Complied with all general and task-specific safety standards, including proper use of any personal protective equipment.	☐
3. Completed the task in an appropriate time frame (Recommendation: 1.5 or 2 times flat rate).	☐
4. Left the work space clean and orderly.	☐
5. Cared for customer property and returned it undamaged.	☐

Student signature _________________________ Date _________________________

Comments:

Have your supervisor/instructor verify satisfactory completion of this procedure, any observations found, and any necessary action(s) recommended.

Evaluation Instructions: The scoring box below is intended to act as a guide for both student and instructor. Each criterion listed will help students understand what is expected from them and evaluators articulate success at a particular task. The scoring is set up to allow a second attempt at each task (see the "Test" and "Retest" columns). Scoring is designed to reward students for correct completion of the task. Points are lost for failure to complete the employability requirements (see "Non-Task-Specific" criteria). When all criteria are evaluated, tally the points for a total at the bottom of each column.

Tasksheet Scoring

	Test		Retest	
Evaluation Items	**Pass**	**Fail**	**Pass**	**Fail**
Task-Specific Evaluation	**(1 pt)**	**(0 pts)**	**(1 pt)**	**(0 pts)**
Research rotor specifications.				
Prepare the on-car lathe for making the initial cut. Compensate the machine.				
Perform a "scratch test", then machine the rotor and compare final measurements to specifications.				
Reassemble the brakes.				
Non-Task-Specific Evaluation	**(0 pts)**	**(−1 pt)**	**(0 pts)**	**(−1 pt)**
Student successfully completed at least three of the non-task-specific steps.				
Student successfully completed all five of the non-task-specific steps.				
Total Score: <total # of points / 4 = %>				

Supervisor:

Supervisor/instructor signature _________________________ Date _______________

Comments:

Retest supervisor/instructor signature _________________________ Date _______________

Comments:

CDX Tasksheet Number: A5024

Student/Intern Information

Name _____________________________________ Date _____________ Class _______________________

Vehicle, Customer, and Service Information

Vehicle used for this activity:

Year _________________ Make _________________________ Model _________________________

Odometer _________________________________ VIN _________________________________

Task-Specific Safety Considerations

- Lifting equipment such as vehicle jacks and stands, vehicle hoists, and engine hoists are important tools that increase productivity and make the job easier. However, they can also cause severe injury or death if used improperly. Make sure you follow the manufacturer's operation procedures. Also, make sure you have your supervisor's/instructor's permission to use any particular type of lifting equipment.
- Comply with personal and environmental safety practices associated with clothing; eye protection; hand tools; power equipment; proper ventilation; and the handling, storage, and disposal of chemicals/materials in accordance with local, state, and federal safety and environmental regulations.
- Extreme caution must be exercised when working around rotating components.

Time off_______________

Time on_______________

Total time_______________

▶ **TASK** Refinish rotor off vehicle; measure final rotor thickness and compare with specification.

MLR
A5D8

Student Instructions: Read through the entire procedure prior to starting. Prepare your workspace and any tools or parts that may be needed to complete the task. When directed by your instructor, begin the procedure to complete the task and check the box as each step is finished. Track your time on this procedure for later comparison to the standard completion time (i.e., "flat rate" or customer pay time).

Procedure:	Step Completed
1. Research rotor specifications.	
a. List the minimum rotor thickness that is acceptable for your vehicles front rotors.	☐
2. Prepare to refinish the rotors.	
a. Lift the vehicle and remove the wheels. Use a micrometer and measure the rotor thickness and compare it to your specification. Is the rotor thick enough to machine?	☐
b. Remove the brake caliper. Do not allow the brake caliper to hang by the brake hose.	☐
c. Remove the rotors. Before machining the rotors, take time to remove any excess rust build up. Make sure that the area of the rotor that contacts the hub flange is completely clean before installing it on the lathe.	☐
3. Refinish the rotor.	
a. Select the correct adapter and install the rotor onto the lathe.	☐
b. Install the silencer band or pads to the rotor before beginning the resurfacing procedure.	☐
c. Perform a "scratch test" by adjusting the cutting tips to the point where they just begin to contact the rotor surface. Continue to lightly cut the rotor as you move the cutting arms in toward the center of the rotor.	☐
d. Adjust the cutting tips and make your first cut. This cut should be a fast pass. How much material should you remove during each pass?	☐
e. Continue making fast passes until the rotor finish is smooth and even. Once this is achieved, perform a slow pass for the final finish.	☐
f. Remove the rotor and take a final measurement. Is the rotor suitable for use after the machining procedure?	☐

4. Install the rotor on the vehicle.	
a. Reassemble the brakes. Torque all fasteners to specification. List those specifications here.	☐
b. Install the wheel and torque the lug nuts to specification. What specification did you use? **Tip:** Use a torque limiting extension bar and an air impact for the initial torque.	☐
c. Lower the vehicle and use a torque wrench to tighten lug nuts to their final specification. **Caution:** You may have placed the vehicle in neutral prior to brake service.	☐
d. ALWAYS pump the brake pedal after brake service, prior to starting the vehicle. Ensure that you have good brake pressure before attempting to move a vehicle. If pistons were collapsed during brake service, it is normal to have to pump the brake pedal two or three times to regain normal pedal feel.	☐

Non-Task-Specific Evaluations:	Step Completed
1. Tools and equipment were used as directed and returned in good working order.	☐
2. Complied with all general and task-specific safety standards, including proper use of any personal protective equipment.	☐
3. Completed the task in an appropriate time frame (Recommendation: 1.5 or 2 times flat rate).	☐
4. Left the work space clean and orderly.	☐
5. Cared for customer property and returned it undamaged.	☐

Student signature _______________________________ Date _______________________________

Comments:

Have your supervisor/instructor verify satisfactory completion of this procedure, any observations found, and any necessary action(s) recommended.

Evaluation Instructions: The scoring box below is intended to act as a guide for both student and instructor. Each criterion listed will help students understand what is expected from them and evaluators articulate success at a particular task. The scoring is set up to allow a second attempt at each task (see the "Test" and "Retest" columns). Scoring is designed to reward students for correct completion of the task. Points are lost for failure to complete the employability requirements (see "Non-Task-Specific" criteria). When all criteria are evaluated, tally the points for a total at the bottom of each column.

Tasksheet Scoring

	Test		Retest	
Evaluation Items	**Pass**	**Fail**	**Pass**	**Fail**
Task-Specific Evaluation	**(1 pt)**	**(O pts)**	**(1 pt)**	**(O pts)**
Research rotor specifications.				
Remove the rotor from the vehicle.				
Use the lathe to machine the rotor. Measure the rotor before and after service.				
Install the rotor.				
Non-Task-Specific Evaluation	**(O pts)**	**(−1 pt)**	**(O pts)**	**(−1 pt)**
Student successfully completed at least three of the non-task-specific steps.				
Student successfully completed all five of the non-task-specific steps.				
Total Score: <total # of points / 4 = %>				

Supervisor:

Supervisor/instructor signature ___________________________ Date ___________________

Comments:

Retest supervisor/instructor signature ___________________________ Date ___________________

Comments:

CDX Tasksheet Number: A5025

Student/Intern Information

Name _________________________________ Date ___________ Class _________________________

Vehicle, Customer, and Service Information

Vehicle used for this activity:

Year _______________ Make _________________________ Model _________________________

Odometer _________________________________ VIN _________________________________

Materials Required

- Blank work order
- Vehicle with available service history records
- Depending on the type of concern, special diagnostic/hand tools may be required. See your supervisor/instructor for instructions to identify what tools may be required.
- Service information database
- Personal protective equipment
- General hand/air tools
- Caliper piston tool

Task-Specific Safety Considerations

- Lifting equipment such as vehicle jacks and stands, vehicle hoists, and engine hoists are important tools that increase productivity and make the job easier. However, they can also cause severe injury or death if used improperly. Make sure you follow the manufacturer's operation procedures. Also, make sure you have your supervisor's/instructor's permission to use any particular type of lifting equipment.
- Comply with personal and environmental safety practices associated with clothing; eye protection; hand tools; power equipment; proper ventilation; and the handling, storage, and disposal of chemicals/materials in accordance with local, state, and federal safety and environmental regulations.

► TASK Retract and re-adjust caliper piston on an integral parking brake system.

MLR
A5D9

Student Instructions: Read through the entire procedure prior to starting. Prepare your workspace and any tools or parts that may be needed to complete the task. When directed by your instructor, begin the procedure to complete the task and check the box as each step is finished. Track your time on this procedure for later comparison to the standard completion time (i.e., "flat rate" or customer pay time).

Time off__________

Time on__________

Total time__________

Procedure:	Step Completed
1. Describe integral parking brake operation.	
a. Explain the difference between a parking brake that is integral to a caliper versus a parking brake that uses brake shoes.	☐
b. Are there any special tools required for this task?	☐
c. Explain how a parking brake is adjusted on an integral brake design.	☐
2. Remove the rear calipers.	
a. Lift the vehicle and remove the rear wheels.	☐
b. Remove the rear brake calipers.	☐
c. Explain what the caliper piston looks like. Is it different from a piston found on a front caliper?	☐
3. Retract the caliper piston.	
a. Use the disc brake piston tool to retract the piston. The tangs of the tool should interlock with the recesses in the piston. Slowly turn the piston clockwise until it is fully seated in its bore. Use care not to tear the piston dust seal.	☐
4. Install the calipers and adjust the caliper pistons.	
a. Install the calipers on the vehicle. Tighten all fasteners to specification. **Tip:** Many times, the rear brake pads will use a locating tab that must be lined up correctly with a recess in the caliper piston. If the piston is not in the correct location, installing the caliper over a new set of pads may be difficult.	☐
b. Following proper procedure, adjust the caliper piston by using the adjusting lever/nut located on the caliper. This action may also be accomplished by using the parking brake several times until the piston is adjusted correctly.	☐

c. Install the wheel and torque the lug nuts to specification. What specification did you use? **Tip:** Use a torque limiting extension bar and an air impact for the initial torque.	☐
d. Lower the vehicle and use a torque wrench to tighten lug nuts to their final specification.	☐
e. ALWAYS pump the brake pedal after brake service, prior to starting the vehicle. Ensure that you have good brake pressure before attempting to move a vehicle. If pistons were collapsed during brake service, it is normal to have to pump the brake pedal two or three times to regain normal pedal feel.	☐

Non-Task-Specific Evaluations:	**Step Completed**
1. Tools and equipment were used as directed and returned in good working order.	☐
2. Complied with all general and task-specific safety standards, including proper use of any personal protective equipment.	☐
3. Completed the task in an appropriate time frame (Recommendation: 1.5 or 2 times flat rate).	☐
4. Left the work space clean and orderly.	☐
5. Cared for customer property and returned it undamaged.	☐

Student signature _________________________________ Date _________________________________

Comments:

Have your supervisor/instructor verify satisfactory completion of this procedure, any observations found, and any necessary action(s) recommended.

Tasksheet Scoring

Evaluation Items	Test		Retest	
	Pass	**Fail**	**Pass**	**Fail**
Task-Specific Evaluation	**(1 pt)**	**(0 pts)**	**(1 pt)**	**(0 pts)**
Research integral parking brake design, function, and adjustment.				
Remove the rear calipers.				
Retract the caliper pistons using the correct tools and methods.				
Install the rear calipers. Adjust the parking brake.				
Non-Task-Specific Evaluation	**(0 pts)**	**(−1 pt)**	**(0 pts)**	**(−1 pt)**
Student successfully completed at least three of the non-task-specific steps.				
Student successfully completed all five of the non-task-specific steps.				
Total Score: <total # of points / 4 = %>				

Supervisor:

Supervisor/instructor signature _________________________________ Date _____________________

Comments:

Retest supervisor/instructor signature _________________________ Date _____________________

Comments:

CDX Tasksheet Number: A5026

Student/Intern Information

Name _______________________________ Date ____________ Class _______________________

Vehicle, Customer, and Service Information

Vehicle used for this activity:

Year _______________ Make _______________________ Model _______________________

Odometer _______________________ VIN _______________________________

Materials Required

- Blank work order
- Vehicle with available service history records
- Depending on the type of concern, special diagnostic/hand tools may be required. See your supervisor/instructor for instructions to identify what tools may be required.
- Service information database
- Personal protective equipment
- General hand/air tools
- Brake pad gauge

Task-Specific Safety Considerations

- Lifting equipment such as vehicle jacks and stands, vehicle hoists, and engine hoists are important tools that increase productivity and make the job easier. However, they can also cause severe injury or death if used improperly. Make sure you follow the manufacturer's operation procedures. Also, make sure you have your supervisor's/instructor's permission to use any particular type of lifting equipment.
- Comply with personal and environmental safety practices associated with clothing; eye protection; hand tools; power equipment; proper ventilation; and the handling, storage, and disposal of chemicals/materials in accordance with local, state, and federal safety and environmental regulations.

Time off___________

Time on___________

Total time___________

▶ **TASK** Check brake pad wear indicator; determine necessary action.

MLR
A5D10

Student Instructions: Read through the entire procedure prior to starting. Prepare your workspace and any tools or parts that may be needed to complete the task. When directed by your instructor, begin the procedure to complete the task and check the box as each step is finished. Track your time on this procedure for later comparison to the standard completion time (i.e., "flat rate" or customer pay time).

Procedure:	Step Completed
1. Inspect for brake pad wear indicators.	
a. Lift the vehicle and remove all wheels.	☐
b. Locate the wear indicators on each pad and inspect for wear.	☐
c. Use a brake pad gauge and measure the thickness of each brake pad. Note the measurements here.	☐
d. Describe how the brake pads are wearing. Is there any irregular wear?	☐
e. Explain what could cause brake pads to wear unevenly.	☐
f. Describe how the driver is notified of a thin brake pad condition. Is the vehicle equipped with sensors that turn on a warning indicator?	☐
g. What is the minimum allowable thickness for brake pads on your vehicle?	☐
2. Based on your inspection, what is your recommendation for service?	☐

Non-Task-Specific Evaluations:	Step Completed
1. Tools and equipment were used as directed and returned in good working order.	☐
2. Complied with all general and task-specific safety standards, including proper use of any personal protective equipment.	☐
3. Completed the task in an appropriate time frame (Recommendation: 1.5 or 2 times flat rate).	☐
4. Left the work space clean and orderly.	☐
5. Cared for customer property and returned it undamaged.	☐

Student signature _____________________________ Date _____________________________

Comments:

Have your supervisor/instructor verify satisfactory completion of this procedure, any observations found, and any necessary action(s) recommended.

Evaluation Instructions: The scoring box below is intended to act as a guide for both student and instructor. Each criterion listed will help students understand what is expected from them and evaluators articulate success at a particular task. The scoring is set up to allow a second attempt at each task (see the "Test" and "Retest" columns). Scoring is designed to reward students for correct completion of the task. Points are lost for failure to complete the employability requirements (see "Non-Task-Specific" criteria). When all criteria are evaluated, tally the points for a total at the bottom of each column.

Tasksheet Scoring

	Test		Retest	
Evaluation Items	**Pass**	**Fail**	**Pass**	**Fail**
Task-Specific Evaluation	**(1 pt)**	**(0 pts)**	**(1 pt)**	**(0 pts)**
Use brake pad wear indicators and a brake pad gauge for inspection.				
Determine whether the vehicle is equipped with a brake pad monitoring system.				
Determine any brake pad wear. Inspect for irregular wear.				
Make recommendations for service based on your inspection.				
Non-Task-Specific Evaluation	**(0 pts)**	**(−1 pt)**	**(0 pts)**	**(−1 pt)**
Student successfully completed at least three of the non-task-specific steps.				
Student successfully completed all five of the non-task-specific steps.				
Total Score: <total # of points / 4 = %>				

Supervisor:

Supervisor/instructor signature ________________________________ Date ________________________

Comments:

Retest supervisor/instructor signature ________________________ Date ________________________

Comments:

CDX Tasksheet Number: A5027

Student/Intern Information

Name _______________________________ Date ___________ Class _______________________

Vehicle, Customer, and Service Information

Vehicle used for this activity:

Year _______________ Make _______________________ Model _______________________

Odometer _______________________ VIN _______________________________

Materials Required

- Blank work order
- Vehicle with available service history records
- Depending on the type of concern, special diagnostic/hand tools may be required. See your supervisor/instructor for instructions to identify what tools may be required.
- Service information database
- Personal protective equipment

Task-Specific Safety Considerations

- Comply with personal and environmental safety practices associated with clothing; eye protection; hand tools; power equipment; proper ventilation; and the handling, storage, and disposal of chemicals/materials in accordance with local, state, and federal safety and environmental regulations.
- Completion of this task may require test-driving the vehicle on the school grounds. Attempt this task only with full permission from your instructor and follow all the guidelines exactly.

▶ TASK Describe the importance of operating the vehicle to burnish/break-in replacement brake pads according to manufacturer's recommendation.

MLR
A5D11

Student Instructions: Read through the entire procedure prior to starting. Prepare your workspace and any tools or parts that may be needed to complete the task. When directed by your instructor, begin the procedure to complete the task and check the box as each step is finished. Track your time on this procedure for later comparison to the standard completion time (i.e., "flat rate" or customer pay time).

Time off______________

Time on______________

Total time______________

Procedure:	Step Completed
1. Research the brake burnishing procedure.	
a. Describe the recommended procedure for burnishing the brakes after brake components have been replaced.	☐
b. Explain how the process is different for a "severe duty" vehicle versus a common passenger vehicle that is driven daily.	☐
c. List three things that a burnishing procedure accomplishes.	☐
d. Explain how the brakes may be affected if a burnishing procedure is not completed.	☐
2. **Optional:** Perform a burnishing procedure.	
a. Following proper procedure, perform a burnishing procedure on your vehicle. **Caution:** Follow all class rules and laws when test driving vehicles. Technicians tend to find areas with very little traffic to perform testing procedures.	☐

Non-Task-Specific Evaluations:	Step Completed
1. Tools and equipment were used as directed and returned in good working order.	☐
2. Complied with all general and task-specific safety standards, including proper use of any personal protective equipment.	☐
3. Completed the task in an appropriate time frame (Recommendation: 1.5 or 2 times flat rate).	☐
4. Left the work space clean and orderly.	☐
5. Cared for customer property and returned it undamaged.	☐

Student signature _________________________ Date _________________________

Comments:

Have your supervisor/instructor verify satisfactory completion of this procedure, any observations found, and any necessary action(s) recommended.

Taksheet Scoring

Evaluation Items	Test		Retest	
	Pass	**Fail**	**Pass**	**Fail**
Task-Specific Evaluation	**(1 pt)**	**(O pts)**	**(1 pt)**	**(O pts)**
Describe the brake burnishing procedure for your vehicle. Explain different processes for this task depending on vehicle type and use.				
Explain the benefits of performing this procedure.				
Explain what may result from not completing a burnishing procedure.				
Complete a burnishing procedure.				
Non-Task-Specific Evaluation	**(O pts)**	**(−1 pt)**	**(O pts)**	**(−1 pt)**
Student successfully completed at least three of the non-task-specific steps.				
Student successfully completed all five of the non-task-specific steps.				
Total Score: <total # of points / 4 = %>				

5E. Power Assist Units

Learning Objective/Task	CDX Tasksheet Number	ASE Education Foundation Reference Number; Priority Level
• Check brake pedal travel with, and without, engine running to verify proper power booster operation.	A5028	A5E1, P-2
• Identify components of the brake power assist system (vacuum and hydraulic); check vacuum supply (manifold or auxiliary pump) to vacuum-type power booster.	A5029	A5E2, P-1

Materials Required

- Blank work order
- Vehicle with available service history records
- Depending on the type of concern, special diagnostic/hand tools may be required. See your supervisor/instructor for instructions to identify what tools may be required.
- Service information database
- Personal protective equipment

Safety Considerations

- When running any vehicles in the shop, make sure you use the shop's exhaust ventilation system to discharge all exhaust gas safely outside.
- Lifting equipment such as vehicle jacks and stands, vehicle hoists, and engine hoists are important tools that increase productivity and make the job easier. However, they can also cause severe injury or death if used improperly. Make sure you follow the manufacturer's operation procedures. Also, make sure you have your supervisor's/instructor's permission to use any particular type of lifting equipment.
- Comply with personal and environmental safety practices associated with clothing; eye protection; hand tools; power equipment; proper ventilation; and the handling, storage, and disposal of chemicals/materials in accordance with local, state, and federal safety and environmental regulations.
- If you need to start the vehicle, you should ensure that the parking brake is firmly applied; if necessary, use wheel chocks to prevent the vehicle from moving when the vehicle is started to verify the completion of these tasks.
- Extreme caution must be exercised when working around rotating components.
- **Caution:** Most types of brake fluid are harmful to painted surfaces. Be sure to prevent brake fluid from coming into contact with a vehicle's paint. Use fender covers to minimize this risk and be sure to wipe up any spilled brake fluid immediately with a wet rag.
- **Caution:** Brake dust may contain asbestos, which has been determined to cause cancer when inhaled or ingested. Treat all brake dust as if it contains asbestos and use Occupational Safety and Health Administration (OSHA)-approved asbestos removal equipment. Do not allow brake dust to become airborne by using anything that would disturb the dust. Also, wear protective gloves during this procedure and dispose of or clean them in an approved manner.

CDX Tasksheet Number: A5028

Student/Intern Information

Name _________________________________ Date ____________ Class _____________________

Vehicle, Customer, and Service Information

Vehicle used for this activity:

Year _______________ Make _______________________ Model ____________________

Odometer _______________________ VIN _______________________________

Materials Required

- Blank work order
- Vehicle with available service history records
- Depending on the type of concern, special diagnostic/hand tools may be required. See your supervisor/instructor for instructions to identify what tools may be required.
- Service information database
- Personal protective equipment

Task-Specific Safety Considerations

- Comply with personal and environmental safety practices associated with clothing; eye protection; hand tools; power equipment; proper ventilation; and the handling, storage, and disposal of chemicals/materials in accordance with local, state, and federal safety and environmental regulations.

▶ TASK Check brake pedal travel with, and without, engine running to verify proper power booster operation.

MLR
A5E1

Time off_______________

Time on_______________

Student Instructions: Read through the entire procedure prior to starting. Prepare your workspace and any tools or parts that may be needed to complete the task. When directed by your instructor, begin the procedure to complete the task and check the box as each step is finished. Track your time on this procedure for later comparison to the standard completion time (i.e., "flat rate" or customer pay time).

Total time_______________

Procedure:	Step Completed
1. Verify proper brake booster operation.	
a. With the engine off, pump the brake pedal three or four times. Describe the changes in the way the brake pedal feels.	☐
b. Press and hold the brake pedal and start the vehicle. Describe the changes in the way the brake pedal feels. Is this a normal condition?	☐
c. With the engine still running, pump the brake pedal. Describe how the pedal feels.	☐
2. Explain the function of the brake booster.	
a. Explain why the brake pedal becomes firm after three or four pumps when the vehicle is not running.	☐
b. How would the brake pedal feel if the booster was not working correctly?	☐
c. How much vacuum should be applied to the booster so that it functions properly? Where does the booster receive its vacuum supply? What type of vehicles require a vacuum pump?	☐
d. What component ensures that vacuum stays stored in the brake booster?	☐

<table>
<thead>
<tr><th>Non-Task-Specific Evaluations:</th><th>Step Completed</th></tr>
</thead>
<tbody>
<tr><td>1. Tools and equipment were used as directed and returned in good working order.</td><td>☐</td></tr>
<tr><td>2. Complied with all general and task-specific safety standards, including proper use of any personal protective equipment.</td><td>☐</td></tr>
<tr><td>3. Completed the task in an appropriate time frame (Recommendation: 1.5 or 2 times flat rate).</td><td>☐</td></tr>
<tr><td>4. Left the work space clean and orderly.</td><td>☐</td></tr>
<tr><td>5. Cared for customer property and returned it undamaged.</td><td>☐</td></tr>
</tbody>
</table>

Student signature _________________________ Date _________________________

Comments:

Have your supervisor/instructor verify satisfactory completion of this procedure, any observations found, and any necessary action(s) recommended.

Evaluation Instructions: The scoring box below is intended to act as a guide for both student and instructor. Each criterion listed will help students understand what is expected from them and evaluators articulate success at a particular task. The scoring is set up to allow a second attempt at each task (see the "Test" and "Retest" columns). Scoring is designed to reward students for correct completion of the task. Points are lost for failure to complete the employability requirements (see "Non-Task-Specific" criteria). When all criteria are evaluated, tally the points for a total at the bottom of each column.

Taksheet Scoring

	Test		Retest	
Evaluation Items	**Pass**	**Fail**	**Pass**	**Fail**
Task-Specific Evaluation	**(1 pt)**	**(0 pts)**	**(1 pt)**	**(0 pts)**
Perform brake testing procedures to verify correct brake booster operation.				
Describe normal pedal feel.				
Describe the pedal feel when the brake booster is malfunctioning.				
Explain a brake booster source vacuum and identify brake booster components.				
Non-Task-Specific Evaluation	**(0 pts)**	**(−1 pt)**	**(0 pts)**	**(−1 pt)**
Student successfully completed at least three of the non-task-specific steps.				
Student successfully completed all five of the non-task-specific steps.				
Total Score: <total # of points / 4 = %>				

Supervisor:

Supervisor/instructor signature _________________________________ Date _______________

Comments:

Retest supervisor/instructor signature _________________________________ Date _______________

Comments:

CDX Tasksheet Number: A5029

Student/Intern Information

Name _________________________________ Date ____________ Class _____________________

Vehicle, Customer, and Service Information

Vehicle used for this activity:

Year _______________ Make _______________________ Model ____________________

Odometer _______________________ VIN _________________________________

> **Materials Required**
> - Blank work order
> - Vehicle with available service history records
> - Depending on the type of concern, special diagnostic/hand tools may be required. See your supervisor/instructor for instructions to identify what tools may be required.
> - Service information database
> - Personal protective equipment
> - Vacuum gauge

Task-Specific Safety Considerations
- Comply with personal and environmental safety practices associated with clothing; eye protection; hand tools; power equipment; proper ventilation; and the handling, storage, and disposal of chemicals/materials in accordance with local, state, and federal safety and environmental regulations.

▶ TASK Identify components of the brake power assist system (vacuum and hydraulic); check vacuum supply (manifold or auxiliary pump) to vacuum-type power booster.

MLR
A5E2

Time off____________

Time on____________

Total time____________

Student Instructions: Read through the entire procedure prior to starting. Prepare your workspace and any tools or parts that may be needed to complete the task. When directed by your instructor, begin the procedure to complete the task and check the box as each step is finished. Track your time on this procedure for later comparison to the standard completion time (i.e., "flat rate" or customer pay time).

Procedure:	Step Completed
1. Identify the brake power assist system on your vehicle.	
a. What type of brake power assist system is used on your vehicle?	☐
b. Explain how a vacuum power assist system operates.	☐
c. Explain how a hydraulic power assist system operates.	☐
2. Locate the components of the brake power assist system.	
a. On the vacuum assist units, locate:	
i. Brake booster	☐
ii. Master cylinder	☐
iii. Brake booster vacuum hose. Where does this hose receive its vacuum?	☐
iv. Check valve. Explain what the check valve does.	☐
b. On the hydraulic assist units locate:	
i. Hydraulic booster	☐
ii. Hydraulic booster lines	☐
iii. Hydraulic boost pump. How is the pump driven? Does this pump operate other systems?	☐
c. What could be the problem if both the brakes and power steering are not operating correctly?	☐

3. Check the vacuum supply.	
a. Remove the vacuum hose from the check valve on the booster. Place the hose from the vacuum pump onto the check valve and draw the booster to 20″ of vacuum.	☐
b. Let the booster sit with a vacuum applied for 5 minutes. If the vacuum does not stay steady at 20″, it is faulty and needs to be replaced. If the vacuum does hold steady at 20″, move to the next step.	☐
c. With 20″ of vacuum in the booster, depress the brake pedal once and release it. The booster should transfer some, but not all, of the vacuum in reserve. Depending on how hard the pedal is depressed, it is normal to see 5–10″ of vacuum depleted from the reserve. The most important thing is to ensure the booster transfers vacuum, but NOT the entire vacuum in its reserve. If the vacuum remains at 20″ or goes to zero, the booster is bad and needs to be replaced.	☐
d. Use a vacuum gauge to measure the supply vacuum to the booster. What readings do you expect? List your vacuum readings.	☐
e. Explain what types of vehicles require an auxiliary pump to supply vacuum to a brake booster and why.	☐

Non-Task-Specific Evaluations:	Step Completed
1. Tools and equipment were used as directed and returned in good working order.	☐
2. Complied with all general and task-specific safety standards, including proper use of any personal protective equipment.	☐
3. Completed the task in an appropriate time frame (Recommendation: 1.5 or 2 times flat rate).	☐
4. Left the work space clean and orderly.	☐
5. Cared for customer property and returned it undamaged.	☐

Student signature _______________________________ Date _______________________________

Comments:

Have your supervisor/instructor verify satisfactory completion of this procedure, any observations found,

and any necessary action(s) recommended.

Evaluation Instructions: The scoring box below is intended to act as a guide for both student and instructor. Each criterion listed will help students understand what is expected from them and evaluators articulate success at a particular task. The scoring is set up to allow a second attempt at each task (see the "Test" and "Retest" columns). Scoring is designed to reward students for correct completion of the task. Points are lost for failure to complete the employability requirements (see "Non-Task-Specific" criteria). When all criteria are evaluated, tally the points for a total at the bottom of each column.

Tasksheet Scoring

	Test		Retest	
Evaluation Items	**Pass**	**Fail**	**Pass**	**Fail**
Task-Specific Evaluation	**(1 pt)**	**(0 pts)**	**(1 pt)**	**(0 pts)**
Identify the brake power assist system on your vehicle.				
Locate the components of the brake power assist system.				
Inspect the brake booster operation. Check the supply vacuum to the brake booster.				
Explain the need for an auxiliary vacuum pump.				
Non-Task-Specific Evaluation	**(0 pts)**	**(−1 pt)**	**(0 pts)**	**(−1 pt)**
Student successfully completed at least three of the non-task-specific steps.				
Student successfully completed all five of the non-task-specific steps.				
Total Score: <total # of points / 4 = %>				

Supervisor:

Supervisor/instructor signature _______________________________ Date _______________

Comments:

Retest supervisor/instructor signature _______________________________ Date _______________

Comments:

5F. Related Systems (i.e., Wheel Bearings, Parking Brakes, Electrical)

Learning Objective/Task	CDX Tasksheet Number	ASE Education Foundation Reference Number; Priority Level
• Remove, clean, inspect, repack, and install wheel bearings; replace seals; install hub and adjust bearings.	A5030	A5F1, P-1
• Check parking brake system components for wear, binding, and corrosion; clean, lubricate, adjust and/or replace as needed.	A5031	A5F2, P-2
• Check parking brake operation and parking brake indicator light system operation; determine necessary action.	A5032	A5F3, P-1
• Check operation of brake stop light system.	A5033	A5F4, P-1
• Replace wheel bearing and race.	A5034	A5F5, P-2
• Inspect and replace wheel studs.	A5035	A5F6, P-1

Materials Required

- Blank work order
- Vehicle with available service history records
- Depending on the type of concern, special diagnostic/hand tools may be required. See your supervisor/instructor for instructions to identify what tools may be required.
- Service information database
- Personal protective equipment

Safety Considerations

- When running any vehicles in the shop, make sure you use the shop's exhaust ventilation system to discharge all exhaust gas safely outside.
- Lifting equipment such as vehicle jacks and stands, vehicle hoists, and engine hoists are important tools that increase productivity and make the job easier. However, they can also cause severe injury or death if used improperly. Make sure you follow the manufacturer's operation procedures. Also, make sure you have your supervisor's/instructor's permission to use any particular type of lifting equipment.
- Comply with personal and environmental safety practices associated with clothing; eye protection; hand tools; power equipment; proper ventilation; and the handling, storage, and disposal of chemicals/materials in accordance with local, state, and federal safety and environmental regulations.
- If you need to start the vehicle, you should ensure that the parking brake is firmly applied; if necessary, use wheel chocks to prevent the vehicle from moving when the vehicle is started to verify the completion of these tasks.
- Extreme caution must be exercised when working around rotating components.
- **Caution:** Most types of brake fluid are harmful to painted surfaces. Be sure to prevent brake fluid from coming into contact with a vehicle's paint. Use fender covers to minimize this risk and be sure to wipe up any spilled brake fluid immediately with a wet rag.

- **Caution:** Brake dust may contain asbestos, which has been determined to cause cancer when inhaled or ingested. Treat all brake dust as if it contains asbestos and use Occupational Safety and Health Administration (OSHA)-approved asbestos removal equipment. Do not allow brake dust to become airborne by using anything that would disturb the dust. Also, wear protective gloves during this procedure and dispose of or clean them in an approved manner.

CDX Tasksheet Number: A5030

Student/Intern Information

Name _________________________________ Date ____________ Class _____________________

Vehicle, Customer, and Service Information

Vehicle used for this activity:

Year _______________ Make _______________________ Model _____________________

Odometer _________________________ VIN _______________________________

Materials Required

- Blank work order
- Vehicle with available service history records
- Depending on the type of concern, special diagnostic/hand tools may be required. See your supervisor/instructor for instructions to identify what tools may be required.
- Service information database
- Personal protective equipment
- General hand/air tools
- Wheel bearing grease

Task-Specific Safety Considerations

- Comply with personal and environmental safety practices associated with clothing; eye protection; hand tools; power equipment; proper ventilation; and the handling, storage, and disposal of chemicals/materials in accordance with local, state, and federal safety and environmental regulations.

▶ TASK Remove, clean, inspect, repack, and install wheel bearings; replace seals; install hub and adjust bearings.

MLR
A5F1

Time off______________

Time on______________

Total time______________

Student Instructions: Read through the entire procedure prior to starting. Prepare your workspace and any tools or parts that may be needed to complete the task. When directed by your instructor, begin the procedure to complete the task and check the box as each step is finished. Track your time on this procedure for later comparison to the standard completion time (i.e., "flat rate" or customer pay time).

Procedure:	Step Completed
1. Research front wheel bearing removal.	
a. List or print off and attach to this sheet the procedure for removing a front wheel bearing from your specific vehicle.	☐
b. List the procedure for preloading the wheel bearing, and list the torque specifications.	☐
c. What type of grease is used for the wheel bearings in your vehicle?	☐
2. Remove the front wheel bearing.	
a. Following proper procedure, remove the serviceable wheel bearing from your vehicle.	☐
b. Using a seal remover tool, remove the inner grease seal from the backside of the rotor. Next, remove the inner bearing.	☐
c. Clean the bearing and hub using the correct methods and solvents.	☐
d. Inspect the bearing rollers, cage, and races for signs of damage. Damage includes pitting, corrosion, spalling, brinelling, or signs of overloading and overheating. Explain the results of your visual inspection. Are you able to reuse the bearing?	☐
3. Install the front wheel bearing.	
a. Apply grease to the wheel bearing. Explain the method you used to pack the bearing full of grease.	☐
b. Install a new inner grease seal.	☐
c. Clean and inspect the spindle before bearing installation. List any action required.	☐
d. Following procedure, install the serviceable bearing and rotor assembly.	☐

	Step Completed
e. Torque the adjusting nut to the proper preload. What torque specification did you use? Why is a preload necessary?	☐
f. Finish the installation. Torque the lug nuts to specification. What torque specification did you use?	☐

Non-Task-Specific Evaluations:	Step Completed
1. Tools and equipment were used as directed and returned in good working order.	☐
2. Complied with all general and task-specific safety standards, including proper use of any personal protective equipment.	☐
3. Completed the task in an appropriate time frame (Recommendation: 1.5 or 2 times flat rate).	☐
4. Left the work space clean and orderly.	☐
5. Cared for customer property and returned it undamaged.	☐

Student signature ______________________________ Date ______________________________

Comments:

Have your supervisor/instructor verify satisfactory completion of this procedure, any observations found,

and any necessary action(s) recommended.

Evaluation Instructions: The scoring box below is intended to act as a guide for both student and instructor. Each criterion listed will help students understand what is expected from them and evaluators articulate success at a particular task. The scoring is set up to allow a second attempt at each task (see the "Test" and "Retest" columns). Scoring is designed to reward students for correct completion of the task. Points are lost for failure to complete the employability requirements (see "Non-Task-Specific" criteria). When all criteria are evaluated, tally the points for a total at the bottom of each column.

Tasksheet Scoring

	Test		Retest	
Evaluation Items	**Pass**	**Fail**	**Pass**	**Fail**
Task-Specific Evaluation	**(1 pt)**	**(0 pts)**	**(1 pt)**	**(0 pts)**
Research wheel bearing removal and installation.				
Remove the front wheel bearing. Clean and inspect the bearing.				
Replace the grease seals and pack the wheel bearings with the correct grease.				
Install the wheel bearing and use the correct procedure to tighten it.				
Non-Task-Specific Evaluation	**(0 pts)**	**(−1 pt)**	**(0 pts)**	**(−1 pt)**
Student successfully completed at least three of the non-task-specific steps.				
Student successfully completed all five of the non-task-specific steps.				
Total Score: <total # of points / 4 = %>				

Supervisor:

Supervisor/instructor signature _______________________ Date _______________

Comments:

Retest supervisor/instructor signature _______________________ Date _______________

Comments:

CDX Tasksheet Number: A5031

Student/Intern Information

Name _________________________________ Date ____________ Class _____________________

Vehicle, Customer, and Service Information

Vehicle used for this activity:

Year _______________ Make ___________________________ Model ____________________

Odometer _________________________________ VIN _________________________________

Task-Specific Safety Considerations
- Comply with personal and environmental safety practices associated with clothing; eye protection; hand tools; power equipment; proper ventilation; and the handling, storage, and disposal of chemicals/materials in accordance with local, state, and federal safety and environmental regulations.

▶ TASK Check parking brake system components for wear, binding, and corrosion; clean, lubricate, adjust and/or replace as needed. **MLR A5F2**

Time off______________

Time on______________

Total time______________

Student Instructions: Read through the entire procedure prior to starting. Prepare your workspace and any tools or parts that may be needed to complete the task. When directed by your instructor, begin the procedure to complete the task and check the box as each step is finished. Track your time on this procedure for later comparison to the standard completion time (i.e., "flat rate" or customer pay time).

Procedure:	Step Completed
1. Perform a parking brake inspection.	
a. With the vehicle on the ground, apply the parking brake. Describe the function of the parking brake pedal/lever. Does it feel tight within three or four clicks? Does it take several clicks to begin to tighten? **Tip:** If possible, when inspecting parking brake operation, place the vehicle in neutral on a slight incline and engage the parking brake. The vehicle should hold steady on a slight incline.	☐
b. Lift the vehicle and remove the rear wheels.	☐
c. Describe what type of parking brake is used (integral, parking shoes, electronic parking brake).	☐
2. Inspect the integral parking brake.	
a. Remove the rear brake calipers.	☐
b. Use a caliper piston tool to retract the piston. Does the piston screw in and retract correctly?	☐
c. Inspect the piston dust seal for cracking, tears, or leaks.	☐
3. Inspect the electronic parking brake.	
a. Use the switch to activate the parking brake.	☐
b. Listen for the parking brake to engage. Does a parking brake indicator illuminate on the dash?	☐
4. Inspect the parking brake shoes.	
a. Remove the rear brakes to expose the parking brake shoes.	☐
b. Inspect the brakes for cracking, delamination, excess wear, and fluid leaks. Describe the condition of the shoes.	☐

c. Have a helper slowly engage the parking brake as you watch the shoes expand outward. Inspect the brake shoes for binding.	☐
d. Using approved methods and solvents, clean the parking brake shoes and adjust the brake shoes and cables to specification. **Tip:** Use service information for brake shoe/cable adjustments.	☐
e. Reinstall all brake components. Install the wheels and tighten the lug nuts to specification.	☐

Non-Task-Specific Evaluations:	**Step Completed**
1. Tools and equipment were used as directed and returned in good working order.	☐
2. Complied with all general and task-specific safety standards, including proper use of any personal protective equipment.	☐
3. Completed the task in an appropriate time frame (Recommendation: 1.5 or 2 times flat rate).	☐
4. Left the work space clean and orderly.	☐
5. Cared for customer property and returned it undamaged.	☐

Student signature _______________________________ Date _____________________________

Comments:

Have your supervisor/instructor verify satisfactory completion of this procedure, any observations found,

and any necessary action(s) recommended.

Evaluation Instructions: The scoring box below is intended to act as a guide for both student and instructor. Each criterion listed will help students understand what is expected from them and evaluators articulate success at a particular task. The scoring is set up to allow a second attempt at each task (see the "Test" and "Retest" columns). Scoring is designed to reward students for correct completion of the task. Points are lost for failure to complete the employability requirements (see "Non-Task-Specific" criteria). When all criteria are evaluated, tally the points for a total at the bottom of each column.

Tasksheet Scoring

	Test		Retest	
Evaluation Items	**Pass**	**Fail**	**Pass**	**Fail**
Task-Specific Evaluation	**(1 pt)**	**(0 pts)**	**(1 pt)**	**(0 pts)**
Inspect the parking brake with the vehicle on the ground and check for proper holding capacity.				
Describe the design of parking brakes.				
Clean the parking brake shoes. Inspect the parking brake and related components for proper function and condition.				
Adjust the parking brake shoes.				
Non-Task-Specific Evaluation	**(0 pts)**	**(−1 pt)**	**(0 pts)**	**(−1 pt)**
Student successfully completed at least three of the non-task-specific steps.				
Student successfully completed all five of the non-task-specific steps.				
Total Score: <total # of points / 4 = %>				

Supervisor:

Supervisor/instructor signature _________________________ Date _______________

Comments:

Retest supervisor/instructor signature _________________________ Date _______________

Comments:

CDX Tasksheet Number: A5032

Student/Intern Information

Name _________________________________ Date _____________ Class _________________________

Vehicle, Customer, and Service Information

Vehicle used for this activity:

Year _______________ Make _______________________________ Model _______________________________

Odometer _________________________________ VIN _______________________________________

Materials Required
- Blank work order
- Vehicle with available service history records
- Depending on the type of concern, special diagnostic/hand tools may be required. See your supervisor/instructor for instructions to identify what tools may be required.
- Service information database
- Personal protective equipment

Task-Specific Safety Considerations
- Comply with personal and environmental safety practices associated with clothing; eye protection; hand tools; power equipment; proper ventilation; and the handling, storage, and disposal of chemicals/materials in accordance with local, state, and federal safety and environmental regulations.

▶ TASK Check parking brake operation and parking brake indicator light system operation; determine necessary action.

MLR A5F3

Time off_____________

Time on_____________

Total time_____________

Student Instructions: Read through the entire procedure prior to starting. Prepare your workspace and any tools or parts that may be needed to complete the task. When directed by your instructor, begin the procedure to complete the task and check the box as each step is finished. Track your time on this procedure for later comparison to the standard completion time (i.e., "flat rate" or customer pay time).

Procedure:	Step Completed
1. Verify parking brake indicator operation.	
a. Cycle the ignition key to the on position and lightly engage the parking brake.	☐
b. With the parking brake engaged, inspect the parking brake indicator light. Does it turn on?	☐
c. Release the parking brake. Does the light turn off?	☐
d. Explain what the light staying on could indicate.	☐
2. Verify parking brake operation.	
a. With the vehicle on the ground, apply the parking brake. Describe the function of the parking brake pedal/lever. Does it feel tight within 3–4 clicks? Does it take several clicks to begin to tighten? **Tip:** If possible, when inspecting parking brake operation, place the vehicle in neutral on a slight incline and engage the parking brake. The vehicle should hold steady on a slight incline.	☐
b. What steps would be next in addressing a parking brake lever that felt loose? (Too many clicks to engage the brake.)	☐
c. If testing the parking brake operation in your service stall, apply the parking brake and lift the vehicle off the ground.	☐
d. Attempt to turn the rear wheels by hand, back and forth. The wheels should hold steady.	☐
3. Based on your inspection, what is your recommendation for service?	☐

Non-Task-Specific Evaluations:	Step Completed
1. Tools and equipment were used as directed and returned in good working order.	☐
2. Complied with all general and task-specific safety standards, including proper use of any personal protective equipment.	☐
3. Completed the task in an appropriate time frame (Recommendation: 1.5 or 2 times flat rate).	☐
4. Left the work space clean and orderly.	☐
5. Cared for customer property and returned it undamaged.	☐

Student signature _________________________ Date _________________________

Comments:

Have your supervisor/instructor verify satisfactory completion of this procedure, any observations found, and any necessary action(s) recommended.

Tasksheet Scoring

Evaluation Items	Test		Retest	
	Pass	**Fail**	**Pass**	**Fail**
Task-Specific Evaluation	**(1 pt)**	**(0 pts)**	**(1 pt)**	**(0 pts)**
Verify parking brake indicator operation. Verify parking brake operation.				
Determine the causes of indicator light malfunction.				
Determine the causes of parking brake malfunction.				
Make recommendations for service based on your inspection.				
Non-Task-Specific Evaluation	**(0 pts)**	**(−1 pt)**	**(0 pts)**	**(−1 pt)**
Student successfully completed at least three of the non-task-specific steps.				
Student successfully completed all five of the non-task-specific steps.				
Total Score: <total # of points / 4 = %>				

Supervisor:

Supervisor/instructor signature _____________________________ Date ________________

Comments:

Retest supervisor/instructor signature _____________________ Date ________________

Comments:

CDX Tasksheet Number: A5033

Student/Intern Information

Name ________________________________ Date ____________ Class ________________________

Vehicle, Customer, and Service Information

Vehicle used for this activity:

Year ______________ Make ________________________ Model ____________________________

Odometer ____________________________ VIN ________________________________

Task-Specific Safety Considerations

- Comply with personal and environmental safety practices associated with clothing; eye protection; hand tools; power equipment; proper ventilation; and the handling, storage, and disposal of chemicals/materials in accordance with local, state, and federal safety and environmental regulations.

▶ **TASK** Check operation of brake stop light system.

MLR
A5F4

Time off____________

Time on____________

Total time____________

Student Instructions: Read through the entire procedure prior to starting. Prepare your workspace and any tools or parts that may be needed to complete the task. When directed by your instructor, begin the procedure to complete the task and check the box as each step is finished. Track your time on this procedure for later comparison to the standard completion time (i.e., "flat rate" or customer pay time).

Procedure:	Step Completed
1. Inspect the operation of the brake light system.	
a. Use a brake pedal depressor to apply the brake.	☐
b. Inspect the brake lights. Did all of the lights illuminate, including the high mounted/third brake light?	☐
c. Some vehicles are equipped with brake lights that will flash during initial brake application. Is your vehicle equipped with this type of light? Why would a flashing brake light be used on a vehicle?	☐
2. Determine the possible causes of brake light failure.	
a. List the possible causes of *one* brake light not working.	☐
b. List the possible causes of *all* brake lights not working.	☐

Non-Task-Specific Evaluations:	Step Completed
1. Tools and equipment were used as directed and returned in good working order.	☐
2. Complied with all general and task-specific safety standards, including proper use of any personal protective equipment.	☐
3. Completed the task in an appropriate time frame (Recommendation: 1.5 or 2 times flat rate).	☐
4. Left the work space clean and orderly.	☐
5. Cared for customer property and returned it undamaged.	☐

Student signature _________________________________ Date _________________________________

Comments:

Have your supervisor/instructor verify satisfactory completion of this procedure, any observations found,

and any necessary action(s) recommended.

Taksheet Scoring

	Test		Retest	
Evaluation Items	Pass	Fail	Pass	Fail
Task-Specific Evaluation	**(1 pt)**	**(0 pts)**	**(1 pt)**	**(0 pts)**
Inspect the operation of the brake light system.				
Inspect the operation of the 3rd brake light.				
List the possible causes of *one* brake light not working.				
List the possible causes of *all* brake lights not working.				
Non-Task-Specific Evaluation	**(0 pts)**	**(−1 pt)**	**(0 pts)**	**(−1 pt)**
Student successfully completed at least three of the non-task-specific steps.				
Student successfully completed all five of the non-task-specific steps.				
Total Score: <total # of points / 4 = %>				

CDX Tasksheet Number: A5034

Student/Intern Information

Name _______________________________ Date ___________ Class _____________________

Vehicle, Customer, and Service Information

Vehicle used for this activity:

Year _______________ Make _________________________________ Model _______________________

Odometer _________________________________ VIN _______________________________

Materials Required

- Blank work order
- Vehicle with available service history records
- Depending on the type of concern, special diagnostic/hand tools may be required. See your supervisor/instructor for instructions to identify what tools may be required.
- Service information database
- Personal protective equipment
- General hand/air tools
- Press and related equipment
- Snap ring pliers
- Torque wrench

Task-Specific Safety Considerations

- Lifting equipment, such as vehicle jacks and stands, vehicle hoists, and engine hoists, are important tools that increase productivity and make the job easier. However, they can also cause severe injury or death if used improperly. Make sure you follow the manufacturer's operation procedures. Also, make sure you have your supervisor's/instructor's permission to use any particular type of lifting equipment.
- Comply with personal and environmental safety practices associated with clothing; eye protection; hand tools; power equipment; proper ventilation; and the handling, storage, and disposal of chemicals/materials in accordance with local, state, and federal safety and environmental regulations.

▶ **TASK** Replace wheel bearing and race.

MLR
A5F5

Time off_____________

Time on_____________

Total time_____________

Student Instructions: Read through the entire procedure prior to starting. Prepare your workspace and any tools or parts that may be needed to complete the task. When directed by your instructor, begin the procedure to complete the task and check the box as each step is finished. Track your time on this procedure for later comparison to the standard completion time (i.e., "flat rate" or customer pay time).

Procedure:	Step Completed
1. Research the procedure for replacing a wheel bearing and race.	
a. List or print off the procedure for replacing a wheel bearing and race on your vehicle.	☐
b. List any torque specifications related to this task.	☐
2. Remove the wheel bearing.	
a. Lift the vehicle and remove the wheel.	☐
b. Remove all brake related components. **Note:** Secure the brake caliper and ensure it does not hang by the brake hose.	☐
c. Remove the steering knuckle. **Tip:** Special care should be taken when removing wheel speed sensors and harnesses. These components are easily damaged.	☐
d. Using the correct adapters and press tools, press the hub from the wheel bearing. Inspect the hub for damage. This part may be reused if in good condition.	☐
e. Remove the bearing retaining snap ring.	☐
f. Press the bearing race from the steering knuckle.	☐
g. Clean and inspect the bearing bore.	☐
3. Install the wheel bearing.	
a. Using the correct adapters, press the bearing into the steering knuckle. **Tip:** Some vehicles use a wheel speed sensor that uses a signal from the wheel bearing. Ensure the bearing is oriented correctly before installing. Describe what would happen to a vehicle if a wheel speed sensor was not functioning correctly.	☐
b. Install the bearing retaining snap ring.	☐
c. Using the correct adapters, press the hub into the bearing.	☐
d. Install the steering knuckle and brake components on the vehicle. Torque all fasteners to specification.	☐
e. Install the wheel and lower the vehicle.	☐

Non-Task-Specific Evaluations:	Step Completed
1. Tools and equipment were used as directed and returned in good working order.	☐
2. Complied with all general and task-specific safety standards, including proper use of any personal protective equipment.	☐
3. Completed the task in an appropriate time frame (Recommendation: 1.5 or 2 times flat rate).	☐
4. Left the work space clean and orderly.	☐
5. Cared for customer property and returned it undamaged.	☐

Student signature _________________________ Date _________________________

Comments:

Have your supervisor/instructor verify satisfactory completion of this procedure, any observations found, and any necessary action(s) recommended.

Evaluation Instructions: The scoring box below is intended to act as a guide for both student and instructor. Each criterion listed will help students understand what is expected from them and evaluators articulate success at a particular task. The scoring is set up to allow a second attempt at each task (see the "Test" and "Retest" columns). Scoring is designed to reward students for correct completion of the task. Points are lost for failure to complete the employability requirements (see "Non-Task-Specific" criteria). When all criteria are evaluated, tally the points for a total at the bottom of each column.

Tasksheet Scoring

	Test		Retest	
Evaluation Items	**Pass**	**Fail**	**Pass**	**Fail**
Task-Specific Evaluation	**(1 pt)**	**(0 pts)**	**(1 pt)**	**(0 pts)**
Research the wheel bearing replacement procedure.				
Remove the brake-related components and the steering knuckle. Press the hub and bearing from the steering knuckle.				
Install the wheel bearing in the correct orientation.				
Install all brake-related components and the steering knuckle.				
Non-Task-Specific Evaluation	**(0 pts)**	**(−1 pt)**	**(0 pts)**	**(−1 pt)**
Student successfully completed at least three of the non-task-specific steps.				
Student successfully completed all five of the non-task-specific steps.				
Total Score: <total # of points / 4 = %>				

Supervisor:

Supervisor/instructor signature ____________________________ Date ________________

Comments:

Retest supervisor/instructor signature ____________________________ Date ________________

Comments:

CDX Tasksheet Number: A5035

Student/Intern Information

Name _________________________ Date __________ Class _________________________

Vehicle, Customer, and Service Information

Vehicle used for this activity:

Year _____________ Make _________________________ Model _________________________

Odometer _________________________ VIN _________________________

> **Materials Required**
> - Blank work order
> - Vehicle with available service history records
> - Depending on the type of concern, special diagnostic/hand tools may be required. See your supervisor/instructor for instructions to identify what tools may be required.
> - Service information database
> - Personal protective equipment

Task-Specific Safety Considerations

- Lifting equipment, such as vehicle jacks and stands, vehicle hoists, and engine hoists, are important tools that increase productivity and make the job easier. However, they can also cause severe injury or death if used improperly. Make sure you follow the manufacturer's operation procedures. Also, make sure you have your supervisor's/instructor's permission to use any particular type of lifting equipment.
- Comply with personal and environmental safety practices associated with clothing; eye protection; hand tools; power equipment; proper ventilation; and the handling, storage, and disposal of chemicals/materials in accordance with local, state, and federal safety and environmental regulations.

▶ **TASK** Inspect and replace wheel studs.

MLR
A5F6

Time off__________

Time on__________

Total time__________

Student Instructions: Read through the entire procedure prior to starting. Prepare your workspace and any tools or parts that may be needed to complete the task. When directed by your instructor, begin the procedure to complete the task and check the box as each step is finished. Track your time on this procedure for later comparison to the standard completion time (i.e., "flat rate" or customer pay time).

Procedure:	Step Completed
1. Remove the wheel stud.	
a. Lift the vehicle and remove the wheel.	☐
b. Remove the brake components, including the brake drum/rotor.	☐
c. Inspect the hub flange and all wheel studs for damage. Explain why a wheel stud might require replacement.	☐
d. Use a hammer to drive out the old wheel stud.	☐
2. Install the wheel stud.	
a. Insert the wheel stud from the rear of the hub flange.	☐
b. Use a stack of washers or a shim to be placed over the wheel stud.	☐
c. Using a standard lug nut, thread the nut onto the new stud and begin to tighten. Continue to tighten until the new wheel stud has been pulled completely though the flange and is sitting flush.	☐
d. Remove the washers. Install the brake components and install the wheel. Torque all fasteners to specification.	☐
e. Lower the vehicle.	☐

Non-Task-Specific Evaluations:	Step Completed
1. Tools and equipment were used as directed and returned in good working order.	☐
2. Complied with all general and task-specific safety standards, including proper use of any personal protective equipment.	☐
3. Completed the task in an appropriate time frame (Recommendation: 1.5 or 2 times flat rate).	☐
4. Left the work space clean and orderly.	☐
5. Cared for customer property and returned it undamaged.	☐

Student signature _____________________________ Date _____________________________

Comments:

Have your supervisor/instructor verify satisfactory completion of this procedure, any observations found, and any necessary action(s) recommended.

Tasksheet Scoring

	Test		Retest	
Evaluation Items	**Pass**	**Fail**	**Pass**	**Fail**
Task-Specific Evaluation	**(1 pt)**	**(O pts)**	**(1 pt)**	**(O pts)**
Remove the brake components to access wheel studs.				
Inspect all wheel studs and the hub flange for damage.				
Remove and install the wheel studs.				
Replace the brake components and torque all fasteners to specification.				
Non-Task-Specific Evaluation	**(O pts)**	**(−1 pt)**	**(O pts)**	**(−1 pt)**
Student successfully completed at least three of the non-task-specific steps.				
Student successfully completed all five of the non-task-specific steps.				
Total Score: <total # of points / 4 = %>				

Supervisor:

Supervisor/instructor signature _______________________________ Date _____________________

Comments:

Retest supervisor/instructor signature _______________________________ Date _____________________

Comments:

5G. Electronic Brake, Traction Control, and Stability Control Systems

Learning Objective/Task	CDX Tasksheet Number	ASE Education Foundation Reference Number; Priority Level
• Identify traction control/vehicle stability control system components.	A5036	A5G1, P-3
• Describe the operation of a regenerative braking system.	A5037	A5G2, P-3

Materials Required

- Blank work order
- Vehicle with available service history records
- Depending on the type of concern, special diagnostic/hand tools may be required. See your supervisor/instructor for instructions to identify what tools may be required.
- Service information database
- Personal protective equipment

Safety Considerations

- When running any vehicles in the shop, make sure you use the shop's exhaust ventilation system to discharge all exhaust gas safely outside.
- Lifting equipment, such as vehicle jacks and stands, vehicle hoists, and engine hoists, are important tools that increase productivity and make the job easier. However, they can also cause severe injury or death if used improperly. Make sure you follow the manufacturer's operation procedures. Also, make sure you have your supervisor's/instructor's permission to use any particular type of lifting equipment.
- Comply with personal and environmental safety practices associated with clothing; eye protection; hand tools; power equipment; proper ventilation; and the handling, storage, and disposal of chemicals/materials in accordance with local, state, and federal safety and environmental regulations.
- If you need to start the vehicle, you should ensure that the parking brake is firmly applied; if necessary, use wheel chocks to prevent the vehicle from moving when the vehicle is started to verify the completion of these tasks.
- Extreme caution must be exercised when working around rotating components.

CDX Tasksheet Number: A5036

Student/Intern Information

Name _________________________ Date __________ Class _________________

Vehicle, Customer, and Service Information

Vehicle used for this activity:

Year _____________ Make __________________ Model ______________

Odometer ________________________ VIN ____________________________

Materials Required

- Blank work order
- Vehicle with available service history records
- Depending on the type of concern, special diagnostic/hand tools may be required. See your supervisor/instructor for instructions to identify what tools may be required.
- Service information database
- Personal protective equipment

Task-Specific Safety Considerations

- Comply with personal and environmental safety practices associated with clothing; eye protection; hand tools; power equipment; proper ventilation; and the handling, storage, and disposal of chemicals/materials in accordance with local, state, and federal safety and environmental regulations.

▶ TASK Identify traction control/vehicle stability control system components.

MLR
A5G1

Time off_____________

Time on_____________

Total time_____________

Student Instructions: Read through the entire procedure prior to starting. Prepare your workspace and any tools or parts that may be needed to complete the task. When directed by your instructor, begin the procedure to complete the task and check the box as each step is finished. Track your time on this procedure for later comparison to the standard completion time (i.e., "flat rate" or customer pay time).

Procedure:	Step Completed
1. Describe the purpose and function of traction control systems.	
a. Explain the purpose of the traction control system. What is it designed to do?	☐
b. Use service information to list the components that make up the traction control system on your vehicle.	☐
c. Describe how the vehicle's computer identifie's wheel slip, what components are used, and how the system works to regain traction/stability.	☐
2. Locate the following traction control components.	
a. Lift the vehicle and locate the wheel speed sensors on each wheel. Inspect the sensor wiring to ensure that they are secured. Inspect for damage.	☐
b. ABS module	☐
c. Throttle body	☐
d. Camshaft actuators (if applicable)	☐
e. Traction control switch	☐
3. Inspect the traction control warning indicator.	
a. Cycle the ignition key to the ON position.	☐
b. Turn the traction control switch on. Does the traction control light illuminate? Turn the switch off. Does the light go out?	☐
c. Start the vehicle. Does the traction control light come on for 3–5 seconds and then go out? What does it mean if the light remains on?	☐

| d. Describe how the light will function in the event of a loss of traction. | ☐ |

Non-Task-Specific Evaluations:	Step Completed
1. Tools and equipment were used as directed and returned in good working order.	☐
2. Complied with all general and task-specific safety standards, including proper use of any personal protective equipment.	☐
3. Completed the task in an appropriate time frame (Recommendation: 1.5 or 2 times flat rate).	☐
4. Left the work space clean and orderly.	☐
5. Cared for customer property and returned it undamaged.	☐

Student signature _______________________________ Date _______________________________

Comments:

Have your supervisor/instructor verify satisfactory completion of this procedure, any observations found, and any necessary action(s) recommended.

Evaluation Instructions: The scoring box below is intended to act as a guide for both student and instructor. Each criterion listed will help students understand what is expected from them and evaluators articulate success at a particular task. The scoring is set up to allow a second attempt at each task (see the "Test" and "Retest" columns). Scoring is designed to reward students for correct completion of the task. Points are lost for failure to complete the employability requirements (see "Non-Task-Specific" criteria). When all criteria are evaluated, tally the points for a total at the bottom of each column.

Tasksheet Scoring

	Test		Retest	
Evaluation Items	**Pass**	**Fail**	**Pass**	**Fail**
Task-Specific Evaluation	**(1 pt)**	**(0 pts)**	**(1 pt)**	**(0 pts)**
Describe the purpose and function of the traction control system.				
List the components that make up the traction control system.				
Locate the components of the traction control system.				
Inspect the traction control warning indicators and describe the function of traction control warning indicators.				
Non-Task-Specific Evaluation	**(0 pts)**	**(−1 pt)**	**(0 pts)**	**(−1 pt)**
Student successfully completed at least three of the non-task-specific steps.				
Student successfully completed all five of the non-task-specific steps.				
Total Score: <total # of points / 4 = %>				

Supervisor:

Supervisor/instructor signature _______________________________ Date _______________

Comments:

Retest supervisor/instructor signature _______________________ Date _______________

Comments:

CDX Tasksheet Number: A5037

Student/Intern Information

Name _________________________________ Date ____________ Class _________________________

Vehicle, Customer, and Service Information

Vehicle used for this activity:

Year _______________ Make _____________________________ Model _____________________________

Odometer _________________________________ VIN _____________________________________

Materials Required

- Blank work order
- Vehicle with available service history records
- Depending on the type of concern, special diagnostic/hand tools may be required. See your supervisor/instructor for instructions to identify what tools may be required.
- Service information database
- Personal protective equipment

Task-Specific Safety Considerations

- Comply with personal and environmental safety practices associated with clothing; eye protection; hand tools; power equipment; proper ventilation; and the handling, storage, and disposal of chemicals/materials in accordance with local, state, and federal safety and environmental regulations.

▶ TASK Describe the operation of a regenerative braking system.

MLR
A5G2

Time off____________

Time on____________

Total time____________

Student Instructions: Read through the entire procedure prior to starting. Prepare your workspace and any tools or parts that may be needed to complete the task. When directed by your instructor, begin the procedure to complete the task and check the box as each step is finished. Track your time on this procedure for later comparison to the standard completion time (i.e., "flat rate" or customer pay time).

Procedure:	Step Completed
1. Research regenerative braking.	
a. Research and describe any service precautions related to servicing the regenerative braking system on a vehicle.	☐
b. What types of vehicles are equipped with regenerative braking systems?	☐
c. How much voltage is used in a regenerative braking system? What color wiring is associated with high-voltages?	☐
d. Describe the main purpose of regenerative braking.	☐
e. Explain how regenerative braking works.	☐
f. List the components that make up a regenerative braking system.	☐
g. Explain what kinetic energy is and how it applies to a regenerative braking system.	☐
h. Describe the conditions when regenerative braking is active.	☐

Non-Task-Specific Evaluations:	**Step Completed**
1. Tools and equipment were used as directed and returned in good working order.	☐
2. Complied with all general and task-specific safety standards, including proper use of any personal protective equipment.	☐
3. Completed the task in an appropriate time frame (Recommendation: 1.5 or 2 times flat rate).	☐
4. Left the work space clean and orderly.	☐
5. Cared for customer property and returned it undamaged.	☐

Student signature ________________________ Date ________________________

Comments:

Have your supervisor/instructor verify satisfactory completion of this procedure, any observations found, and any necessary action(s) recommended.

Taksheet Scoring

		Test		Retest	
Evaluation Items	**Pass**	**Fail**	**Pass**	**Fail**	
Task-Specific Evaluation	**(1 pt)**	**(0 pts)**	**(1 pt)**	**(0 pts)**	
Identify service precautions related to regenerative braking systems.					
Identify voltage requirements and high-voltage wiring colors.					
Describe the purpose and function of regenerative braking systems.					
Define kinetic energy. Identify components related to regenerative braking systems and describe the conditions when regenerative braking is active.					
Non-Task-Specific Evaluation	**(0 pts)**	**(−1 pt)**	**(0 pts)**	**(−1 pt)**	
Student successfully completed at least three of the non-task-specific steps.					
Student successfully completed all five of the non-task-specific steps.					
Total Score: <total # of points / 4 = %>					

Supervisor:

Supervisor/instructor signature _________________________________ Date _______________________

Comments:

Retest supervisor/instructor signature _________________________________ Date _______________________

Comments:

6A. Electrical/Electronic Systems

Learning Objective/Task	CDX Tasksheet Number	ASE Education Foundation Reference Number; Priority Level
• Research vehicle service information, including vehicle service history, service precautions, and technical service bulletins (TSBs).	A6001	A6A1, P-1
• Demonstrate knowledge of electrical/electronic series, parallel, and series-parallel circuits using principles of electricity (Ohm's law).	A6002	A6A2, P-1
• Use wiring diagrams to trace electrical/electronic circuits.	A6003	A6A3, P-1
• Demonstrate proper use of a digital multimeter when measuring source voltage, voltage drop (including grounds), current flow, and resistance.	A6004	A6A4, P-1
• Demonstrate knowledge of the causes and effects from shorts, grounds, opens, and resistance problems in electrical/electronic circuits.	A6005	A6A5, P-1
• Use a test light to check operation of electrical circuits.	A6006	A6A6, P-2
• Use fused jumper wires to check operation of electrical circuits.	A6007	A6A7, P-2
• Measure key-off battery drain (parasitic draw).	A6008	A6A8, P-1
• Inspect and test fusible links, circuit breakers, and fuses, and determine necessary action.	A6009	A6A9, P-1
• Repair and/or replace connectors, terminal ends, and wiring of electrical/electronic systems (including solder repair).	A6010	A6A10, P-1
• Identify electrical/electronic system components and configuration.	A6011	A6A11, P-1

Materials Required

- Blank work order
- Vehicle with available service history records
- Depending on the type of concern, special diagnostic/hand tools may be required. See your supervisor/instructor for instructions to identify what tools may be required.
- Service information database
- Personal protective equipment

Safety Considerations

- When running any vehicles in the shop, make sure you use the shop's exhaust ventilation system to discharge all exhaust gas safely outside.
- Lifting equipment, such as vehicle jacks and stands, vehicle hoists, and engine hoists, are important tools that increase productivity and make the job easier. However, they can also cause severe injury or death if used improperly. Make sure you follow the manufacturer's operation procedures. Also, make sure you have your supervisor's/instructor's permission to use any particular type of lifting equipment.
- Comply with personal and environmental safety practices associated with clothing; eye protection; hand tools; power equipment; proper ventilation; and the handling, storage, and disposal of chemicals/materials in accordance with local, state, and federal safety and environmental regulations.
- If you need to start the vehicle, you should ensure that the parking brake is firmly applied; if necessary, use wheel chocks to prevent the vehicle from moving when the vehicle is started to verify the completion of these tasks.

CDX Tasksheet Number: A6001

Student/Intern Information

Name _______________________________________ Date _____________ Class _________________________

Vehicle, Customer, and Service Information

Vehicle used for this activity:

Year _________________ Make _________________________________ Model _________________________

Odometer _________________________________ VIN ____

Materials Required

- Blank work order
- Vehicle with available service history records
- Depending on the type of concern, special diagnostic/hand tools may be required. See your supervisor/instructor for instructions to identify what tools may be required.
- Service information database
- Personal protective equipment

Task-Specific Safety Considerations

- Comply with personal and environmental safety practices associated with clothing; eye protection; hand tools; power equipment; proper ventilation; and the handling, storage, and disposal of chemicals/materials in accordance with local, state, and federal safety and environmental regulations.

▶ **TASK** Research vehicle service information, including vehicle service history, service precautions, and technical service bulletins (TSBs).

MLR
A6A1

Time off________________

Time on________________

Total time____________

Student Instructions: Read through the entire procedure prior to starting. Prepare your workspace and any tools or parts that may be needed to complete the task. When directed by your instructor, begin the procedure to complete the task and check the box as each step is finished. Track your time on this procedure for later comparison to the standard completion time (i.e., "flat rate" or customer pay time).

Procedure:	Step Completed
1. Research the vehicle's service information.	
a. Use service information to identify any service precautions related to working on your vehicle's electrical system.	☐

2. Research the vehicle's service history.	
a. Determine any outstanding or completed recall work.	☐
b. Determine if any repairs were made in the service history.	☐
3. Determine if any TSBs have been issued for your specific vehicle.	
a. Identify one electrical/electronics-related TSB and list the TSB number, vehicle concern, and repair procedure.	☐

Non-Task-Specific Evaluations:	Step Completed
1. Tools and equipment were used as directed and returned in good working order.	☐
2. Complied with all general and task-specific safety standards, including proper use of any personal protective equipment.	☐
3. Completed the task in an appropriate time frame (Recommendation: 1.5 or 2 times flat rate).	☐
4. Left the work space clean and orderly.	☐
5. Cared for customer property and returned it undamaged.	☐

Student signature ________________________________ Date ________________________________

Comments:

Have your supervisor/instructor verify satisfactory completion of this procedure, any observations found, and any necessary action(s) recommended.

Tasksheet Scoring

Evaluation Items	Test		Retest	
	Pass	Fail	Pass	Fail
Task-Specific Evaluation	**(1 pt)**	**(0 pts)**	**(1 pt)**	**(0 pts)**
Research the vehicle's service history.				
Identify service precautions.				
Identify any completed/outstanding recalls.				
Identify any brake-related TSBs.				
Non-Task-Specific Evaluation	**(0 pts)**	**(−1 pt)**	**(0 pts)**	**(−1 pt)**
Student successfully completed at least three of the non-task-specific steps.				
Student successfully completed all five of the non-task-specific steps.				
Total Score: <total # of points / 4 = %>				

Supervisor:

Supervisor/instructor signature _________________________________ Date _________________________

Comments:

Retest supervisor/instructor signature _________________________________ Date _________________________

Comments:

CDX Tasksheet Number: A6002

Student/Intern Information

Name _________________________________ Date ____________ Class _____________________

Vehicle, Customer, and Service Information

Vehicle used for this activity:

Year ______________ Make _________________________ Model _____________________

Odometer ___________________________ VIN _______________________________

Task-Specific Safety Considerations

- Comply with personal and environmental safety practices associated with clothing; eye protection; hand tools; power equipment; proper ventilation; and the handling, storage, and disposal of chemicals/materials in accordance with local, state, and federal safety and environmental regulations.

Time off______________

Time on______________

Total time______________

▶ **TASK** Demonstrate knowledge of electrical/electronic series, parallel, and series-parallel circuits using principles of electricity (Ohm's law).

MLR
A6A2

Student Instructions: Read through the entire procedure prior to starting. Prepare your workspace and any tools or parts that may be needed to complete the task. When directed by your instructor, begin the procedure to complete the task and check the box as each step is finished. Track your time on this procedure for later comparison to the standard completion time (i.e., "flat rate" or customer pay time).

Procedure:	Step Completed
1. Define Ohm's law.	
a. Define Ohm's law and explain how voltage, amperage, and resistance are related and how they affect each other.	☐

b. Write the equation for Ohm's law.	☐
2. Understand series circuits.	
a. Explain what is unique about amperage in a series circuit.	☐
b. Explain how you calculate the total resistance in a series circuit with more than one resistor	☐
c. How is voltage drop calculated and how does it affect a circuit with more than one load?	☐
d. How does an open circuit condition affect a series circuit?	☐
e. List one example of a series circuit that can be found in a vehicle.	☐
3. Understand parallel circuits.	
a. Explain how a parallel circuit is different than a series circuit.	☐
b. Write the formula for calculating the total resistance in a parallel circuit.	☐
c. As more branches are added to a parallel circuit, explain what happens to the total resistance. Why does this happen?	☐

d. How is source voltage distributed throughout a parallel circuit?	☐
e. How is total amperage calculated in a parallel circuit?	☐
f. How does an open circuit condition affect a parallel circuit?	☐
g. List one example of a parallel circuit that can be found in a vehicle.	☐
4. Understand series-parallel circuits.	
a. Explain how total resistance is calculated in a series-parallel circuit.	☐
b. Explain how amperage is distributed throughout this type of circuit.	☐
c. Explain voltage drop in a series-parallel circuit.	☐
d. List one example of a series-parallel circuit that can be found in a vehicle.	☐

Non-Task-Specific Evaluations:	**Step Completed**
1. Tools and equipment were used as directed and returned in good working order.	☐
2. Complied with all general and task-specific safety standards, including proper use of any personal protective equipment.	☐
3. Completed the task in an appropriate time frame (Recommendation: 1.5 or 2 times flat rate).	☐
4. Left the work space clean and orderly.	☐
5. Cared for customer property and returned it undamaged.	☐

Student signature _________________________________ Date _________________________________

Comments:

Have your supervisor/instructor verify satisfactory completion of this procedure, any observations found, and any necessary action(s) recommended.

Evaluation Instructions: The scoring box below is intended to act as a guide for both student and instructor. Each criterion listed will help students understand what is expected from them and evaluators articulate success at a particular task. The scoring is set up to allow a second attempt at each task (see the "Test" and "Retest" columns). Scoring is designed to reward students for correct completion of the task. Points are lost for failure to complete the employability requirements (see "Non-Task-Specific" criteria). When all criteria are evaluated, tally the points for a total at the bottom of each column.

Tasksheet Scoring

	Test		Retest	
Evaluation Items	**Pass**	**Fail**	**Pass**	**Fail**
Task-Specific Evaluation	**(1 pt)**	**(0 pts)**	**(1 pt)**	**(0 pts)**
Define Ohm's law and write the equation for calculating values.				
Describe the principles of series circuits and list an example of one.				
Describe the principles of parallel circuits and list an example of one.				
Describe the principles of series–parallel circuits and list an example of one.				
Non-Task-Specific Evaluation	**(0 pts)**	**(−1 pt)**	**(0 pts)**	**(−1 pt)**
Student successfully completed at least three of the non-task-specific steps.				
Student successfully completed all five of the non-task-specific steps.				
Total Score: <total # of points / 4 = %>				

Supervisor:

Supervisor/instructor signature _________________________________ Date _________________

Comments:

Retest supervisor/instructor signature _________________________________ Date _________________

Comments:

CDX Tasksheet Number: A6003

Student/Intern Information

Name ___________________________________ Date ____________ Class _______________________

Vehicle, Customer, and Service Information

Vehicle used for this activity:

Year _______________ Make _____________________________ Model _______________________

Odometer _________________________________ VIN _____________________________________

Materials Required

- Blank work order
- Vehicle with available service history records
- Depending on the type of concern, special diagnostic/hand tools may be required. See your supervisor/instructor for instructions to identify what tools may be required.
- Service information database
- Personal protective equipment
- Wiring diagram

Task-Specific Safety Considerations

- Comply with personal and environmental safety practices associated with clothing; eye protection; hand tools; power equipment; proper ventilation; and the handling, storage, and disposal of chemicals/materials in accordance with local, state, and federal safety and environmental regulations.

▶ TASK Use wiring diagrams to trace electrical/electronic circuits. **MLR A6A3**

Time off _______________

Time on _______________

Student Instructions: Read through the entire procedure prior to starting. Prepare your workspace and any tools or parts that may be needed to complete the task. When directed by your instructor, begin the procedure to complete the task and check the box as each step is finished. Track your time on this procedure for later comparison to the standard completion time (i.e., "flat rate" or customer pay time).

Total time _______________

Procedure:	Step Completed
1. Draw the symbol for the following wiring diagram components:	
a. Ground	☐

b. Fuse	☐
c. Circuit breaker	☐
d. Battery	☐
e. Resistor	☐
f. Diode	☐
2. Use the wiring diagram attached to this sheet to answer the following questions:	
a. What fuse (number) feeds the horn circuit? What is the amperage rating of the horn fuse? Where is the fuse located?	☐
b. When is power supplied to the horn fuse?	☐
c. What component is used to ground the coil side of Horn Relay 21 to activate the horn?	☐
d. How many horns is this vehicle equipped with?	☐

e. What color wire supplies voltage to the horns?	☐
f. What is the ground number and location of the horn ground?	☐

Non-Task-Specific Evaluations:	**Step Completed**
1. Tools and equipment were used as directed and returned in good working order.	☐
2. Complied with all general and task-specific safety standards, including proper use of any personal protective equipment.	☐
3. Completed the task in an appropriate time frame (Recommendation: 1.5 or 2 times flat rate).	☐
4. Left the work space clean and orderly.	☐
5. Cared for customer property and returned it undamaged.	☐

Student signature _________________________ Date _________________________

Comments:

Have your supervisor/instructor verify satisfactory completion of this procedure, any observations found,

and any necessary action(s) recommended.

Evaluation Instructions: The scoring box below is intended to act as a guide for both student and instructor. Each criterion listed will help students understand what is expected from them and evaluators articulate success at a particular task. The scoring is set up to allow a second attempt at each task (see the "Test" and "Retest" columns). Scoring is designed to reward students for correct completion of the task. Points are lost for failure to complete the employability requirements (see "Non-Task-Specific" criteria). When all criteria are evaluated, tally the points for a total at the bottom of each column.

Tasksheet Scoring

	Test		Retest	
Evaluation Items	Pass	Fail	Pass	Fail
Task-Specific Evaluation	**(1 pt)**	**(0 pts)**	**(1 pt)**	**(0 pts)**
Draw wiring diagram symbols.				
Determine fuse names, numbers, locations, and amperage ratings.				
Trace power in the horn circuit. Identify when power is supplied in the circuit. Determine the wiring color.				
Determine the ground number and location.				
Non-Task-Specific Evaluation	**(0 pts)**	**(−1 pt)**	**(0 pts)**	**(−1 pt)**
Student successfully completed at least three of the non-task-specific steps.				
Student successfully completed all five of the non-task-specific steps.				
Total Score: <total # of points / 4 = %>				

Supervisor:

Supervisor/instructor signature _________________________ Date _________________

Comments:

Retest supervisor/instructor signature _________________________ Date _________________

Comments:

CDX Tasksheet Number: A6004

Student/Intern Information

Name _________________________________ Date ___________ Class _________________________

Vehicle, Customer, and Service Information

Vehicle used for this activity:

Year _______________ Make _________________________ Model _____________________

Odometer _________________________ VIN _________________________________

> **Materials Required**
> - Blank work order
> - Vehicle with available service history records
> - Depending on the type of concern, special diagnostic/hand tools may be required. See your supervisor/instructor for instructions to identify what tools may be required.
> - Service information database
> - Personal protective equipment
> - Digital multimeter (DMM)
> - General hand tools
> - Inductive clamp

Task-Specific Safety Considerations
- Comply with personal and environmental safety practices associated with clothing; eye protection; hand tools; power equipment; proper ventilation; and the handling, storage, and disposal of chemicals/materials in accordance with local, state, and federal safety and environmental regulations.
- Special care should be taken when working with electronic components. Keep open flames away from batteries at all times.

▶ **TASK** Demonstrate proper use of a digital multimeter when measuring source voltage, voltage drop (including grounds), current flow, and resistance. **MLR A6A4**

Time off________

Time on________

Total time________

Student Instructions: Read through the entire procedure prior to starting. Prepare your workspace and any tools or parts that may be needed to complete the task. When directed by your instructor, begin the procedure to complete the task and check the box as each step is finished. Track your time on this procedure for later comparison to the standard completion time (i.e., "flat rate" or customer pay time).

Procedure:	Step Completed
1. Measure the source voltage.	
a. Obtain a wiring diagram for the starting/charging system on your vehicle.	☐

b. Measure the source voltage of the battery in the vehicle. Describe what meter setting and meter lead location you will use to perform this measurement.	☐
c. What is the voltage of the battery? What is the battery state of health based on this measurement?	☐
2. Measure the voltage drop.	
a. Describe how you should measure voltage drop in a circuit. Describe what meter setting and lead location you will use to perform this measurement.	☐
b. How much voltage drop is acceptable through switches and wiring?	☐
c. How much voltage drop do you expect to see on the power cable from the battery to the alternator?	☐
d. Describe how you would measure voltage drop on the ground side of the charging system. How much voltage drop do you expect to see on the ground side?	☐
e. Use a DMM to measure the voltage drop of the power wire from the battery to the alternator. What is your reading? Is this an acceptable voltage drop?	☐

f. Use a DMM to measure the voltage drop of the ground side of the charging system. What is your reading? Is this an acceptable voltage drop?	☐
g. Explain what could cause excessive voltage drop in a circuit.	☐
3. Measure the current.	
a. Describe how you should measure amperage in a circuit. Describe what meter setting and meter lead location you will use to perform this measurement.	☐
b. Use an inductive clamp and measure the amperage of the charging system with no accessories turned on. What is your reading?	☐
c. Turn on the rear defrost, radio, high-beam headlights. What is your amperage reading now? Did your reading change?	☐
d. Use service information to determine the rated output of the alternator. How do your readings compare to this specification?	☐
4. Measure the resistance.	
a. Describe how you should measure the resistance of an electrical component. Describe what meter setting and meter lead location you will use to perform this measurement.	☐
b. Disconnect the battery ground terminal.	☐
c. Trace the body ground cable to where it connects to the frame/engine block and disconnect it.	☐

| d. Use a DMM to measure the resistance of the negative battery cable. What is your reading? Is this an acceptable resistance value? | ☐ |
| e. Explain what symptoms may be caused by a battery cable with excessive resistance. | ☐ |

Non-Task-Specific Evaluations:	Step Completed
1. Tools and equipment were used as directed and returned in good working order.	☐
2. Complied with all general and task-specific safety standards, including proper use of any personal protective equipment.	☐
3. Completed the task in an appropriate time frame (Recommendation: 1.5 or 2 times flat rate).	☐
4. Left the work space clean and orderly.	☐
5. Cared for customer property and returned it undamaged.	☐

Student signature _________________________ Date _______________________

Comments:

Have your supervisor/instructor verify satisfactory completion of this procedure, any observations found, and any necessary action(s) recommended.

Evaluation Instructions: The scoring box below is intended to act as a guide for both student and instructor. Each criterion listed will help students understand what is expected from them and evaluators articulate success at a particular task. The scoring is set up to allow a second attempt at each task (see the "Test" and "Retest" columns). Scoring is designed to reward students for correct completion of the task. Points are lost for failure to complete the employability requirements (see "Non-Task-Specific" criteria). When all criteria are evaluated, tally the points for a total at the bottom of each column.

Tasksheet Scoring

		Test		Retest	
Evaluation Items	**Pass**	**Fail**	**Pass**	**Fail**	
Task-Specific Evaluation	**(1 pt)**	**(O pts)**	**(1 pt)**	**(O pts)**	
Measure the source voltage.					
Measure the voltage drop.					
Measure the amperage.					
Measure the resistance.					
Non-Task-Specific Evaluation	**(O pts)**	**(−1 pt)**	**(O pts)**	**(−1 pt)**	
Student successfully completed at least three of the non-task-specific steps.					
Student successfully completed all five of the non-task-specific steps.					
Total Score: <total # of points / 4 = %>					

CDX Tasksheet Number: A6005

Student/Intern Information

Name ___________________________ Date ___________ Class ___________________

Vehicle, Customer, and Service Information

Vehicle used for this activity:

Year ___________ Make ___________________ Model ___________________

Odometer ___________________ VIN ___________________

Materials Required

- Blank work order
- Vehicle with available service history records
- Depending on the type of concern, special diagnostic/hand tools may be required. See your supervisor/instructor for instructions to identify what tools may be required.
- Service information database
- Personal protective equipment
- Digital multimeter
- Fuse/circuit breaker tool

Task-Specific Safety Considerations

- Comply with personal and environmental safety practices associated with clothing; eye protection; hand tools; power equipment; proper ventilation; and the handling, storage, and disposal of chemicals/materials in accordance with local, state, and federal safety and environmental regulations.

▶ TASK Demonstrate knowledge of the causes and effects from shorts, grounds, opens, and resistance problems in electrical/electronic circuits.
Note: Refer to your instructor in regard to what circuit on your vehicle has a fault in need of diagnosis.

MLR
A6A5

Time off___________

Time on___________

Total time___________

Student Instructions: Read through the entire procedure prior to starting. Prepare your workspace and any tools or parts that may be needed to complete the task. When directed by your instructor, begin the procedure to complete the task and check the box as each step is finished. Track your time on this procedure for later comparison to the standard completion time (i.e., "flat rate" or customer pay time).

Procedure:	Step Completed
1. Test for a short circuit.	
a. What device is installed in a circuit to protect against circuit overload?	☐

b. Obtain a wiring diagram for the circuit on your vehicle that has been affected by a short circuit condition.	☐
c. Describe how the circuit has been affected by this condition.	☐
d. Describe how power flows through the circuit. Is the circuit powered on all the time? List the fuses, relays, switches, and components that make up the circuit.	☐
e. Inspect the circuit protection device and describe its condition.	☐
f. Install the fuse/circuit breaker tool in place of the blown fuse. Activate the circuit and observe the tool. Did the tool trip and open the circuit or did the load in the circuit activate?	☐
g. If the circuit breaker tool trips immediately, what does this tell you about your fault? If the load in the circuit powers up and begins working, what does this tell you about the fault?	☐
h. Describe how you would continue testing to determine the cause/location of a short to ground.	☐
i. Describe how you would continue testing to determine the cause/location of an intermittent short to ground.	☐
j. Explain what it means to have a short-to-power condition and how it affects circuits in a vehicle.	☐

2. Test for an open circuit.	
a. Obtain a wiring diagram for the circuit on your vehicle that has been affected by an open-circuit condition.	☐
b. Describe how the circuit has been affected by this condition.	☐
c. Describe how power flows through the circuit. Is the circuit powered on all the time? List the fuses, relays, switches, and components that make up the circuit.	☐
d. Begin diagnosis by checking the open circuit voltage of the battery. What is the voltage?	☐
e. Measure the voltage at the circuit protection device. Is the device in working condition with power available?	☐
f. Measure voltage at the component. What is your voltage reading? If voltage is available to this point, where might the open-circuit condition be located?	☐
g. Measure voltage drop on the ground side of the circuit and record your findings.	☐
h. According to your diagnosis, where did current stop flowing in the circuit?	☐
3. Test for excessive resistance in a circuit.	
a. Describe how excessive resistance will affect a circuit.	☐

b. Explain how you should perform a test for resistance using a digital multimeter.	☐
c. Obtain a wiring diagram for the circuit on your vehicle that has been affected by a high-resistance condition.	☐
d. Describe how performing voltage drop tests on a circuit could help to pinpoint the location of a high-resistance condition.	☐
e. Perform voltage drop tests to determine the source of the excessive resistance. List your results.	☐
f. Verify the source of the high resistance by disconnecting the component from the circuit and using a multimeter to measure the resistance. Did the results of this test verify that this is the cause of the high resistance in the circuit?	☐

Non-Task-Specific Evaluations:	Step Completed
1. Tools and equipment were used as directed and returned in good working order.	☐
2. Complied with all general and task-specific safety standards, including proper use of any personal protective equipment.	☐
3. Completed the task in an appropriate time frame (Recommendation: 1.5 or 2 times flat rate).	☐
4. Left the work space clean and orderly.	☐
5. Cared for customer property and returned it undamaged.	☐

Student signature _______________________________ Date _______________________________

Comments:

Have your supervisor/instructor verify satisfactory completion of this procedure, any observations found, and any necessary action(s) recommended.

Tasksheet Scoring

		Test		Retest	
Evaluation Items	**Pass**	**Fail**	**Pass**	**Fail**
Task-Specific Evaluation	**(1 pt)**	**(0 pts)**	**(1 pt)**	**(0 pts)**
Obtain a wiring diagram for circuits. List important components that makeup the circuit. Describe cause and effect of shorts, opens, and circuits with excessive resistance.				
Perform tests to determine the source of an open circuit.				
Perform tests to determine the source of a short circuit.				
Perform tests to determine the source of excessive resistance.				
Non-Task-Specific Evaluation	**(0 pts)**	**(−1 pt)**	**(0 pts)**	**(−1 pt)**
Student successfully completed at least three of the non-task-specific steps.				
Student successfully completed all five of the non-task-specific steps.				
Total Score: <total # of points / 4 = %>				

Supervisor:

Supervisor/instructor signature _________________________ Date _______________

Comments:

Retest supervisor/instructor signature _________________________ Date _______________

Comments:

CDX Tasksheet Number: A6006

Student/Intern Information

Name ________________________________ Date ____________ Class ________________________

Vehicle, Customer, and Service Information

Vehicle used for this activity:

Year ______________ Make ______________________ Model ____________________________

Odometer __________________________ VIN ____

Materials Required
• Blank work order
• Vehicle with available service history records
• Depending on the type of concern, special diagnostic/hand tools may be required. See your supervisor/instructor for instructions to identify what tools may be required.
• Service information database
• Personal protective equipment
• Test light
• Back probe

Task-Specific Safety Considerations
- Comply with personal and environmental safety practices associated with clothing; eye protection; hand tools; power equipment; proper ventilation; and the handling, storage, and disposal of chemicals/materials in accordance with local, state, and federal safety and environmental regulations.

Time off________________

Time on________________

Total time________________

▶ **TASK** Use a test light to check operation of electrical circuits.

MLR
A6A6

Student Instructions: Read through the entire procedure prior to starting. Prepare your workspace and any tools or parts that may be needed to complete the task. When directed by your instructor, begin the procedure to complete the task and check the box as each step is finished. Track your time on this procedure for later comparison to the standard completion time (i.e., "flat rate" or customer pay time).

Procedure:	Step Completed
1. Obtain a wiring diagram for the headlights on your vehicle.	
a. What color is the wiring that feeds the low-beam headlight?	☐

2. Inspect for voltage at the headlight connector.	
a. Remove any objects obstructing access to the headlight connector.	☐
b. Back probe the power wire on the low-beam headlight connector.	☐
c. Install the test light clamping end to battery negative post. Apply the tip of the test light to the battery positive post. Did the test light illuminate? Explain why this is an important first step to any electrical diagnosis using a test light.	☐
d. With the test light clamping end still connected to the negative battery post, apply the test light tip to the back probe at the headlight connector. Turn the low-beam headlights on. Did the test light illuminate?	☐
e. Explain what your next steps in the diagnostic process would be if the test light does not illuminate.	☐
3. Explain how a test light could be misleading when diagnosing an electrical concern.	☐

Non-Task-Specific Evaluations:	Step Completed
1. Tools and equipment were used as directed and returned in good working order.	☐
2. Complied with all general and task-specific safety standards, including proper use of any personal protective equipment.	☐
3. Completed the task in an appropriate time frame (Recommendation: 1.5 or 2 times flat rate).	☐
4. Left the work space clean and orderly.	☐
5. Cared for customer property and returned it undamaged.	☐

Student signature _________________________________ Date _________________________________

Comments:

Have your supervisor/instructor verify satisfactory completion of this procedure, any observations found, and any necessary action(s) recommended.

Evaluation Instructions: The scoring box below is intended to act as a guide for both student and instructor. Each criterion listed will help students understand what is expected from them and evaluators articulate success at a particular task. The scoring is set up to allow a second attempt at each task (see the "Test" and "Retest" columns). Scoring is designed to reward students for correct completion of the task. Points are lost for failure to complete the employability requirements (see "Non-Task-Specific" criteria). When all criteria are evaluated, tally the points for a total at the bottom of each column.

Tasksheet Scoring

		Test		Retest	
Evaluation Items	**Pass**	**Fail**	**Pass**	**Fail**	
Task-Specific Evaluation	**(1 pt)**	**(0 pts)**	**(1 pt)**	**(0 pts)**	
Obtain a wiring diagram for the headlight circuit on your vehicle. Describe the wiring color.					
Back probe the headlight connector.					
Use a test light to inspect for voltage at the headlight connector.					
Explain how a test light can be misleading when testing.					
Non-Task-Specific Evaluation	**(0 pts)**	**(−1 pt)**	**(0 pts)**	**(−1 pt)**	
Student successfully completed at least three of the non-task-specific steps.					
Student successfully completed all five of the non-task-specific steps.					
Total Score: <total # of points / 4 = %>					

Supervisor:

Supervisor/instructor signature ___________________________________ Date ___________________

Comments:

Retest supervisor/instructor signature ___________________________________ Date ___________________

Comments:

CDX Tasksheet Number: A6007

Student/Intern Information

Name _________________________________ Date _____________ Class _______________________

Vehicle, Customer, and Service Information

Vehicle used for this activity:

Year _____________ Make _____________________________ Model _______________________

Odometer _________________________________ VIN _______________________________

Task-Specific Safety Considerations

- Comply with personal and environmental safety practices associated with clothing; eye protection; hand tools; power equipment; proper ventilation; and the handling, storage, and disposal of chemicals/materials in accordance with local, state, and federal safety and environmental regulations.

Time off___________

Time on___________

Total time___________

▶ **TASK** Use fused jumper wires to check operation of electrical circuits.

Note: Refer to your instructor in regard to which circuit on your vehicle requires testing.

MLR
A6A7

Student Instructions: Read through the entire procedure prior to starting. Prepare your workspace and any tools or parts that may be needed to complete the task. When directed by your instructor, begin the procedure to complete the task and check the box as each step is finished. Track your time on this procedure for later comparison to the standard completion time (i.e., "flat rate" or customer pay time).

Procedure:	Step Completed
1. Obtain a wiring diagram for the affected circuit on your vehicle.	
a. What is the amperage rating of the fuse used to protect this circuit?	☐

b. What color wire feeds power to load in the circuit?	☐
c. Explain when a fused jumper wire would be used during diagnosis of an electrical fault.	☐
d. What is the electrical concern on your vehicle? How do you plan to install the fused jumper wire?	☐

2. Install the fused jumper wire.	
a. Ensure that the fuse you are using in your jumper wire has the same rating as the fuse used in the circuit originally.	☐
b. Locate a point in the circuit to install the fused jumper wire. **Tip:** The less invasive you can be during this procedure, the better. Do not cut wiring open to install a fused jumper wire.	☐
c. Install the fused jumper wire and activate the circuit. Does the load in the circuit begin to work?	☐
d. If the load in the circuit begins to work, explain where the problem may be located.	☐
e. If the load in the circuit does *not* begin to work, explain where the problem may be located.	☐

Non-Task-Specific Evaluations:	**Step Completed**
1. Tools and equipment were used as directed and returned in good working order.	☐
2. Complied with all general and task-specific safety standards, including proper use of any personal protective equipment.	☐
3. Completed the task in an appropriate time frame (Recommendation: 1.5 or 2 times flat rate).	☐
4. Left the work space clean and orderly.	☐
5. Cared for customer property and returned it undamaged.	☐

Student signature _________________________________ Date _________________________________

Comments:

Have your supervisor/instructor verify satisfactory completion of this procedure, any observations found, and any necessary action(s) recommended.

Taksheet Scoring

Evaluation Items	Test		Retest	
	Pass	**Fail**	**Pass**	**Fail**
Task-Specific Evaluation	**(1 pt)**	**(O pts)**	**(1 pt)**	**(O pts)**
Obtain a wiring diagram for the affected circuit.				
Describe the use of a fused jumper wire.				
Install a fused jumper wire and activate the circuit.				
Determine if further action is required during the diagnostic process.				
Non-Task-Specific Evaluation	**(O pts)**	**(−1 pt)**	**(O pts)**	**(−1 pt)**
Student successfully completed at least three of the non-task-specific steps.				
Student successfully completed all five of the non-task-specific steps.				
Total Score: <total # of points / 4 = %>				

Supervisor:

Supervisor/instructor signature ________________________________ Date ________________

Comments:

Retest supervisor/instructor signature ________________________ Date ________________

Comments:

CDX Tasksheet Number: A6008

Student/Intern Information

Name ________________________________ Date ____________ Class ________________________

Vehicle, Customer, and Service Information

Vehicle used for this activity:

Year ______________ Make ____________________________ Model ________________________

Odometer ________________________________ VIN ________________________________

Materials Required

- Blank work order
- Vehicle with available service history records
- Depending on the type of concern, special diagnostic/hand tools may be required. See your supervisor/instructor for instructions to identify what tools may be required.
- Service information database
- Personal protective equipment
- Digital multimeter
- Parasitic drain battery disconnect tool

Task-Specific Safety Considerations

- Comply with personal and environmental safety practices associated with clothing; eye protection; hand tools; power equipment; proper ventilation; and the handling, storage, and disposal of chemicals/materials in accordance with local, state, and federal safety and environmental regulations.

▶ TASK Measure key-off battery drain (parasitic draw).

MLR
A6A8

Time off____________

Time on____________

Total time____________

Student Instructions: Read through the entire procedure prior to starting. Prepare your workspace and any tools or parts that may be needed to complete the task. When directed by your instructor, begin the procedure to complete the task and check the box as each step is finished. Track your time on this procedure for later comparison to the standard completion time (i.e., "flat rate" or customer pay time).

Procedure:	Step Completed
1. Research parasitic drain.	
a. Explain what causes parasitic drain.	☐

b. Explain how can you pinpoint the source of a parasitic drain. What tools and procedures would you use? Are there any special precautions when performing this task?	☐
c. What is an acceptable amount of key-off battery drain for your vehicle?	☐
2. Measure parasitic drain.	
a. Open the hood. Unplug the underhood light if one is equipped. If needed, lower the driver window to gain access to the ignition switch. Ensure that all lights are off and doors are closed before beginning this procedure.	☐
b. Using a battery disconnect tool or similar method, install a multimeter in series with the negative battery cable.	☐
c. Explain what setting your meter is on and where the meter leads are located on the meter itself.	☐
d. What is your initial battery drain reading?	☐
e. Wait 10 minutes and check the reading once more. Did the reading change? If so, why? **Tip:** Be sure not to open any doors or activate any circuits as this will skew your drain readings.	☐
3. Explain the next steps in this procedure to diagnose a higher-than-normal drain reading. How would you continue testing to pinpoint the source of the parasitic drain?	☐

Non-Task-Specific Evaluations:	Step Completed
1. Tools and equipment were used as directed and returned in good working order.	☐
2. Complied with all general and task-specific safety standards, including proper use of any personal protective equipment.	☐
3. Completed the task in an appropriate time frame (Recommendation: 1.5 or 2 times flat rate).	☐
4. Left the work space clean and orderly.	☐
5. Cared for customer property and returned it undamaged.	☐

Student signature _________________________ Date _________________________

Comments:

Have your supervisor/instructor verify satisfactory completion of this procedure, any observations found, and any necessary action(s) recommended.

Tasksheet Scoring

	Test		Retest	
Evaluation Items	**Pass**	**Fail**	**Pass**	**Fail**
Task-Specific Evaluation	**(1 pt)**	**(0 pts)**	**(1 pt)**	**(0 pts)**
Research the procedure for performing a parasitic drain test and the acceptable values. Explain the steps to pinpoint the source of battery drain.				
Explain the causes of parasitic drain and the reasons that drain values may change over time.				
Use the correct procedure to measure the initial drain values.				
Measure the parasitic drain after a 10-minute wait period.				
Non-Task-Specific Evaluation	**(0 pts)**	**(−1 pt)**	**(0 pts)**	**(−1 pt)**
Student successfully completed at least three of the non-task-specific steps.				
Student successfully completed all five of the non-task-specific steps.				
Total Score: <total # of points / 4 = %>				

Supervisor:

Supervisor/instructor signature _________________________ Date _________________

Comments:

Retest supervisor/instructor signature _________________________ Date _________________

Comments:

CDX Tasksheet Number: A6009

Student/Intern Information

Name _______________________________ Date _____________ Class _______________________

Vehicle, Customer, and Service Information

Vehicle used for this activity:

Year _______________ Make _____________________________ Model _______________________

Odometer _______________________________ VIN _______________________________

Materials Required

- Blank work order
- Vehicle with available service history records
- Depending on the type of concern, special diagnostic/hand tools may be required. See your supervisor/instructor for instructions to identify what tools may be required.
- Service information database
- Personal protective equipment
- Digital multimeter (DMM)
- Test light

Task-Specific Safety Considerations

- Comply with personal and environmental safety practices associated with clothing; eye protection; hand tools; power equipment; proper ventilation; and the handling, storage, and disposal of chemicals/materials in accordance with local, state, and federal safety and environmental regulations.

Time off_______________

Time on_______________

Total time_______________

▶ TASK Inspect and test fusible links, circuit breakers, and fuses, and determine necessary action.

MLR
A6A9

Student Instructions: Read through the entire procedure prior to starting. Prepare your workspace and any tools or parts that may be needed to complete the task. When directed by your instructor, begin the procedure to complete the task and check the box as each step is finished. Track your time on this procedure for later comparison to the standard completion time (i.e., "flat rate" or customer pay time).

Procedure:	Step Completed
1. Inspect and test the underhood fuses.	
a. Open the hood and locate the fuse box cover.	☐
b. Remove the cover and locate the fuses. List the fuses and amperage ratings contained in the fuse box.	☐

© 2020 Jones & Bartlett Learning, LLC, an Ascend Learning Company

c. Install the clamping end of the test light to the negative battery terminal. Touch the tip of the test light to the positive battery post. Does the test light illuminate? Explain why this is an important first step when using this tool.	☐
d. Use the test light and test the fuses. List the fuses that do not have power at this time. Are these fuses supposed to have power or does the circuit has to be turned on? Explain how you can figure out this information.	☐
e. Are there any fuses that are supposed to have power but do not? Remove any fuses in question for further inspection.	☐
f. Inspect the fuse legs for damage and correct pin fit. Ensure that fuses fit tightly in place. Is the center portion of the fuse blown?	☐
g. Another method of inspecting fuses is to use a multimeter. Use the meter set to direct current (DC) voltage and install the negative meter lead to the negative battery post. Use the positive test lead to check for 12 volts across the fuse in question. A fuse in good working condition should read battery voltage across the fuse. You may also remove a fuse and check the resistance across the fuse. Resistance should be minimal.	☐
h. **Optional:** Inspect all fuses using a multimeter set to DC voltage.	☐
2. Inspect fusible links.	
a. Explain where the fusible links are located on your vehicle. **Tip:** A wiring diagram can be helpful in determining fusible link location.	☐
b. Begin by locating the fusible links on your vehicle. Perform your first check by gently pulling on each end of the link. A melted link will stretch since the copper wire core is melted.	☐
c. Use a multimeter or test light to check for available voltage at both ends of the link.	☐

d. A third way of testing a fusible link would be to use a multimeter and check the resistance of the link. If a link is melted, what will the meter display?	☐
3. Inspect the circuit breakers.	
a. Explain how a circuit breaker is different than a fuse or fusible link. Explain how a circuit breaker works.	☐
b. Which electrical circuits use a circuit breaker on your vehicle?	☐
c. Remove a circuit breaker from your vehicle.	☐
d. With the circuit breaker out of the vehicle, use a multimeter to check the resistance across the terminals. If the circuit breaker is cooled off, there should be continuity between the two terminals.	☐

Non-Task-Specific Evaluations:	Step Completed
1. Tools and equipment were used as directed and returned in good working order.	☐
2. Complied with all general and task-specific safety standards, including proper use of any personal protective equipment.	☐
3. Completed the task in an appropriate time frame (Recommendation: 1.5 or 2 times flat rate).	☐
4. Left the work space clean and orderly.	☐
5. Cared for customer property and returned it undamaged.	☐

Student signature _________________________ Date _________________________

Comments:

Have your supervisor/instructor verify satisfactory completion of this procedure, any observations found, and any necessary action(s) recommended.

Tasksheet Scoring

Evaluation Items	Test		Retest	
	Pass	Fail	Pass	Fail
Task-Specific Evaluation	**(1 pt)**	**(0 pts)**	**(1 pt)**	**(0 pts)**
Test the functionality of the test light prior to using the tool for diagnostic purposes.				
Use a test light to test the fuses. Use a wiring diagram to decipher the fuses that supply power at all times and the fuses that require a circuit to be activated.				
Use a DMM to check the voltage across the fuse.				
Locate and inspect the fusible links.				
Non-Task-Specific Evaluation	**(0 pts)**	**(−1 pt)**	**(0 pts)**	**(−1 pt)**
Student successfully completed at least three of the non-task-specific steps.				
Student successfully completed all five of the non-task-specific steps.				
Total Score: <total # of points / 4 = %>				

Supervisor:

Supervisor/instructor signature ___________________________________ Date _______________________

Comments:

Retest supervisor/instructor signature ___________________________________ Date _______________________

Comments:

CDX Tasksheet Number: A6010

Student/Intern Information

Name ________________________________ Date ____________ Class ________________________

Vehicle, Customer, and Service Information

Vehicle used for this activity:

Year ________________ Make ________________________ Model ________________________

Odometer ________________________ VIN ________________________

Materials Required

- Blank work order
- Vehicle with available service history records
- Depending on the type of concern, special diagnostic/hand tools may be required. See your supervisor/instructor for instructions to identify what tools may be required.
- Service information database
- Personal protective equipment
- Wire crimp/strip tools
- Crimp connectors and shrink wrap
- Soldering iron and solder
- Wire harness
- Terminal repair kit
- Heat gun

Task-Specific Safety Considerations

- Comply with personal and environmental safety practices associated with clothing; eye protection; hand tools; power equipment; proper ventilation; and the handling, storage, and disposal of chemicals/materials in accordance with local, state, and federal safety and environmental regulations.
- *Soldering Tools* produce heat that can cause bodily harm if used incorrectly.

Time off________

Time on________

Total time________

▶ TASK Repair and/or replace connectors, terminal ends, and wiring of electrical/electronic systems (including solder repair).

MLR
A6A10

NOTE: The preferred method of repairing wiring is to use solder, shrink wrap, and electrical tape. A section of this task will focus on an acceptable method of using crimp connectors.

Student Instructions: Read through the entire procedure prior to starting. Prepare your workspace and any tools or parts that may be needed to complete the task. When directed by your instructor, begin the procedure to complete the task and check the box as each step is finished. Track your time on this procedure for later comparison to the standard completion time (i.e., "flat rate" or customer pay time).

Procedure:	Step Completed
1. Research acceptable wiring repair practices.	
a. What is the preferred method of making wiring repair?	☐
b. Explain what types of wiring harness should be replaced rather than repaired when damaged.	☐
2. Perform a wiring repair using crimp connectors.	
a. Cut a 1-foot-long section of wire from the bulk wire harness using the wire cutters.	☐
b. Cut the 1-foot section of wire in half.	☐
c. Use wire strippers to remove approximately $\frac{1}{4}''$ of insulation from one side of each wire. What gauge wiring are you using?	☐
d. Select the proper gauge crimp connector and install each stripped section of wire into the crimp connector. **Tip:** If possible, use a crimp connector that has the ability to be heated and shrunk to make a secure, water proof connection.	☐
e. Using the crimping tool to crimp the wires together. Lightly pull on the wiring to ensure that the connection is tight.	☐
f. Use a heat gun to shrink the crimp connector until it fits securely around the wire. It is also acceptable to use a piece of shrink wrap over top of the crimp connector to eliminate moisture intrusion.	☐
3. Perform a wiring repair using solder.	
a. Using the same section of wire from the previous task, use wire strippers to remove approximately $\frac{1}{2}''$ of insulation from both ends of the wire.	☐
b. Slide a piece of shrink tubing over the length of wire. This will be heated in later steps.	☐
c. Connect the ends of the wire together by twisting the exposed copper portion of the wire together. This should form a circle of wire at this point. What type of solder is needed for making this repair?	☐

d. Use a soldering iron to heat the exposed section of copper wire. Once the wire is warm enough, apply solder to the wire. Solder should flow through the copper wire making a secure connection. **Tip:** It may be necessary to clean and tin the soldering iron before use.	☐
e. Slide the section of shrink wrap tubing over the freshly soldered section of wire. Use a heat gun to shrink the tubing making a secure, moisture-proof connection.	☐
4. Repair the connector's terminals.	
a. Select a connector from the bulk wiring harness and remove it. Leave approximately 6″ of wiring on the connector when it is cut off.	☐
b. Determine the type of terminal locking mechanism that locks the terminals in place inside the connector. Select the correct terminal tool to release this locking mechanism.	☐
c. With the locking mechanism removed, use the appropriate terminal tool to release the tab that locks the terminal in place. At the same time, gently pull on the wire that you are working with and remove it from the connector.	☐
d. At this time, select a new terminal that is identical to the one that you just removed.	☐
e. Use wire cutters to remove the terminal from the section of wire. Strip $\frac{1}{4}$″ of insulation from the wire and twist the newly exposed copper section of wire.	☐
f. Crimp the new terminal onto the exposed section of copper wiring. **Tip:** Connectors will use some type of insulation near the terminal where it is inserted into the connector housing. Be sure to install this insulation before crimping a new terminal onto the wire.	☐
g. Once the new terminal and insulation is in place, insert the terminal back into the connector housing. You should hear/feel the terminal lock into place. Gently pull on the wire to ensure that it is secure within the connector.	☐
h. Reinstall the terminal locking mechanism.	☐

Non-Task-Specific Evaluations:	**Step Completed**
1. Tools and equipment were used as directed and returned in good working order.	☐
2. Complied with all general and task-specific safety standards, including proper use of any personal protective equipment.	☐
3. Completed the task in an appropriate time frame (Recommendation: 1.5 or 2 times flat rate).	☐
4. Left the work space clean and orderly.	☐
5. Cared for customer property and returned it undamaged.	☐

Student signature _______________________________ Date _______________________________

Comments:

Have your supervisor/instructor verify satisfactory completion of this procedure, any observations found, and any necessary action(s) recommended.

Evaluation Instructions: The scoring box below is intended to act as a guide for both student and instructor. Each criterion listed will help students understand what is expected from them and evaluators articulate success at a particular task. The scoring is set up to allow a second attempt at each task (see the "Test" and "Retest" columns). Scoring is designed to reward students for correct completion of the task. Points are lost for failure to complete the employability requirements (see "Non-Task-Specific" criteria). When all criteria are evaluated, tally the points for a total at the bottom of each column.

Taksheet Scoring

	Test		Retest	
Evaluation Items	Pass	Fail	Pass	Fail
Task-Specific Evaluation	**(1 pt)**	**(O pts)**	**(1 pt)**	**(O pts)**
Research acceptable wiring repair practices. Explain the types of wiring that should not be repaired.				
Use crimp connectors to repair wiring.				
Use solder to repair wiring.				
Install new terminals inside a connector.				
Non-Task-Specific Evaluation	**(O pts)**	**(–1 pt)**	**(O pts)**	**(–1 pt)**
Student successfully completed at least three of the non-task-specific steps.				
Student successfully completed all five of the non-task-specific steps.				
Total Score: <total # of points / 4 = %>				

Supervisor:

Supervisor/instructor signature _________________________________ Date _________________

Comments:

Retest supervisor/instructor signature _________________________________ Date _________________

Comments:

CDX Tasksheet Number: A6011

Student/Intern Information

Name ________________________________ Date ____________ Class ________________________

Vehicle, Customer, and Service Information

Vehicle used for this activity:

Year ______________ Make ____________________________ Model ____________________________

Odometer ________________________________ VIN ________________________________

Materials Required

- Blank work order
- Vehicle with available service history records
- Depending on the type of concern, special diagnostic/hand tools may be required. See your supervisor/instructor for instructions to identify what tools may be required.
- Service information database
- Personal protective equipment
- General hand tools
- Scan tool

Task-Specific Safety Considerations

- Comply with personal and environmental safety practices associated with clothing; eye protection; hand tools; power equipment; proper ventilation; and the handling, storage, and disposal of chemicals/materials in accordance with local, state, and federal safety and environmental regulations.

▶ TASK Identify electrical/electronic system components and configuration. **MLR A6A11**

Time off________________

Time on________________

Student Instructions: Read through the entire procedure prior to starting. Prepare your workspace and any tools or parts that may be needed to complete the task. When directed by your instructor, begin the procedure to complete the task and check the box as each step is finished. Track your time on this procedure for later comparison to the standard completion time (i.e., "flat rate" or customer pay time).

Total time________________

Procedure:	Step Completed
1. Research network communications on your vehicle.	
a. Using service information, research and describe the type of communication networking used in your vehicle.	☐

b. What classification of network is used for systems, such as powertrain and vehicle dynamic control?	☐
c. Is the CAN protocol used on your vehicle?	☐
d. List the modules that are used on your vehicle and describe what each module controls.	☐
e. Describe the configuration of the modules. How are the modules linked together and how do they communicate? If one module malfunctions, how does it affect the other modules in the vehicle?	☐
2. Describe the location of the data link connector (DLC).	
a. What is the purpose of pin 4 of the DLC?	☐
b. What is the purpose of pin 5 of the DLC?	☐
c. What is the purpose of pin 16 of the DLC?	☐
d. What pin is used for CAN H (high)?	☐
e. What pin is used for CAN L (low)?	☐

f. What resistance value should be present between pins 6 and 14?	☐
3. Locate the engine control module (ECM)/powertrain control module (PCM).	
a. Using service information, determine the location of the ECM/PCM.	☐
b. Locate the module on the vehicle and describe what steps were needed to locate the module. Did any panels need to be removed?	☐
c. Describe the location of the ECM/PCM.	☐

Non-Task-Specific Evaluations:	Step Completed
1. Tools and equipment were used as directed and returned in good working order.	☐
2. Complied with all general and task-specific safety standards, including proper use of any personal protective equipment.	☐
3. Completed the task in an appropriate time frame (Recommendation: 1.5 or 2 times flat rate).	☐
4. Left the work space clean and orderly.	☐
5. Cared for customer property and returned it undamaged.	☐

Student signature _______________________________ Date _______________________________

Comments:

Have your supervisor/instructor verify satisfactory completion of this procedure, any observations found, and any necessary action(s) recommended.

Evaluation Instructions: The scoring box below is intended to act as a guide for both student and instructor. Each criterion listed will help students understand what is expected from them and evaluators articulate success at a particular task. The scoring is set up to allow a second attempt at each task (see the "Test" and "Retest" columns). Scoring is designed to reward students for correct completion of the task. Points are lost for failure to complete the employability requirements (see "Non-Task-Specific" criteria). When all criteria are evaluated, tally the points for a total at the bottom of each column.

Tasksheet Scoring

	Test		Retest	
Evaluation Items	**Pass**	**Fail**	**Pass**	**Fail**
Task-Specific Evaluation	**(1 pt)**	**(0 pts)**	**(1 pt)**	**(0 pts)**
Research and describe the type of communication networking used in your vehicle.				
List the modules that are used on your vehicle and describe what each module controls. Describe the configuration of the modules on your vehicle.				
Describe the pin layout of the DLC.				
Locate the ECM/PCM.				
Non-Task-Specific Evaluation	**(0 pts)**	**(−1 pt)**	**(0 pts)**	**(−1 pt)**
Student successfully completed at least three of the non-task-specific steps.				
Student successfully completed all five of the non-task-specific steps.				
Total Score: <total # of points / 4 = %>				

Supervisor:

Supervisor/instructor signature ______________________ Date _____________

Comments:

Retest supervisor/instructor signature _________________ Date ____________

Comments:

6B. Battery Service

Learning Objective/Task	CDX Tasksheet Number	ASE Education Foundation Reference Number; Priority Level
• Perform battery state-of-charge test, and determine necessary action.	A6012	A6B1, P-1
• Confirm proper battery capacity for vehicle application; perform battery capacity and load test, and determine necessary action.	A6013	A6B2, P-1
• Maintain or restore electronic memory functions.	A6014	A6B3, P-1
• Inspect and clean battery; fill battery cells; and check battery cables, connectors, clamps, and hold-downs.	A6015	A6B4, P-1
• Perform slow/fast battery charge according to manufacturer's recommendations.	A6016	A6B5, P-1
• Jump-start vehicle using jumper cables and a booster battery or an auxiliary power supply.	A6017	A6B6, P-1
• Identify safety precautions for high voltage systems on electric, hybrid-electric, and diesel vehicles.	A6018	A6B7, P-2
• Identify electrical/electronic modules, security systems, radios, and other accessories that require reinitialization or code entry after reconnecting vehicle battery.	A6019	A6B8, P-1
• Identify hybrid vehicle auxiliary (12V) battery service, repair, and test procedures.	A6020	A6B9, P-2

Materials Required

- Blank work order
- Vehicle with available service history records
- Depending on the type of concern, special diagnostic/hand tools may be required. See your supervisor/instructor for instructions to identify what tools may be required.
- Service information database
- Personal protective equipment

Safety Considerations

- When running any vehicles in the shop, make sure you use the shop's exhaust ventilation system to discharge all exhaust gas safely outside.
- Lifting equipment, such as vehicle jacks and stands, vehicle hoists, and engine hoists, are important tools that increase productivity and make the job easier. However, they can also cause severe injury or death if used improperly. Make sure you follow the manufacturer's operation procedures. Also, make sure you have your supervisor's/instructor's permission to use any particular type of lifting equipment.
- Comply with personal and environmental safety practices associated with clothing; eye protection; hand tools; power equipment; proper ventilation; and the handling, storage, and disposal of chemicals/materials in accordance with local, state, and federal safety and environmental regulations.
- If you need to start the vehicle, you should ensure that the parking brake is firmly applied; if necessary, use wheel chocks to prevent the vehicle from moving when the vehicle is started to verify the completion of these tasks.

CDX Tasksheet Number: A6012

Student/Intern Information

Name _________________________________ Date ____________ Class _____________________

Vehicle, Customer, and Service Information

Vehicle used for this activity:

Year ________________ Make ____________________________ Model __________________________

Odometer _______________________________ VIN _______________________________________

Materials Required

- Blank work order
- Vehicle with available service history records
- Depending on the type of concern, special diagnostic/hand tools may be required. See your supervisor/instructor for instructions to identify what tools may be required.
- Service information database
- Personal protective equipment
- Refractometer
- Infrared temperature tool
- Digital multimeter

Task-Specific Safety Considerations

- Comply with personal and environmental safety practices associated with clothing; eye protection; hand tools; power equipment; proper ventilation; and the handling, storage, and disposal of chemicals/materials in accordance with local, state, and federal safety and environmental regulations.
- **Caution:** Keep open flames away from batteries at all times.

▶ TASK Perform battery state-of-charge test, and determine necessary action.

MLR
A6B1

Student Instructions: Read through the entire procedure prior to starting. Prepare your workspace and any tools or parts that may be needed to complete the task. When directed by your instructor, begin the procedure to complete the task and check the box as each step is finished. Track your time on this procedure for later comparison to the standard completion time (i.e., "flat rate" or customer pay time).

Procedure:	Step Completed
1. Perform a battery state-of-charge test.	
a. Does the battery have a built-in hydrometer? If so, describe what you see in the sight glass. What color is the glass and what does this say about the battery's state-of-charge?	☐
b. Remove the battery caps and check the fluid level. What would you use to top off the battery?	☐
c. Use the infrared temperature tool and measure the temperature of the battery case. Record your findings. Describe how temperature can affect a battery's state-of-charge.	☐
d. Measure the specific gravity of each battery cell and record your findings.	☐
e. Use a multimeter set to direct current volts and measure the open circuit voltage of the battery. Record your findings.	☐
2. On the basis of your inspections, determine the state-of-charge of the battery. Are you able to figure a percentage of charge level based on your inspection?	☐

Non-Task-Specific Evaluations:	Step Completed
1. Tools and equipment were used as directed and returned in good working order.	☐
2. Complied with all general and task-specific safety standards, including proper use of any personal protective equipment.	☐
3. Completed the task in an appropriate time frame (Recommendation: 1.5 or 2 times flat rate).	☐
4. Left the work space clean and orderly.	☐
5. Cared for customer property and returned it undamaged.	☐

Student signature _________________________________ Date _________________

Comments:

Have your supervisor/instructor verify satisfactory completion of this procedure, any observations found, and any necessary action(s) recommended.

Evaluation Instructions: The scoring box below is intended to act as a guide for both student and instructor. Each criterion listed will help students understand what is expected from them and evaluators articulate success at a particular task. The scoring is set up to allow a second attempt at each task (see the "Test" and "Retest" columns). Scoring is designed to reward students for correct completion of the task. Points are lost for failure to complete the employability requirements (see "Non-Task-Specific" criteria). When all criteria are evaluated, tally the points for a total at the bottom of each column.

Tasksheet Scoring

Evaluation Items	Test		Retest	
	Pass	Fail	Pass	Fail
Task-Specific Evaluation	**(1 pt)**	**(0 pts)**	**(1 pt)**	**(0 pts)**
Inspect the electrolyte solution level.				
Measure the battery case's temperature and explain how temperature corresponds with state-of-charge.				
Measure the specific gravity on each battery cell. Measure the open circuit voltage of the battery.				
Make recommendations for service based on your inspection.				
Non-Task-Specific Evaluation	**(0 pts)**	**(−1 pt)**	**(0 pts)**	**(−1 pt)**
Student successfully completed at least three of the non-task-specific steps.				
Student successfully completed all five of the non-task-specific steps.				
Total Score: <total # of points / 4 = %>				

Supervisor:

Supervisor/instructor signature _______________________________ Date _______________

Comments:

Retest supervisor/instructor signature _______________________________ Date _______________

Comments:

CDX Tasksheet Number: A6013

Student/Intern Information

Name _________________________________ Date ____________ Class _____________________

Vehicle, Customer, and Service Information

Vehicle used for this activity:

Year _______________ Make ____________________________ Model _____________________

Odometer _______________________________ VIN _________________________________

Materials Required

- Blank work order
- Vehicle with available service history records
- Depending on the type of concern, special diagnostic/hand tools may be required. See your supervisor/instructor for instructions to identify what tools may be required.
- Service information database
- Personal protective equipment
- Capacity tester
- Conductance tester

Task-Specific Safety Considerations

- Comply with personal and environmental safety practices associated with clothing; eye protection; hand tools; power equipment; proper ventilation; and the handling, storage, and disposal of chemicals/materials in accordance with local, state, and federal safety and environmental regulations.

▶ **TASK** Confirm proper battery capacity for vehicle application; perform battery capacity and load test, and determine necessary action.

MLR
A6B2

Student Instructions: Read through the entire procedure prior to starting. Prepare your workspace and any tools or parts that may be needed to complete the task. When directed by your instructor, begin the procedure to complete the task and check the box as each step is finished. Track your time on this procedure for later comparison to the standard completion time (i.e., "flat rate" or customer pay time).

Time off________________

Time on________________

Total time________________

Procedure:	Step Completed
1. Research battery capacity specifications.	
a. Using service information, determine the battery capacity required for your vehicle. List the following: cold-cranking amps (CCAs), CAs, and reserve capacity.	☐

b. Compare the battery that is currently in the vehicle to the specifications found in Step 1a.	☐
c. Explain the difference between performing battery tests with a load tester versus a conductance tester.	☐
2. Load test the battery.	
a. Determine the CCA of your battery. Install the positive and negative cables to the battery.	☐
b. Install the inductive clamp over the positive battery cable.	☐
c. Apply a load (half the value of the CCA rating) to the battery and observe the voltmeter on the machine.	☐
d. List the voltage reading. What is the minimum voltage that should be seen on the display?	☐
e. Wait 30 seconds and repeat the load test one more time. Record the voltage readings.	☐
f. On the basis of your inspection, did the battery pass the load test?	☐
3. Perform a conductance test on the battery.	
a. Install the conductance tester leads to the appropriate battery terminals.	☐
b. Follow the prompts on the tester unit and select the appropriate CCA/CA ratings and temperature readings.	☐
c. The tester unit may prompt you to remove the surface charge. If so, turn the headlights on for 10 seconds and retry the test.	☐
d. Test the battery. What are the results of the test according to the conductance tester unit?	☐

Non-Task-Specific Evaluations:	Step Completed
1. Tools and equipment were used as directed and returned in good working order.	☐
2. Complied with all general and task-specific safety standards, including proper use of any personal protective equipment.	☐
3. Completed the task in an appropriate time frame (Recommendation: 1.5 or 2 times flat rate).	☐
4. Left the work space clean and orderly.	☐
5. Cared for customer property and returned it undamaged.	☐

Student signature _______________________________ Date _________________________________

Comments:

Have your supervisor/instructor verify satisfactory completion of this procedure, any observations found, and any necessary action(s) recommended.

Tasksheet Scoring

	Test		Retest	
Evaluation Items	**Pass**	**Fail**	**Pass**	**Fail**
Task-Specific Evaluation	**(1 pt)**	**(0 pts)**	**(1 pt)**	**(0 pts)**
Research battery specifications for your vehicle. Explain the different types of battery testing procedures and how they measure different aspects of battery performance.				
Install the load tester and test using the correct specifications.				
Determine battery performance based on the voltage displayed during load testing.				
Use a conductance tester to measure battery performance.				
Non-Task-Specific Evaluation	**(0 pts)**	**(−1 pt)**	**(0 pts)**	**(−1 pt)**
Student successfully completed at least three of the non-task-specific steps.				
Student successfully completed all five of the non-task-specific steps.				
Total Score: <total # of points / 4 = %>				

Supervisor:

Supervisor/instructor signature _________________________________ Date _______________________

Comments:

Retest supervisor/instructor signature _________________________________ Date _______________________

Comments:

CDX Tasksheet Number: A6014

Student/Intern Information

Name _________________________________ Date _____________ Class _________________________

Vehicle, Customer, and Service Information

Vehicle used for this activity:

Year _______________ Make _________________________ Model _________________________

Odometer _________________________ VIN _________________________________

> **Materials Required**
> - Blank work order
> - Vehicle with available service history records
> - Depending on the type of concern, special diagnostic/hand tools may be required. See your supervisor/instructor for instructions to identify what tools may be required.
> - Service information database
> - Personal protective equipment
> - Memory saver
> - Auxiliary power source
> - General hand tools

Task-Specific Safety Considerations
- Comply with personal and environmental safety practices associated with clothing; eye protection; hand tools; power equipment; proper ventilation; and the handling, storage, and disposal of chemicals/materials in accordance with local, state, and federal safety and environmental regulations.

▶ TASK Maintain or restore electronic memory functions.

MLR
A6B3

Student Instructions: Read through the entire procedure prior to starting. Prepare your workspace and any tools or parts that may be needed to complete the task. When directed by your instructor, begin the procedure to complete the task and check the box as each step is finished. Track your time on this procedure for later comparison to the standard completion time (i.e., "flat rate" or customer pay time).

Time off_____________

Time on_____________

Total time_____________

Procedure:	Step Completed
1. Maintain electronic memory functions.	
a. Explain the importance of saving electronic memory functions in a vehicle when disconnecting the battery. Other than saving clock and radio presets, why is this procedure important?	☐

b. Take note of the time display and radio presets. After this procedure is finished, you will check to make sure that radio presets and time were saved correctly.	☐
c. Install the memory saver into the data link connector or cigar lighter.	☐
d. Connect the memory saver cable to an auxiliary power source.	
e. Disconnect the negative battery cable and wait 1 minute. Reconnect the negative battery cable.	☐
f. Remove the auxiliary power source and memory saver device.	☐
g. Turn the ignition to the ON position and verify that the clock and radio settings were maintained during the battery disconnect procedure.	☐
2. Use service information to determine what other systems on the vehicle would require initialization after the battery has been disconnected. List those systems and explain how the initialization procedure is performed.	☐

Non-Task-Specific Evaluations:	Step Completed
1. Tools and equipment were used as directed and returned in good working order.	☐
2. Complied with all general and task-specific safety standards, including proper use of any personal protective equipment.	☐
3. Completed the task in an appropriate time frame (Recommendation: 1.5 or 2 times flat rate).	☐
4. Left the work space clean and orderly.	☐
5. Cared for customer property and returned it undamaged.	☐

Student signature _______________________________ Date _______________________________

Comments:

Have your supervisor/instructor verify satisfactory completion of this procedure, any observations found, and any necessary action(s) recommended.

Evaluation Instructions: The scoring box below is intended to act as a guide for both student and instructor. Each criterion listed will help students understand what is expected from them and evaluators articulate success at a particular task. The scoring is set up to allow a second attempt at each task (see the "Test" and "Retest" columns). Scoring is designed to reward students for correct completion of the task. Points are lost for failure to complete the employability requirements (see "Non-Task-Specific" criteria). When all criteria are evaluated, tally the points for a total at the bottom of each column.

Tasksheet Scoring

	Test		Retest	
Evaluation Items	**Pass**	**Fail**	**Pass**	**Fail**
Task-Specific Evaluation	**(1 pt)**	**(0 pts)**	**(1 pt)**	**(0 pts)**
Explain the importance of the saving electronic memory function.				
Compare the clock and radio presets before and after the battery disconnect procedure to confirm that the memory saver function was successful.				
Install the memory saver. Disconnect the battery and wait 1 minute. Reconnect the battery.				
Use service information to determine what systems would require initialization after the battery has been disconnected.				
Non-Task-Specific Evaluation	**(0 pts)**	**(−1 pt)**	**(0 pts)**	**(−1 pt)**
Student successfully completed at least three of the non-task-specific steps.				
Student successfully completed all five of the non-task-specific steps.				
Total Score: <total # of points / 4 = %>				

Supervisor:

Supervisor/instructor signature _________________________________ Date _______________________

Comments:

Retest supervisor/instructor signature _________________________________ Date _______________________

Comments:

CDX Tasksheet Number: A6015

Student/Intern Information

Name _________________________________ Date ____________ Class _____________________

Vehicle, Customer, and Service Information

Vehicle used for this activity:

Year ________________ Make ______________________________ Model ______________________

Odometer _________________________________ VIN ____

Materials Required
- Blank work order
- Vehicle with available service history records
- Depending on the type of concern, special diagnostic/hand tools may be required. See your supervisor/instructor for instructions to identify what tools may be required.
- Service information database
- Personal protective equipment
- Battery terminal brush
- Distilled water
- Baking soda, battery terminal cleaner, and battery terminal protector

Task-Specific Safety Considerations
- Comply with personal and environmental safety practices associated with clothing; eye protection; hand tools; power equipment; proper ventilation; and the handling, storage, and disposal of chemicals/materials in accordance with local, state, and federal safety and environmental regulations.
- **Caution:** Keep open flames away from the battery at all times.

▶ **TASK** Inspect and clean battery; fill battery cells; and check battery cables, connectors, clamps, and hold-downs.

MLR
A6B4

Time off_______________

Time on_______________

Student Instructions: Read through the entire procedure prior to starting. Prepare your workspace and any tools or parts that may be needed to complete the task. When directed by your instructor, begin the procedure to complete the task and check the box as each step is finished. Track your time on this procedure for later comparison to the standard completion time (i.e., "flat rate" or customer pay time).

Total time____________

Procedure:	Step Completed
1. Inspect and clean the battery, cables, and hold-downs.	
a. Describe the general appearance of the battery terminals and cables.	☐

b. Check that the cables are tight around the battery terminals by twisting them back and forth.	☐
c. Is a battery hold-down present? Is it holding the battery in place securely?	☐
d. Before removing battery caps, use baking soda or a battery cleaner solution to remove built-up acid or dirt.	☐
e. Remove the battery caps and inspect the level of electrolyte solution. If the battery needs to be topped off, explain what you would use. Top off as necessary.	☐
f. If the battery terminals and cables require cleaning, refer to instructions from tasksheet A6014 and install a memory saver before removing the cables.	☐
g. Remove the battery cables. Explain the order in which you should remove the cables. Why is this order important?	☐
h. Remove the battery hold-down and remove the battery and place it on the work bench.	☐
i. Use the battery terminal brush to clean the battery cables. All corrosion must be removed. Next, clean the terminals of the battery.	☐
j. Inspect the battery tray and describe its condition.	☐
k. Inspect the battery case for leaks, cracks, or the appearance of a swollen case.	☐
l. Once the battery and cables are inspected and cleaned, install the battery, battery hold-down, and cables. A battery protector spray may be applied to the terminals at this time. What order should the battery cables be installed?	☐
m. Remove the memory saver and verify that the memory functions have been maintained and the vehicle starts.	☐

Non-Task-Specific Evaluations:	**Step Completed**
1. Tools and equipment were used as directed and returned in good working order.	☐
2. Complied with all general and task-specific safety standards, including proper use of any personal protective equipment.	☐
3. Completed the task in an appropriate time frame (Recommendation: 1.5 or 2 times flat rate).	☐
4. Left the work space clean and orderly.	☐
5. Cared for customer property and returned it undamaged.	☐

Student signature _______________________________ Date _______________________________

Comments:

Have your supervisor/instructor verify satisfactory completion of this procedure, any observations found, and any necessary action(s) recommended.

Tasksheet Scoring

		Test		Retest	
Evaluation Items	**Pass**	**Fail**	**Pass**	**Fail**
Task-Specific Evaluation	**(1 pt)**	**(0 pts)**	**(1 pt)**	**(0 pts)**
Perform a visual inspection of the battery case, cables, and hold-downs. Inspect the battery level and top off as needed.				
Install a memory saver unit prior to battery removal.				
Remove the battery. Clean the terminals and cables. Inspect the battery case and tray.				
Install the battery. Apply protectant. Remove the memory saver unit and verify the vehicle starts and memory functions have been maintained.				
Non-Task-Specific Evaluation	**(0 pts)**	**(−1 pt)**	**(0 pts)**	**(−1 pt)**
Student successfully completed at least three of the non-task-specific steps.				
Student successfully completed all five of the non-task-specific steps.				
Total Score: <total # of points / 4 = %>				

Supervisor:

Supervisor/instructor signature _______________________________ Date _________________

Comments:

Retest supervisor/instructor signature _______________________________ Date _________________

Comments:

CDX Tasksheet Number: A6016

Student/Intern Information

Name _________________________________ Date ____________ Class _________________________

Vehicle, Customer, and Service Information

Vehicle used for this activity:

Year _______________ Make _______________________ Model _____________________

Odometer _______________________ VIN _________________________________

Materials Required

- Blank work order
- Vehicle with available service history records
- Depending on the type of concern, special diagnostic/hand tools may be required. See your supervisor/instructor for instructions to identify what tools may be required.
- Service information database
- Personal protective equipment
- Battery charger

Task-Specific Safety Considerations

- Comply with personal and environmental safety practices associated with clothing; eye protection; hand tools; power equipment; proper ventilation; and the handling, storage, and disposal of chemicals/materials in accordance with local, state, and federal safety and environmental regulations.

▶ **TASK** Perform slow/fast battery charge according to manufacturer's recommendations.

MLR
A6B5

Time off_______________

Time on_______________

Total time_______________

Student Instructions: Read through the entire procedure prior to starting. Prepare your workspace and any tools or parts that may be needed to complete the task. When directed by your instructor, begin the procedure to complete the task and check the box as each step is finished. Track your time on this procedure for later comparison to the standard completion time (i.e., "flat rate" or customer pay time).

Procedure:	Step Completed
1. Research battery charging procedures.	
a. Use service information to determine the recommended battery charging procedure. **Tip:** Depending on the battery design and condition, the procedure for battery charging is subject to change. List the procedure for charging the battery in your vehicle.	☐

2. Prepare to charge the battery.	
a. Clean the area surrounding the battery caps. Remove the caps and inspect the electrolyte level. Top off as needed. **Tip:** Do not charge a battery if the fluid inside is frozen.	☐
b. Install a memory saver and disconnect the battery cables.	☐
c. Measure the open circuit voltage of the battery and list your results. On the basis of this inspection, what is the percentage of charge of your battery?	☐
3. Charge the battery.	
a. With the charger unplugged, install the charger cables to the correct battery terminals.	☐
b. Turn the charger on and select the correct settings depending on your battery design and condition.	☐
c. What amperage setting are you using to charge the battery? How long will the battery charge for?	☐
d. After the charging is complete, turn the charger off and remove the cables.	☐
e. Install the battery cables in the correct sequence. Remove the memory saver unit and attempt to start the vehicle.	☐

Non-Task-Specific Evaluations:	Step Completed
1. Tools and equipment were used as directed and returned in good working order.	☐
2. Complied with all general and task-specific safety standards, including proper use of any personal protective equipment.	☐
3. Completed the task in an appropriate time frame (Recommendation: 1.5 or 2 times flat rate).	☐
4. Left the work space clean and orderly.	☐
5. Cared for customer property and returned it undamaged.	☐

Student signature _________________________________ Date _____________________________

Comments:

Have your supervisor/instructor verify satisfactory completion of this procedure, any observations found,

and any necessary action(s) recommended.

Evaluation Instructions: The scoring box below is intended to act as a guide for both student and instructor. Each criterion listed will help students understand what is expected from them and evaluators articulate success at a particular task. The scoring is set up to allow a second attempt at each task (see the "Test" and "Retest" columns). Scoring is designed to reward students for correct completion of the task. Points are lost for failure to complete the employability requirements (see "Non-Task-Specific" criteria). When all criteria are evaluated, tally the points for a total at the bottom of each column.

Tasksheet Scoring

	Test		Retest	
Evaluation Items	**Pass**	**Fail**	**Pass**	**Fail**
Task-Specific Evaluation	**(1 pt)**	**(0 pts)**	**(1 pt)**	**(0 pts)**
Research the procedure and recommended settings for charging your battery.				
Inspect the battery fluid level. Measure the open circuit voltage and determine the percentage of charge.				
Install a memory saver prior to disconnecting the battery cables. Disconnect the battery cables. Install the battery charger and select the correct amperage and time settings.				
Charge the battery. Reassemble the battery cables and attempt to start the vehicle.				
Non-Task-Specific Evaluation	**(0 pts)**	**(−1 pt)**	**(0 pts)**	**(−1 pt)**
Student successfully completed at least three of the non-task-specific steps.				
Student successfully completed all five of the non-task-specific steps.				
Total Score: <total # of points / 4 = %>				

Supervisor:

Supervisor/instructor signature _______________________________ Date _______________

Comments:

Retest supervisor/instructor signature _______________________________ Date _______________

Comments:

CDX Tasksheet Number: A6017

Student/Intern Information

Name _________________________________ Date ____________ Class _________________________

Vehicle, Customer, and Service Information

Vehicle used for this activity:

Year ______________ Make ______________________________ Model _________________________

Odometer ___________________________________ VIN ____

Materials Required
- Blank work order
- Vehicle with available service history records
- Depending on the type of concern, special diagnostic/hand tools may be required. See your supervisor/instructor for instructions to identify what tools may be required.
- Service information database
- Personal protective equipment
- Jumper cables
- Auxiliary power supply

Task-Specific Safety Considerations
- Comply with personal and environmental safety practices associated with clothing; eye protection; hand tools; power equipment; proper ventilation; and the handling, storage, and disposal of chemicals/materials in accordance with local, state, and federal safety and environmental regulations.
- **Caution:** Be sure to install jumper cables in correct orientation. Discharged batteries emit a flammable hydrogen gas that can cause an explosion if a spark or open flame is present.

▶**TASK** Jump-start vehicle using jumper cables and a booster battery or an auxiliary power supply.

MLR
A6B6

Time off_____________

Time on_____________

Total time_____________

Student Instructions: Read through the entire procedure prior to starting. Prepare your workspace and any tools or parts that may be needed to complete the task. When directed by your instructor, begin the procedure to complete the task and check the box as each step is finished. Track your time on this procedure for later comparison to the standard completion time (i.e., "flat rate" or customer pay time).

Procedure:	Step Completed
1. Research the procedure for jump starting a vehicle.	
a. Explain the correct sequence of installing jumper cables onto a vehicle with a dead battery.	☐

2. Jump start a vehicle with a dead battery using jumper cables.	
a. Open the hood of both vehicles and ensure that both vehicles are off before installing the jumper cables.	☐
b. Install the jumper cables in the correct order. **Caution:** This step is critical to eliminate the chance of a battery explosion.	☐
c. Start the vehicle with the good battery. Allow the vehicle to run for 2 minutes before attempting to start the vehicle with the dead battery.	☐
d. Attempt to start the vehicle with the dead battery. If the vehicle starts, remove the jumper cables in reverse order. If the vehicle still does not start, allow 2 more minutes of charging before attempting to start again. **Caution:** Be careful working around an engine that is running.	☐
3. Jump start a vehicle with a dead battery using an auxiliary power supply/ boost pack.	
a. Ensure that the boost pack is in the off position. Install the cables of the boost pack to the correct battery terminals.	☐
b. Turn the boost pack on. Some boosters require a short wait period before attempting to start the vehicle.	☐
c. Attempt to start the vehicle. If the vehicle starts, turn the boost pack off and disconnect the cables.	☐

Non-Task-Specific Evaluations:	Step Completed
1. Tools and equipment were used as directed and returned in good working order.	☐
2. Complied with all general and task-specific safety standards, including proper use of any personal protective equipment.	☐
3. Completed the task in an appropriate time frame (Recommendation: 1.5 or 2 times flat rate).	☐
4. Left the work space clean and orderly.	☐
5. Cared for customer property and returned it undamaged.	☐

Student signature _______________________________ Date _______________________________

Comments:

Have your supervisor/instructor verify satisfactory completion of this procedure, any observations found, and any necessary action(s) recommended.

Evaluation Instructions: The scoring box below is intended to act as a guide for both student and instructor. Each criterion listed will help students understand what is expected from them and evaluators articulate success at a particular task. The scoring is set up to allow a second attempt at each task (see the "Test" and "Retest" columns). Scoring is designed to reward students for correct completion of the task. Points are lost for failure to complete the employability requirements (see "Non-Task-Specific" criteria). When all criteria are evaluated, tally the points for a total at the bottom of each column.

Tasksheet Scoring

	Test		Retest	
Evaluation Items	**Pass**	**Fail**	**Pass**	**Fail**
Task-Specific Evaluation	**(1 pt)**	**(0 pts)**	**(1 pt)**	**(0 pts)**
Research the procedure for jump starting a vehicle. Install jumper cables in the correct sequence and allow the donor vehicle to charge the dead vehicle.				
Attempt to start the vehicle with the dead battery.				
Remove the jumper cables in the correct sequence.				
Use a boost pack to jump start a vehicle.				
Non-Task-Specific Evaluation	**(0 pts)**	**(–1 pt)**	**(0 pts)**	**(–1 pt)**
Student successfully completed at least three of the non-task-specific steps.				
Student successfully completed all five of the non-task-specific steps.				
Total Score: <total # of points / 4 = %>				

Supervisor:

Supervisor/instructor signature _________________________ Date _________________

Comments:

Retest supervisor/instructor signature _________________________ Date _________________

Comments:

CDX Tasksheet Number: A6018

Student/Intern Information

Name _________________________________ Date _____________ Class _______________________

Vehicle, Customer, and Service Information

Vehicle used for this activity:

Year _______________ Make _______________________ Model _______________________

Odometer _________________________________ VIN _______________________________

<table>
<tr><td>

Materials Required
- Blank work order
- Vehicle with available service history records
- Depending on the type of concern, special diagnostic/hand tools may be required. See your supervisor/instructor for instructions to identify what tools may be required.
- Service information database
- Personal protective equipment

</td></tr>
</table>

Task-Specific Safety Considerations
- Comply with personal and environmental safety practices associated with clothing; eye protection; hand tools; power equipment; proper ventilation; and the handling, storage, and disposal of chemicals/materials in accordance with local, state, and federal safety and environmental regulations.

Time off______________

Time on______________

Total time______________

▶ TASK Identify safety precautions for high voltage systems on electric, hybrid-electric, and diesel vehicles.

MLR
A6B7

Student Instructions: Read through the entire procedure prior to starting. Prepare your workspace and any tools or parts that may be needed to complete the task. When directed by your instructor, begin the procedure to complete the task and check the box as each step is finished. Track your time on this procedure for later comparison to the standard completion time (i.e., "flat rate" or customer pay time).

Procedure:	Step Completed
1. Determine safety precautions for high voltage systems.	
a. List the personal protective equipment and tools that are required for servicing high voltage systems.	☐

b. Research the wiring colors used on electric or hybrid-electric vehicles. What colors are used to warn you of high voltages? What amount of voltage is present in these systems?	☐
c. Research the wiring colors used on diesel vehicles. What colors are used to warn you of high voltages? What amount of voltage is present in these systems?	☐
d. Explain how high voltage systems can be disabled on the vehicles. Are there special precautions used when servicing these systems?	☐
e. Describe what type of warning indicators may illuminate to warn the driver of a vehicle that there is a fault with a high voltage system.	☐

Non-Task-Specific Evaluations:	Step Completed
1. Tools and equipment were used as directed and returned in good working order.	☐
2. Complied with all general and task-specific safety standards, including proper use of any personal protective equipment.	☐
3. Completed the task in an appropriate time frame (Recommendation: 1.5 or 2 times flat rate).	☐
4. Left the work space clean and orderly.	☐
5. Cared for customer property and returned it undamaged.	☐

Student signature _________________________________ Date _________________________________

Comments:

Have your supervisor/instructor verify satisfactory completion of this procedure, any observations found, and any necessary action(s) recommended.

Evaluation Instructions: The scoring box below is intended to act as a guide for both student and instructor. Each criterion listed will help students understand what is expected from them and evaluators articulate success at a particular task. The scoring is set up to allow a second attempt at each task (see the "Test" and "Retest" columns). Scoring is designed to reward students for correct completion of the task. Points are lost for failure to complete the employability requirements (see "Non-Task-Specific" criteria). When all criteria are evaluated, tally the points for a total at the bottom of each column.

Tasksheet Scoring

	Test		Retest	
Evaluation Items	**Pass**	**Fail**	**Pass**	**Fail**
Task-Specific Evaluation	**(1 pt)**	**(0 pts)**	**(1 pt)**	**(0 pts)**
List the personal protective equipment and tools needed to service high voltage systems.				
Research the wiring colors and voltage capacities of electric and hybrid-electric vehicles.				
Research the wiring color and voltage capacity of diesel vehicles.				
Describe the procedures for disabling high voltage systems and any special service precautions. Describe the warning indicators for faults with high voltage systems.				
Non-Task-Specific Evaluation	**(0 pts)**	**(−1 pt)**	**(0 pts)**	**(−1 pt)**
Student successfully completed at least three of the non-task-specific steps.				
Student successfully completed all five of the non-task-specific steps.				
Total Score: <total # of points / 4 = %>				

Supervisor:

Supervisor/instructor signature ________________________________ Date ________________

Comments:

Retest supervisor/instructor signature ________________________________ Date ________________

Comments:

CDX Tasksheet Number: A6019

Student/Intern Information

Name _________________________________ Date _____________ Class _________________________

Vehicle, Customer, and Service Information

Vehicle used for this activity:

Year _______________ Make _________________________ Model _____________________

Odometer _________________________________ VIN _________________________________

Task-Specific Safety Considerations

- Comply with personal and environmental safety practices associated with clothing; eye protection; hand tools; power equipment; proper ventilation; and the handling, storage, and disposal of chemicals/materials in accordance with local, state, and federal safety and environmental regulations.
- When running any vehicles in the shop, make sure you use the shop's exhaust ventilation system to discharge all exhaust gas safely outside.

▶ **TASK** Identify electrical/electronic modules, security systems, radios, and other accessories that require reinitialization or code entry after reconnecting vehicle battery.

MLR
A6B8

Time off_______________

Time on_______________

Total time_______________

Student Instructions: Read through the entire procedure prior to starting. Prepare your workspace and any tools or parts that may be needed to complete the task. When directed by your instructor, begin the procedure to complete the task and check the box as each step is finished. Track your time on this procedure for later comparison to the standard completion time (i.e., "flat rate" or customer pay time).

Procedure:	Step Completed
1. Research the electronic systems that require reinitialization after a battery has been disconnected/reconnected.	
a. Use service information to determine what electronic systems on your vehicle require initialization after a battery has been reconnected. List those systems and the initialization procedure.	☐
b. Inspect the underhood and in-car labels for systems that require initialization. Many times, labels can be found on the vehicle explaining what systems require initialization. These systems may include, but are not limited to: power windows, sunroof, radio, clock, memory seats, engine/transmission relearn.	☐
2. Disconnect the vehicle's battery.	
a. Disconnect the negative cable from the battery and allow 5 minutes before reconnecting the cable.	☐
3. Perform initialization procedures on designated electronic systems.	
a. Following procedure from your service information, perform the relearn procedures for all systems that require initialization.	☐
b. Confirm that all systems work correctly. Are there any systems that proved difficult to relearn?	☐
c. Start the vehicle and confirm that it starts and idles correctly. **Note:** Sometimes transmission shift quality can be affected by a battery disconnect procedure. A relearn and/or drive pattern would be required to obtain normal shift feel once more. Also, idle quality can be affected. A relearn and/or throttle service may be required.	☐

<table>
<tr><td>Non-Task-Specific Evaluations:</td><td>Step Completed</td></tr>
<tr><td>1. Tools and equipment were used as directed and returned in good working order.</td><td>☐</td></tr>
<tr><td>2. Complied with all general and task-specific safety standards, including proper use of any personal protective equipment.</td><td>☐</td></tr>
<tr><td>3. Completed the task in an appropriate time frame (Recommendation: 1.5 or 2 times flat rate).</td><td>☐</td></tr>
<tr><td>4. Left the work space clean and orderly.</td><td>☐</td></tr>
<tr><td>5. Cared for customer property and returned it undamaged.</td><td>☐</td></tr>
</table>

Student signature _______________________________ Date _______________________________

Comments:

Have your supervisor/instructor verify satisfactory completion of this procedure, any observations found, and any necessary action(s) recommended.

Evaluation Instructions: The scoring box below is intended to act as a guide for both student and instructor. Each criterion listed will help students understand what is expected from them and evaluators articulate success at a particular task. The scoring is set up to allow a second attempt at each task (see the "Test" and "Retest" columns). Scoring is designed to reward students for correct completion of the task. Points are lost for failure to complete the employability requirements (see "Non-Task-Specific" criteria). When all criteria are evaluated, tally the points for a total at the bottom of each column.

Tasksheet Scoring

	Test		Retest	
Evaluation Items	**Pass**	**Fail**	**Pass**	**Fail**
Task-Specific Evaluation	**(1 pt)**	**(0 pts)**	**(1 pt)**	**(0 pts)**
Use service information to determine the systems that require reinitialization after a battery disconnect. Inspect the labels on your vehicle to determine the systems that require reinitialization after a battery disconnect.				
Describe the procedure for initializing electronic systems.				
Disconnect the battery's negative cable and clear the memory from electronic systems.				
Perform relearn procedures on all electronic systems affected on your vehicle.				
Non-Task-Specific Evaluation	**(0 pts)**	**(−1 pt)**	**(0 pts)**	**(−1 pt)**
Student successfully completed at least three of the non-task-specific steps.				
Student successfully completed all five of the non-task-specific steps.				
Total Score: <total # of points / 4 = %>				

Supervisor:

Supervisor/instructor signature _________________________ Date _________________

Comments:

Retest supervisor/instructor signature _________________________ Date _________________

Comments:

CDX Tasksheet Number: A6020

Student/Intern Information

Name _________________________________ Date _____________ Class _____________________

Vehicle, Customer, and Service Information

Vehicle used for this activity:

Year _______________ Make _____________________________ Model _____________________

Odometer _________________________________ VIN _____________________________

Materials Required
- Blank work order
- Vehicle with available service history records
- Depending on the type of concern, special diagnostic/hand tools may be required. See your supervisor/instructor for instructions to identify what tools may be required.
- Service information database
- Personal protective equipment

Task-Specific Safety Considerations
- Comply with personal and environmental safety practices associated with clothing; eye protection; hand tools; power equipment; proper ventilation; and the handling, storage, and disposal of chemicals/materials in accordance with local, state, and federal safety and environmental regulations.

▶ TASK Identify hybrid vehicle auxiliary (12V) battery service, repair, and test procedures.

MLR
A6B9

Time off_____________

Time on_____________

Total time_____________

Student Instructions: Read through the entire procedure prior to starting. Prepare your workspace and any tools or parts that may be needed to complete the task. When directed by your instructor, begin the procedure to complete the task and check the box as each step is finished. Track your time on this procedure for later comparison to the standard completion time (i.e., "flat rate" or customer pay time).

Procedure:	Step Completed
1. Identify hybrid vehicle auxiliary battery service procedures.	
a. Use service information to determine the location of the 12-volt battery in your vehicle.	☐
b. Explain the purpose of the 12-volt battery on your hybrid vehicle.	☐

c. Determine any precautions when servicing the 12-volt battery. Are there any special procedures or tools needed to test, remove, or replace the 12-volt battery?	☐
d. Explain the procedure for jump starting a hybrid vehicle. Are there jump starting terminals located away from the 12-volt battery?	☐
e. Explain how you would test the 12-volt battery in a hybrid vehicle. What symptoms may a customer notice if the 12-volt battery was under performing?	☐

Non-Task-Specific Evaluations:	Step Completed
1. Tools and equipment were used as directed and returned in good working order.	☐
2. Complied with all general and task-specific safety standards, including proper use of any personal protective equipment.	☐
3. Completed the task in an appropriate time frame (Recommendation: 1.5 or 2 times flat rate).	☐
4. Left the work space clean and orderly.	☐
5. Cared for customer property and returned it undamaged.	☐

Student signature _________________________________ Date _________________________

Comments:

Have your supervisor/instructor verify satisfactory completion of this procedure, any observations found, and any necessary action(s) recommended.

Evaluation Instructions: The scoring box below is intended to act as a guide for both student and instructor. Each criterion listed will help students understand what is expected from them and evaluators articulate success at a particular task. The scoring is set up to allow a second attempt at each task (see the "Test" and "Retest" columns). Scoring is designed to reward students for correct completion of the task. Points are lost for failure to complete the employability requirements (see "Non-Task-Specific" criteria). When all criteria are evaluated, tally the points for a total at the bottom of each column.

Tasksheet Scoring

Evaluation Items	Test		Retest	
	Pass	**Fail**	**Pass**	**Fail**
Task-Specific Evaluation	**(1 pt)**	**(0 pts)**	**(1 pt)**	**(0 pts)**
Determine the location of the 12-volt battery in your hybrid vehicle. Explain the purpose of the 12-volt battery in a hybrid vehicle.				
Determine any special service precautions or tools required for servicing the 12-volt battery.				
Explain the procedure for jump starting a hybrid vehicle.				
Explain the procedure for testing a 12-volt battery on a hybrid vehicle.				
Non-Task-Specific Evaluation	**(0 pts)**	**(−1 pt)**	**(0 pts)**	**(−1 pt)**
Student successfully completed at least three of the non-task-specific steps.				
Student successfully completed all five of the non-task-specific steps.				
Total Score: <total # of points / 4 = %>				

Supervisor:

Supervisor/instructor signature _________________________ Date _________________

Comments:

Retest supervisor/instructor signature _________________________ Date _________________

Comments:

6C. Starting Systems

Learning Objective/Task	CDX Tasksheet Number	ASE Education Foundation Reference Number; Priority Level
• Perform starter current draw test, and determine necessary action.	A6021	A6C1, P-1
• Perform starter circuit voltage drop tests, and determine necessary action.	A6022	A6C2, P-1
• Inspect and test starter relays and solenoids, and determine necessary action.	A6023	A6C3, P-2
• Remove and install starter in a vehicle.	A6024	A6C4, P-1
• Inspect and test switches, connectors, and wires of starter control circuits, and determine necessary action.	A6025	A6C5, P-2
• Demonstrate knowledge of an automatic idle-stop/start-stop system.	A6026	A6C6, P-3

Materials Required

- Blank work order
- Vehicle with available service history records
- Depending on the type of concern, special diagnostic/hand tools may be required. See your supervisor/instructor for instructions to identify what tools may be required.
- Service information database
- Personal protective equipment

Safety Considerations

- When running any vehicles in the shop, make sure you use the shop's exhaust ventilation system to discharge all exhaust gas safely outside.
- Lifting equipment, such as vehicle jacks and stands, vehicle hoists, and engine hoists, are important tools that increase productivity and make the job easier. However, they can also cause severe injury or death if used improperly. Make sure you follow the manufacturer's operation procedures. Also, make sure you have your supervisor's/instructor's permission to use any particular type of lifting equipment.
- Comply with personal and environmental safety practices associated with clothing; eye protection; hand tools; power equipment; proper ventilation; and the handling, storage, and disposal of chemicals/materials in accordance with local, state, and federal safety and environmental regulations.
- If you need to start the vehicle, you should ensure that the parking brake is firmly applied; if necessary, use wheel chocks to prevent the vehicle from moving when the vehicle is started to verify the completion of these tasks.

CDX Tasksheet Number: A6021

Student/Intern Information

Name _________________________________ Date _____________ Class _________________________

Vehicle, Customer, and Service Information

Vehicle used for this activity:

Year _______________ Make _______________________________ Model _______________________

Odometer _________________________________ VIN _________________________________

Materials Required

- Blank work order
- Vehicle with available service history records
- Depending on the type of concern, special diagnostic/hand tools may be required. See your supervisor/instructor for instructions to identify what tools may be required.
- Service information database
- Personal protective equipment
- Digital multimeter
- Inductive clamp
- VAT machine

Task-Specific Safety Considerations

- Comply with personal and environmental safety practices associated with clothing; eye protection; hand tools; power equipment; proper ventilation; and the handling, storage, and disposal of chemicals/materials in accordance with local, state, and federal safety and environmental regulations.

▶ **TASK** Perform starter current draw test, and determine necessary action. **MLR A6C1**

Time off________________

Student Instructions: Read through the entire procedure prior to starting. Prepare your workspace and any tools or parts that may be needed to complete the task. When directed by your instructor, begin the procedure to complete the task and check the box as each step is finished. Track your time on this procedure for later comparison to the standard completion time (i.e., "flat rate" or customer pay time).

Time on________________

Total time________________

Procedure:	Step Completed
1. Test the battery.	
a. Before performing a starter current draw test, confirm that the battery is in good condition. Refer to tasksheet A6013 for battery testing procedure.	☐
b. Describe the results from this test.	☐

c. How would a weak battery affect the starter amperage draw test?	☐
2. Perform a starter current draw test.	
a. Disable the fuel and ignition systems to prevent the engine from starting while testing. Explain how you accomplished this task.	☐
b. Use service information to identify acceptable current draw and minimum battery voltage. Record the specifications.	☐
c. Several tools may be available to perform this test. Connect whichever tool specified by your instructor and be sure to zero the ammeter.	☐
d. If a remote starter is available, install the tool at this time. If no tool is available, have a helper crank the engine as you observe the ammeter display.	☐
e. Crank the engine for 5–10 seconds and observe and list the following readings: initial amperage draw, continued amperage draw, and battery voltage after 10 seconds.	☐
f. On the basis of the test results, is the starting system in proper working condition?	☐

Non-Task-Specific Evaluations:	Step Completed
1. Tools and equipment were used as directed and returned in good working order.	☐
2. Complied with all general and task-specific safety standards, including proper use of any personal protective equipment.	☐
3. Completed the task in an appropriate time frame (Recommendation: 1.5 or 2 times flat rate).	☐
4. Left the work space clean and orderly.	☐
5. Cared for customer property and returned it undamaged.	☐

Student signature _________________________ Date _____________________

Comments:

Have your supervisor/instructor verify satisfactory completion of this procedure, any observations found, and any necessary action(s) recommended.

Evaluation Instructions: The scoring box below is intended to act as a guide for both student and instructor. Each criterion listed will help students understand what is expected from them and evaluators articulate success at a particular task. The scoring is set up to allow a second attempt at each task (see the "Test" and "Retest" columns). Scoring is designed to reward students for correct completion of the task. Points are lost for failure to complete the employability requirements (see "Non-Task-Specific" criteria). When all criteria are evaluated, tally the points for a total at the bottom of each column.

Tasksheet Scoring

	Test		Retest	
Evaluation Items	**Pass**	**Fail**	**Pass**	**Fail**
Task-Specific Evaluation	**(1 pt)**	**(0 pts)**	**(1 pt)**	**(0 pts)**
Perform a battery test.				
Disable the fuel and ignition systems.				
Determine amperage draw specifications. Perform a starter current draw test.				
Make recommendations for service based on your inspection.				
Non-Task-Specific Evaluation	**(0 pts)**	**(−1 pt)**	**(0 pts)**	**(−1 pt)**
Student successfully completed at least three of the non-task-specific steps.				
Student successfully completed all five of the non-task-specific steps.				
Total Score: <total # of points / 4 = %>				

Supervisor:

Supervisor/instructor signature _______________________________ Date _______________

Comments:

Retest supervisor/instructor signature _______________________________ Date _______________

Comments:

CDX Tasksheet Number: A6022

Student/Intern Information

Name ________________________________ Date ____________ Class ____________________

Vehicle, Customer, and Service Information

Vehicle used for this activity:

Year ______________ Make ________________________ Model ____________________

Odometer ________________________ VIN ________________________________

Materials Required

- Blank work order
- Vehicle with available service history records
- Depending on the type of concern, special diagnostic/hand tools may be required. See your supervisor/instructor for instructions to identify what tools may be required.
- Service information database
- Personal protective equipment
- Digital multimeter (DMM)

Task-Specific Safety Considerations

- Comply with personal and environmental safety practices associated with clothing; eye protection; hand tools; power equipment; proper ventilation; and the handling, storage, and disposal of chemicals/materials in accordance with local, state, and federal safety and environmental regulations.

▶ **TASK** Perform starter circuit voltage drop tests, and determine necessary action.

MLR
A6C2

Time off________________

Time on________________

Student Instructions: Read through the entire procedure prior to starting. Prepare your workspace and any tools or parts that may be needed to complete the task. When directed by your instructor, begin the procedure to complete the task and check the box as each step is finished. Track your time on this procedure for later comparison to the standard completion time (i.e., "flat rate" or customer pay time).

Total time________________

Procedure:	Step Completed
1. Test the battery.	
a. Before performing a starter voltage drop test, confirm that the battery is in good condition. Refer to tasksheet A6013 for battery testing procedure.	☐
b. Describe the results from this test.	☐

c. How would a weak battery affect the starter voltage drop test?	☐
2. Perform a voltage drop test of the starter.	
a. Explain what components make up the starting system.	☐
b. Perform a visual inspection of the battery cables and starter connections. Inspect for loose or corroded connections.	☐
c. Describe how much voltage drop is acceptable on both the positive and negative side of the starter circuit. Where should the majority of the voltage in the circuit be consumed?	☐
d. Disable the fuel and ignition systems to prevent the engine from starting while testing. Explain how you accomplished this task.	☐
e. Connect the DMM to the battery's positive and negative posts and crank the engine for 5 seconds. Record the voltage observed during this test. Is this an acceptable battery voltage when cranking the engine?	☐
f. Test the voltage drop of the positive battery cable. Connect the positive meter lead to the positive battery cable near the battery terminal. Connect the negative meter lead to the other end of the positive cable that attaches to the starter. Crank the engine and observe the meter reading. How much voltage drop is on the positive cable? Is this an acceptable value?	☐
g. Test the voltage drop of the negative side of the starter circuit. Connect one DMM lead to the negative battery cable near the negative battery terminal and the other lead to the starter case. Crank the engine and observe the meter reading. How much voltage drop is on the positive cable? Is this an acceptable value?	☐

3. Explain how excessive voltage drop can affect engine cranking speed.	
a. What causes excessive voltage drop?	☐

Non-Task-Specific Evaluations:	**Step Completed**
1. Tools and equipment were used as directed and returned in good working order.	☐
2. Complied with all general and task-specific safety standards, including proper use of any personal protective equipment.	☐
3. Completed the task in an appropriate time frame (Recommendation: 1.5 or 2 times flat rate).	☐
4. Left the work space clean and orderly.	☐
5. Cared for customer property and returned it undamaged.	☐

Student signature ___________________________ Date ___________________________

Comments:

Have your supervisor/instructor verify satisfactory completion of this procedure, any observations found, and any necessary action(s) recommended.

Evaluation Instructions: The scoring box below is intended to act as a guide for both student and instructor. Each criterion listed will help students understand what is expected from them and evaluators articulate success at a particular task. The scoring is set up to allow a second attempt at each task (see the "Test" and "Retest" columns). Scoring is designed to reward students for correct completion of the task. Points are lost for failure to complete the employability requirements (see "Non-Task-Specific" criteria). When all criteria are evaluated, tally the points for a total at the bottom of each column.

Tasksheet Scoring

Evaluation Items	Test		Retest	
	Pass	**Fail**	**Pass**	**Fail**
Task-Specific Evaluation	**(1 pt)**	**(0 pts)**	**(1 pt)**	**(0 pts)**
Test the battery and explain how a weak battery can affect starter voltage drop testing. Perform a visual inspection of the starter circuit wiring and connections.				
Disable the fuel and ignition systems.				
Perform a voltage drop test of the positive side of the starter circuit.				
Perform a voltage drop test of the negative side of the starter circuit.				
Non-Task-Specific Evaluation	**(0 pts)**	**(−1 pt)**	**(0 pts)**	**(−1 pt)**
Student successfully completed at least three of the non-task-specific steps.				
Student successfully completed all five of the non-task-specific steps.				
Total Score: <total # of points / 4 = %>				

Supervisor:

Supervisor/instructor signature ______________________________ Date ________________________

Comments:

Retest supervisor/instructor signature ______________________________ Date ____________________

Comments:

CDX Tasksheet Number: A6023

Student/Intern Information

Name _________________________________ Date ___________ Class _________________________

Vehicle, Customer, and Service Information

Vehicle used for this activity:

Year _______________ Make _______________________ Model _____________________

Odometer _______________________________ VIN _____________________________

Materials Required

- Blank work order
- Vehicle with available service history records
- Depending on the type of concern, special diagnostic/hand tools may be required. See your supervisor/instructor for instructions to identify what tools may be required.
- Service information database
- Personal protective equipment
- Digital multimeter (DMM)

Task-Specific Safety Considerations

- Comply with personal and environmental safety practices associated with clothing; eye protection; hand tools; power equipment; proper ventilation; and the handling, storage, and disposal of chemicals/materials in accordance with local, state, and federal safety and environmental regulations.

▶ **TASK** Inspect and test starter relays and solenoids, and determine necessary action.

MLR
A6C3

Time off____________

Time on____________

Total time____________

Student Instructions: Read through the entire procedure prior to starting. Prepare your workspace and any tools or parts that may be needed to complete the task. When directed by your instructor, begin the procedure to complete the task and check the box as each step is finished. Track your time on this procedure for later comparison to the standard completion time (i.e., "flat rate" or customer pay time).

Procedure:	Step Completed
1. Determine the starter relay and solenoid location.	
a. Research the starting system design and operation on your vehicle. Is the starting system computer controlled?	☐

b. Determine the location of the starting system fuses, relays, and solenoids. Explain where these components are located.	☐
2. Inspect and test the starter relay and solenoid.	
a. Begin by performing a visual inspection of the starter relay and solenoid. Are they properly installed and free of corrosion?	☐
b. Perform a test of the battery to verify that it is fully charged.	☐
c. Disable the fuel and ignition systems on the vehicle.	☐
d. How much voltage should be present at the starter solenoid/relay ignition terminal?	☐
e. Connect the DMM negative test lead to the negative battery terminal and the positive lead to the solenoid/relay ignition terminal. What voltage should be present when cranking the vehicle?	☐
f. Crank the engine and record the voltage.	☐
g. Now move the positive meter lead to the ground side of the solenoid/relay. Crank the engine and record the voltage drop.	☐
h. Re-enable the fuel and ignition systems. Attempt to start the vehicle. Did the vehicle start?	☐
3. On the basis of your inspection, does the starter solenoid/relay appear to be working correctly?	☐

Non-Task-Specific Evaluations:	**Step Completed**
1. Tools and equipment were used as directed and returned in good working order.	☐
2. Complied with all general and task-specific safety standards, including proper use of any personal protective equipment.	☐
3. Completed the task in an appropriate time frame (Recommendation: 1.5 or 2 times flat rate).	☐
4. Left the work space clean and orderly.	☐
5. Cared for customer property and returned it undamaged.	☐

Student signature _______________________________ Date _______________________________

Comments:

Have your supervisor/instructor verify satisfactory completion of this procedure, any observations found, and any necessary action(s) recommended.

Tasksheet Scoring

Evaluation Items	Test		Retest	
	Pass	**Fail**	**Pass**	**Fail**
Task-Specific Evaluation	**(1 pt)**	**(0 pts)**	**(1 pt)**	**(0 pts)**
Research your vehicle's starting system design and operation. Determine the location of the starting system fuses, relays, and solenoids.				
Perform a visual inspection. Disable the fuel and ignition systems.				
Inspect the starter solenoid/relay for correct function.				
Make recommendations for service based on your inspection.				
Non-Task-Specific Evaluation	**(0 pts)**	**(−1 pt)**	**(0 pts)**	**(−1 pt)**
Student successfully completed at least three of the non-task-specific steps.				
Student successfully completed all five of the non-task-specific steps.				
Total Score: <total # of points / 4 = %>				

Supervisor:

Supervisor/instructor signature ________________________________ Date ________________

Comments:

Retest supervisor/instructor signature ________________________________ Date ________________

Comments:

CDX Tasksheet Number: A6024

Student/Intern Information

Name _______________________________ Date ____________ Class __________________________

Vehicle, Customer, and Service Information

Vehicle used for this activity:

Year ______________ Make ______________________ Model ____________________________

Odometer ______________________________ VIN ____

Materials Required
- Blank work order
- Vehicle with available service history records
- Depending on the type of concern, special diagnostic/hand tools may be required. See your supervisor/instructor for instructions to identify what tools may be required.
- Service information database
- Personal protective equipment
- General hand/air tools

Task-Specific Safety Considerations
- Comply with personal and environmental safety practices associated with clothing; eye protection; hand tools; power equipment; proper ventilation; and the handling, storage, and disposal of chemicals/materials in accordance with local, state, and federal safety and environmental regulations.

▶ **TASK** Remove and install starter in a vehicle.

MLR
A6C4

Time off______________

Time on______________

Total time______________

Student Instructions: Read through the entire procedure prior to starting. Prepare your workspace and any tools or parts that may be needed to complete the task. When directed by your instructor, begin the procedure to complete the task and check the box as each step is finished. Track your time on this procedure for later comparison to the standard completion time (i.e., "flat rate" or customer pay time).

Procedure:	Step Completed
1. Research the starter replacement procedure.	
a. List or print off the procedure for replacing the starter. Include any torque specifications related to this task.	☐

b. Are there any special service precautions related to this task?	☐
2. Remove the starter.	
a. Disconnect the negative battery cable.	☐
b. Depending on the starter location, remove as necessary any components hindering access to the starter connections.	☐
c. Remove the starter connections. **Tip:** Take a picture or label the connections to avoid confusion during installation.	☐
d. Remove the starter bolts and the starter. There may be a shim in between the starter and the starter mounting surface. Do not lose this shim, as it will most likely be required during replacement.	☐
e. With the starter removed, inspect the flexplate/flywheel ring gear teeth for excessive wear. Inspect the starter housing for signs of damage. Inspect the starter wiring and connections.	☐
3. Install the starter.	
a. Following procedure install the starter. If a shim was removed, it will most likely need reinstalled. Refer to your service information for inspecting and adjusting starter pinion depth using shims.	☐
b. Torque all starter bolts to specification.	☐
c. Install the starter connections.	☐
d. Install any components that were removed to access the starter.	☐
e. Connect the negative battery cable and attempt to start the vehicle. Does the vehicle start normally? Make sure that there is no grinding noise or binding of the starter pinion gear.	☐

<table>
<tr><td>Non-Task-Specific Evaluations:</td><td>Step Completed</td></tr>
<tr><td>1. Tools and equipment were used as directed and returned in good working order.</td><td>☐</td></tr>
<tr><td>2. Complied with all general and task-specific safety standards, including proper use of any personal protective equipment.</td><td>☐</td></tr>
<tr><td>3. Completed the task in an appropriate time frame (Recommendation: 1.5 or 2 times flat rate).</td><td>☐</td></tr>
<tr><td>4. Left the work space clean and orderly.</td><td>☐</td></tr>
<tr><td>5. Cared for customer property and returned it undamaged.</td><td>☐</td></tr>
</table>

Student signature _______________________________ Date _______________________________

Comments:

Have your supervisor/instructor verify satisfactory completion of this procedure, any observations found, and any necessary action(s) recommended.

Tasksheet Scoring

Evaluation Items	Test		Retest	
	Pass	**Fail**	**Pass**	**Fail**
Task-Specific Evaluation	**(1 pt)**	**(0 pts)**	**(1 pt)**	**(0 pts)**
Research the starter replacement procedure.				
Disconnect the negative battery cables and remove necessary components to access the starter. Remove the starter connections and the starter.				
Inspect the starter, all wiring and connections, and ring gear.				
Install the starter and torque fasteners to specification.				
Non-Task-Specific Evaluation	**(0 pts)**	**(−1 pt)**	**(0 pts)**	**(−1 pt)**
Student successfully completed at least three of the non-task-specific steps.				
Student successfully completed all five of the non-task-specific steps.				
Total Score: <total # of points / 4 = %>				

Supervisor:

Supervisor/instructor signature ________________________ Date ________________

Comments:

Retest supervisor/instructor signature ________________________ Date ________________

Comments:

CDX Tasksheet Number: A6025

Student/Intern Information

Name _________________________________ Date _____________ Class _____________________

Vehicle, Customer, and Service Information

Vehicle used for this activity:

Year _______________ Make _________________________ Model _____________________

Odometer _________________________ VIN _________________________________

Materials Required

- Blank work order
- Vehicle with available service history records
- Depending on the type of concern, special diagnostic/hand tools may be required. See your supervisor/instructor for instructions to identify what tools may be required.
- Service information database
- Personal protective equipment
- Battery tester
- Digital multimeter (DMM)

Task-Specific Safety Considerations

- Comply with personal and environmental safety practices associated with clothing; eye protection; hand tools; power equipment; proper ventilation; and the handling, storage, and disposal of chemicals/materials in accordance with local, state, and federal safety and environmental regulations.

▶ **TASK** Inspect and test switches, connectors, and wires of starter control circuits, and determine necessary action.

MLR
A6C5

Time off_____________

Time on_____________

Total time_____________

Student Instructions: Read through the entire procedure prior to starting. Prepare your workspace and any tools or parts that may be needed to complete the task. When directed by your instructor, begin the procedure to complete the task and check the box as each step is finished. Track your time on this procedure for later comparison to the standard completion time (i.e., "flat rate" or customer pay time).

Procedure:	Step Completed
1. Describe how a security system or engine immobilizer malfunction can prohibit the vehicle from starting.	

2. Inspect the starter control circuits for functionality.	
a. Test the battery and ensure that it is fully charged. **Tip:** When diagnosing starting system concerns, it is best practice to start by verifying that the battery is in proper operating condition.	☐
b. Inspect the connections at the battery for corrosion or loose-fitting connections.	☐
c. Use a DMM to verify the presence of voltage at the starter solenoid connection at the starter. When should voltage be present?	☐
d. If no voltage is present at this connection, list what components could be at fault.	☐
3. Obtain a wiring diagram for the starting system for your vehicle.	
a. Where is the fuse for the starting system? What is its amperage rating?	☐
b. Is a starter relay used? Explain how you would check to see if the relay is receiving voltage from the ignition switch.	☐
c. Describe how a park/neutral switch is used in a starting system.	☐
d. Using your wiring diagram, determine the location of the ignition switch connector.	☐

Non-Task-Specific Evaluations:	Step Completed
1. Tools and equipment were used as directed and returned in good working order.	☐
2. Complied with all general and task-specific safety standards, including proper use of any personal protective equipment.	☐
3. Completed the task in an appropriate time frame (Recommendation: 1.5 or 2 times flat rate).	☐
4. Left the work space clean and orderly.	☐
5. Cared for customer property and returned it undamaged.	☐

Student signature ______________________________ Date ______________________________

Comments:

Have your supervisor/instructor verify satisfactory completion of this procedure, any observations found, and any necessary action(s) recommended.

Evaluation Instructions: The scoring box below is intended to act as a guide for both student and instructor. Each criterion listed will help students understand what is expected from them and evaluators articulate success at a particular task. The scoring is set up to allow a second attempt at each task (see the "Test" and "Retest" columns). Scoring is designed to reward students for correct completion of the task. Points are lost for failure to complete the employability requirements (see "Non-Task-Specific" criteria). When all criteria are evaluated, tally the points for a total at the bottom of each column.

Tasksheet Scoring

	Test		Retest	
Evaluation Items	Pass	Fail	Pass	Fail
Task-Specific Evaluation	**(1 pt)**	**(0 pts)**	**(1 pt)**	**(0 pts)**
Obtain the starting system wiring diagram and locate various components. Describe the function of the starter interruption/immobilizer systems.				
Perform a battery test and check the cable condition.				
Verify the voltage at the starter solenoid.				
Describe the starter relay testing procedure.				
Non-Task-Specific Evaluation	**(0 pts)**	**(−1 pt)**	**(0 pts)**	**(−1 pt)**
Student successfully completed at least three of the non-task-specific steps.				
Student successfully completed all five of the non-task-specific steps.				
Total Score: <total # of points / 4 = %>				

Supervisor:

Supervisor/instructor signature _______________________________ Date _______________

Comments:

Retest supervisor/instructor signature _______________________________ Date _______________

Comments:

CDX Tasksheet Number: A6026

Student/Intern Information

Name _________________________________ Date ____________ Class _________________________

Vehicle, Customer, and Service Information

Vehicle used for this activity:

Year _______________ Make _______________________ Model _____________________

Odometer _____________________________ VIN _________________________________

Materials Required
- Blank work order
- Vehicle with available service history records
- Depending on the type of concern, special diagnostic/hand tools may be required. See your supervisor/instructor for instructions to identify what tools may be required.
- Service information database
- Personal protective equipment

Task-Specific Safety Considerations
- Comply with personal and environmental safety practices associated with clothing; eye protection; hand tools; power equipment; proper ventilation; and the handling, storage, and disposal of chemicals/materials in accordance with local, state, and federal safety and environmental regulations.

▶ TASK Demonstrate knowledge of an automatic idle-stop/start–stop system.

MLR A6C6

Time off_____________

Time on_____________

Total time_____________

Student Instructions: Read through the entire procedure prior to starting. Prepare your workspace and any tools or parts that may be needed to complete the task. When directed by your instructor, begin the procedure to complete the task and check the box as each step is finished. Track your time on this procedure for later comparison to the standard completion time (i.e., "flat rate" or customer pay time).

Procedure:	Step Completed
1. Describe the purpose of idle-stop/start systems.	☐

2. Research and describe what conditions must be met before the idle-stop/start function is enabled.	☐
3. Research and explain how a conventional starter can be used on an idle-stop/start system.	☐
4. Research and explain how a motor generator can be used on an idle-stop/start system.	☐
5. Describe how vehicle systems continue to operate even with the engine off.	☐

Non-Task-Specific Evaluations:	Step Completed
1. Tools and equipment were used as directed and returned in good working order.	☐
2. Complied with all general and task-specific safety standards, including proper use of any personal protective equipment.	☐
3. Completed the task in an appropriate time frame (Recommendation: 1.5 or 2 times flat rate).	☐
4. Left the work space clean and orderly.	☐
5. Cared for customer property and returned it undamaged.	☐

Student signature ___________________________ Date ___________________________

Comments:

Have your supervisor/instructor verify satisfactory completion of this procedure, any observations found, and any necessary action(s) recommended.

Evaluation Instructions: The scoring box below is intended to act as a guide for both student and instructor. Each criterion listed will help students understand what is expected from them and evaluators articulate success at a particular task. The scoring is set up to allow a second attempt at each task (see the "Test" and "Retest" columns). Scoring is designed to reward students for correct completion of the task. Points are lost for failure to complete the employability requirements (see "Non-Task-Specific" criteria). When all criteria are evaluated, tally the points for a total at the bottom of each column.

Tasksheet Scoring

		Test		Retest	
Evaluation Items	**Pass**	**Fail**	**Pass**	**Fail**
Task-Specific Evaluation	**(1 pt)**	**(0 pts)**	**(1 pt)**	**(0 pts)**
Describe the purpose of the idle-stop/start system.				
Research the enabling criteria for stop/start activation.				
Research different starter designs and configurations.				
Describe the function of various vehicle systems when the engine is not running.				
Non-Task-Specific Evaluation	**(0 pts)**	**(−1 pt)**	**(0 pts)**	**(−1 pt)**
Student successfully completed at least three of the non-task-specific steps.				
Student successfully completed all five of the non-task-specific steps.				
Total Score: <total # of points / 4 = %>				

Supervisor:

Supervisor/instructor signature _______________________ Date _______________

Comments:

Retest supervisor/instructor signature _______________________ Date _______________

Comments:

6D. Charging System

Learning Objective/Task	CDX Tasksheet Number	ASE Education Foundation Reference Number; Priority Level
• Perform charging system output test, and determine necessary action.	A6027	A6D1, P-1
• Inspect, adjust, and/or replace generator (alternator) drive belts; check pulleys and tensioners for wear; and check pulley and belt alignment.	A6028	A6D2, P-1
• Remove, inspect, and/or replace generator (alternator).	A6029	A6D3, P-2
• Perform charging circuit voltage drop tests, and determine necessary action.	A6030	A6D4, P-2

Materials Required

- Blank work order
- Vehicle with available service history records
- Depending on the type of concern, special diagnostic/hand tools may be required. See your supervisor/instructor for instructions to identify what tools may be required.
- Service information database
- Personal protective equipment

Safety Considerations

- When running any vehicles in the shop, make sure you use the shop's exhaust ventilation system to discharge all exhaust gas safely outside.
- Lifting equipment, such as vehicle jacks and stands, vehicle hoists, and engine hoists, are important tools that increase productivity and make the job easier. However, they can also cause severe injury or death if used improperly. Make sure you follow the manufacturer's operation procedures. Also, make sure you have your supervisor's/instructor's permission to use any particular type of lifting equipment.
- Comply with personal and environmental safety practices associated with clothing; eye protection; hand tools; power equipment; proper ventilation; and the handling, storage, and disposal of chemicals/materials in accordance with local, state, and federal safety and environmental regulations.
- If you need to start the vehicle, you should ensure that the parking brake is firmly applied; if necessary, use wheel chocks to prevent the vehicle from moving when the vehicle is started to verify the completion of these tasks.

CDX Tasksheet Number: A6027

Student/Intern Information

Name _________________________________ Date ___________ Class _________________________

Vehicle, Customer, and Service Information

Vehicle used for this activity:

Year _______________ Make _____________________________ Model _______________________

Odometer _________________________________ VIN _____________________________________

Materials Required
- Blank work order
- Vehicle with available service history records
- Depending on the type of concern, special diagnostic/hand tools may be required. See your supervisor/instructor for instructions to identify what tools may be required.
- Service information database
- Personal protective equipment
- VAT machine, digital multimeter, and inductive clamp

Task-Specific Safety Considerations
- Comply with personal and environmental safety practices associated with clothing; eye protection; hand tools; power equipment; proper ventilation; and the handling, storage, and disposal of chemicals/materials in accordance with local, state, and federal safety and environmental regulations.

▶ **TASK** Perform charging system output test, and determine necessary action.

MLR
A6D1

Time off_____________

Time on_____________

Total time_____________

Student Instructions: Read through the entire procedure prior to starting. Prepare your workspace and any tools or parts that may be needed to complete the task. When directed by your instructor, begin the procedure to complete the task and check the box as each step is finished. Track your time on this procedure for later comparison to the standard completion time (i.e., "flat rate" or customer pay time).

Procedure:	Step Completed
1. Research charging system specifications.	
a. Use service information to determine the rated voltage and amperage output of the alternator on your vehicle.	☐
b. How many revolutions per minute (RPM) should the engine be running at while perform a charging system output test?	☐

2. Perform a charging system output test.	
a. Perform a battery test. Confirm that the battery is fully charged.	☐
b. Several tools can be used to perform this test. Refer to your instructor for directions regarding the tool you will be using. Install the tester to the alternator output wire.	☐
c. Ensure all testing materials are clear of rotating engine components and start the vehicle.	☐
d. What is the amperage and voltage output at idle? Raise the engine RPM and record the readings once more. How do these values compare to your specification?	☐
e. Turn on multiple electrical accessories, such as high-beam headlights, blower fan, rear defrost, and radio. Observe the amperage and voltage output. How do these readings compare to your specification?	☐
3. On the basis of your inspection, what is your recommendation for service?	
a. Explain what might cause an under-charging situation.	☐
b. Explain what might cause an over-charging situation.	☐

Non-Task-Specific Evaluations:	**Step Completed**
1. Tools and equipment were used as directed and returned in good working order.	☐
2. Complied with all general and task-specific safety standards, including proper use of any personal protective equipment.	☐
3. Completed the task in an appropriate time frame (Recommendation: 1.5 or 2 times flat rate).	☐
4. Left the work space clean and orderly.	☐
5. Cared for customer property and returned it undamaged.	☐

Student signature _________________________ Date _________________________

Comments:

Have your supervisor/instructor verify satisfactory completion of this procedure, any observations found, and any necessary action(s) recommended.

Evaluation Instructions: The scoring box below is intended to act as a guide for both student and instructor. Each criterion listed will help students understand what is expected from them and evaluators articulate success at a particular task. The scoring is set up to allow a second attempt at each task (see the "Test" and "Retest" columns). Scoring is designed to reward students for correct completion of the task. Points are lost for failure to complete the employability requirements (see "Non-Task-Specific" criteria). When all criteria are evaluated, tally the points for a total at the bottom of each column.

Taksheet Scoring

	Test		Retest	
Evaluation Items	**Pass**	**Fail**	**Pass**	**Fail**
Task-Specific Evaluation	**(1 pt)**	**(0 pts)**	**(1 pt)**	**(0 pts)**
Research charging system specifications.				
Install the testing equipment.				
Perform a battery test and a charging system output test.				
Make recommendations for service based on your inspection.				
Non-Task-Specific Evaluation	**(0 pts)**	**(−1 pt)**	**(0 pts)**	**(−1 pt)**
Student successfully completed at least three of the non-task-specific steps.				
Student successfully completed all five of the non-task-specific steps.				
Total Score: <total # of points / 4 = %>				

Supervisor:

Supervisor/instructor signature _________________________________ Date _________________

Comments:

Retest supervisor/instructor signature _________________________ Date _________________

Comments:

CDX Tasksheet Number: A6028

Student/Intern Information

Name _________________________________ Date ____________ Class _____________________

Vehicle, Customer, and Service Information

Vehicle used for this activity:

Year ______________ Make _______________________ Model _____________________

Odometer _______________________ VIN _______________________________

Materials Required

- Blank work order
- Vehicle with available service history records
- Depending on the type of concern, special diagnostic tools may be required. See your supervisor/instructor for instructions to identify what tools may be required.
- Service information
- Serpentine belt removal tool set
- Belt tension tester
- Pulley alignment laser (if available) or straight edge

Task-Specific Safety Considerations

- Comply with personal and environmental safety practices associated with clothing; eye protection; hand tools; power equipment; proper ventilation; and the handling, storage, and disposal of chemicals/materials in accordance with local, state, and federal safety and environmental regulations.

▶ TASK Inspect, adjust, and/or replace generator (alternator) drive belts; check pulleys and tensioners for wear; and check pulley and belt alignment.

MLR A6D2

Time off_____________

Time on_____________

Total time_____________

Student Instructions: Read through the entire procedure prior to starting. Prepare your workspace and any tools or parts that may be needed to complete the task. When directed by your instructor, begin the procedure to complete the task and check the box as each step is finished. Track your time on this procedure for later comparison to the standard completion time (i.e., "flat rate" or customer pay time).

Procedure:	Step Completed
1. Locate the belt routing diagram.	
a. Draw or print off and attach to this sheet the belt routing diagram for your specific vehicle.	☐

2. Inspect the alternator drive belt.	
a. What type of belt is used to drive the alternator?	☐
b. Does your vehicle use a manual or automatic belt tensioner?	☐
c. Inspect the belt for cracks, fraying, or looseness. Attempt to rotate the alternator pulley by hand with the belt still attached. Are you able to spin the pulley? On the basis of your observation, what is your recommendation for service?	☐
d. Inspect the tensioner for wear. Some tensioners are equipped with a wear indicator.	☐
3. Adjust, remove, and replace the alternator drive belt.	
a. Using the proper procedure, use the serpentine belt removal tool to remove the belt.	☐
b. Inspect all engine accessory drive pulleys for binding or looseness. List any further necessary action.	☐
c. Check the alignment of the pulley. Mount the pulley alignment laser tool to the crankshaft. A straight-edge may be used in place of a laser tool. Check the alignment of the pulley. List any further necessary action.	☐
d. Use the belt routing diagram and install the drive belt. Ensure that the belt is tensioned properly.	☐
e. Start the vehicle and inspect belt for correct operation. If a manual tensioner is being used, refer to service information for the tensioning procedure.	☐

Non-Task-Specific Evaluations:	Step Completed
1. Tools and equipment were used as directed and returned in good working order.	☐
2. Complied with all general and task-specific safety standards, including proper use of any personal protective equipment.	☐
3. Completed the task in an appropriate time frame (Recommendation: 1.5 or 2 times flat rate).	☐
4. Left the work space clean and orderly.	☐
5. Cared for customer property and returned it undamaged.	☐

Student signature _________________________________ Date _________________________________

Comments:

Have your supervisor/instructor verify satisfactory completion of this procedure, any observations found, and any necessary action(s) recommended.

Tasksheet Scoring

Evaluation Items	Test		Retest	
	Pass	Fail	Pass	Fail
Task-Specific Evaluation	**(1 pt)**	**(0 pts)**	**(1 pt)**	**(0 pts)**
Obtain a belt routing diagram. Identify the type of drive belt and tensioner used on the vehicle.				
Inspect the belt's condition and tension prior to removal.				
Remove the belt and inspect all pulleys for correct operation and alignment.				
Install belt and tension as required.				
Non-Task-Specific Evaluation	**(0 pts)**	**(−1 pt)**	**(0 pts)**	**(−1 pt)**
Student successfully completed at least three of the non-task-specific steps.				
Student successfully completed all five of the non-task-specific steps.				
Total Score: <total # of points / 4 = %>				

Supervisor:

Supervisor/instructor signature _________________________________ Date _______________________

Comments:

Retest supervisor/instructor signature _________________________________ Date _______________________

Comments:

CDX Tasksheet Number: A6029

Student/Intern Information

Name _________________________________ Date ____________ Class _________________________

Vehicle, Customer, and Service Information

Vehicle used for this activity:

Year _______________ Make _____________________________ Model _______________________

Odometer _________________________________ VIN _______________________________________

> **Materials Required**
> - Blank work order
> - Vehicle with available service history records
> - Depending on the type of concern, special diagnostic/hand tools may be required. See your supervisor/instructor for instructions to identify what tools may be required.
> - Service information database
> - Personal protective equipment
> - General hand tools
> - Serpentine belt tools

Task-Specific Safety Considerations
- Comply with personal and environmental safety practices associated with clothing; eye protection; hand tools; power equipment; proper ventilation; and the handling, storage, and disposal of chemicals/materials in accordance with local, state, and federal safety and environmental regulations.

▶ TASK Remove, inspect, and/or replace generator (alternator). **MLR** **A6D3**

Time off____________

Student Instructions: Read through the entire procedure prior to starting. Prepare your workspace and any tools or parts that may be needed to complete the task. When directed by your instructor, begin the procedure to complete the task and check the box as each step is finished. Track your time on this procedure for later comparison to the standard completion time (i.e., "flat rate" or customer pay time).

Time on____________

Total time____________

Procedure:	Step Completed
1. Research the procedure for replacing the alternator.	
a. Obtain the belt routing diagram for your vehicle.	☐
b. Determine any special service precautions and torque specifications needed for this task.	☐

c. List the voltage and amperage output ratings specified for the alternator on your vehicle.	☐
2. Remove the alternator.	
a. Disconnect the battery cables.	☐
b. Use the serpentine belt removal tools to remove the belt. Describe how the belt is tensioned.	☐
c. Describe the location of the alternator. Remove any electrical connections from the alternator.	☐
d. Remove the alternator mounting bolts, then remove the alternator from the vehicle.	☐
e. Inspect the alternator for damage. Are there any signs of foreign debris entering the alternator? Inspect the pulley and bearings for wear.	☐
3. Install the alternator.	
a. Install the alternator into the vehicle and torque the fasteners to specification. **Tip:** If installing a new alternator, compare the voltage and amperage ratings with the specifications that you researched. Inspect the mounting areas and electrical connections. Inspect the alternator pulley to ensure that it is the same size with the same number of grooves for the belt to fit correctly.	☐
b. Install the electrical connections at the alternator.	☐
c. Use a belt routing diagram and install the belt. Ensure that the belt is tensioned correctly.	☐
d. Reconnect the battery. Start the vehicle and inspect the belt and alternator for correct operation. Inspect the instrument cluster to ensure that no charging system warning indicators are illuminated.	☐

Non-Task-Specific Evaluations:	Step Completed
1. Tools and equipment were used as directed and returned in good working order.	☐
2. Complied with all general and task-specific safety standards, including proper use of any personal protective equipment.	☐
3. Completed the task in an appropriate time frame (Recommendation: 1.5 or 2 times flat rate).	☐
4. Left the work space clean and orderly.	☐
5. Cared for customer property and returned it undamaged.	☐

Student signature _________________________ Date _________________________

Comments:

Have your supervisor/instructor verify satisfactory completion of this procedure, any observations found, and any necessary action(s) recommended.

Evaluation Instructions: The scoring box below is intended to act as a guide for both student and instructor. Each criterion listed will help students understand what is expected from them and evaluators articulate success at a particular task. The scoring is set up to allow a second attempt at each task (see the "Test" and "Retest" columns). Scoring is designed to reward students for correct completion of the task. Points are lost for failure to complete the employability requirements (see "Non-Task-Specific" criteria). When all criteria are evaluated, tally the points for a total at the bottom of each column.

Tasksheet Scoring

	Test		Retest	
Evaluation Items	**Pass**	**Fail**	**Pass**	**Fail**
Task-Specific Evaluation	**(1 pt)**	**(0 pts)**	**(1 pt)**	**(0 pts)**
Obtain a belt routing diagram and research the alternator removal procedure and output specifications.				
Remove the serpentine belt using the correct tools. Remove the alternator.				
Inspect the alternator.				
Install the alternator and drive belt. Verify correct installation and operation of both components.				
Non-Task-Specific Evaluation	**(0 pts)**	**(−1 pt)**	**(0 pts)**	**(−1 pt)**
Student successfully completed at least three of the non-task-specific steps.				
Student successfully completed all five of the non-task-specific steps.				
Total Score: <total # of points / 4 = %>				

Supervisor:

Supervisor/instructor signature ________________________ Date ________________

Comments:

Retest supervisor/instructor signature ________________________ Date ________________

Comments:

CDX Tasksheet Number: A6030

Student/Intern Information

Name _________________________________ Date _____________ Class _________________________

Vehicle, Customer, and Service Information

Vehicle used for this activity:

Year _______________ Make _________________________ Model _____________________________

Odometer _________________________ VIN ____

Materials Required

- Blank work order
- Vehicle with available service history records
- Depending on the type of concern, special diagnostic/hand tools may be required. See your supervisor/instructor for instructions to identify what tools may be required.
- Service information database
- Personal protective equipment
- Digital multimeter
- Battery tester

Task-Specific Safety Considerations

- Comply with personal and environmental safety practices associated with clothing; eye protection; hand tools; power equipment; proper ventilation; and the handling, storage, and disposal of chemicals/materials in accordance with local, state, and federal safety and environmental regulations.
- When running any vehicles in the shop, make sure you use the shop's exhaust ventilation system to discharge all exhaust gas safely outside.
- Special care must be taken when working around rotating components.

▶ TASK Perform charging circuit voltage drop tests, and determine necessary action.

MLR
A6D4

Time off_______________

Time on_______________

Total time_______________

Student Instructions: Read through the entire procedure prior to starting. Prepare your workspace and any tools or parts that may be needed to complete the task. When directed by your instructor, begin the procedure to complete the task and check the box as each step is finished. Track your time on this procedure for later comparison to the standard completion time (i.e., "flat rate" or customer pay time).

Procedure:	Step Completed
1. Perform a voltage drop test of the charging system.	
a. Measure the open circuit voltage of the battery and record your readings.	☐
b. Perform a battery test to ensure that the vehicle has a fully charged battery. Is the battery fully charged?	☐
c. Start the vehicle and measure the voltage at the battery. What is the voltage at the battery? Is your reading normal and within specifications?	☐
d. Measure the voltage drop of the positive side of the charging system. Connect the positive meter lead to the output terminal of the alternator. Connect the negative meter lead to the positive battery terminal. Record your results. Are these values acceptable?	☐
e. Measure voltage drop of the negative side of the charging system. Connect the positive meter lead to the alternator case. Connect the negative meter lead to the negative battery cable. Record your results. Are these values acceptable?	☐
f. Explain what could cause an excessive voltage drop on the positive side.	☐
g. Explain what could cause an excessive voltage drop on the negative side.	☐

<table>
<tr><td>h. On the basis of your inspection, what are your recommendations for service?</td><td>☐</td></tr>
</table>

Non-Task-Specific Evaluations:	**Step Completed**
1. Tools and equipment were used as directed and returned in good working order.	☐
2. Complied with all general and task-specific safety standards, including proper use of any personal protective equipment.	☐
3. Completed the task in an appropriate time frame (Recommendation: 1.5 or 2 times flat rate).	☐
4. Left the work space clean and orderly.	☐
5. Cared for customer property and returned it undamaged.	☐

Student signature _______________________ Date _______________________

Comments:

Have your supervisor/instructor verify satisfactory completion of this procedure, any observations found,

and any necessary action(s) recommended.

Tasksheet Scoring

Evaluation Items	Test		Retest	
	Pass	**Fail**	**Pass**	**Fail**
Task-Specific Evaluation	**(1 pt)**	**(O pts)**	**(1 pt)**	**(O pts)**
Measure the open circuit voltage of the battery. Perform a battery test.				
Test for excessive voltage drop on the positive side of the charging circuit.				
Test for excessive voltage drop on the negative side of the charging circuit.				
Describe the cause of an excessive voltage drop. Make recommendations for service based on your inspection.				
Non-Task-Specific Evaluation	**(O pts)**	**(−1 pt)**	**(O pts)**	**(−1 pt)**
Student successfully completed at least three of the non-task-specific steps.				
Student successfully completed all five of the non-task-specific steps.				
Total Score: <total # of points / 4 = %>				

Supervisor:

Supervisor/instructor signature _______________________ Date _______________

Comments:

Retest supervisor/instructor signature _______________________ Date _______________

Comments:

6E. Lighting, Instrument Cluster, Driver Information, and Body Electrical Systems

Learning Objective/Task	CDX Tasksheet Number	ASE Education Foundation Reference Number; Priority Level
• Inspect interior and exterior lamps and sockets, including headlights and auxiliary lights (fog lights/driving lights), and replace as needed.	A6031	A6E1, P-1
• Aim headlights.	A6032	A6E2, P-2
• Identify system voltage and safety precautions associated with high-intensity discharge headlights.	A6033	A6E3, P-2
• Disable and enable supplemental restraint system, and verify indicator lamp operation.	A6034	A6E4, P-1
• Remove and reinstall door panel.	A6035	A6E5, P-1
• Describe the operation of keyless entry/remote-start systems.	A6036	A6E6, P-3
• Verify operation of instrument panel gauges and warning/indicator lights, and reset maintenance indicators.	A6037	A6E7, P-1
• Verify windshield wiper and washer operation, and replace wiper blades.	A6038	A6E8, P-1

Materials Required

- Blank work order
- Vehicle with available service history records
- Depending on the type of concern, special diagnostic/hand tools may be required. See your supervisor/instructor for instructions to identify what tools may be required.
- Service information database
- Personal protective equipment

Safety Considerations

- When running any vehicles in the shop, make sure you use the shop's exhaust ventilation system to discharge all exhaust gas safely outside.
- Lifting equipment, such as vehicle jacks and stands, vehicle hoists, and engine hoists, are important tools that increase productivity and make the job easier. However, they can also cause severe injury or death if used improperly. Make sure you follow the manufacturer's operation procedures. Also, make sure you have your supervisor's/instructor's permission to use any particular type of lifting equipment.
- Comply with personal and environmental safety practices associated with clothing; eye protection; hand tools; power equipment; proper ventilation; and the handling, storage, and disposal of chemicals/materials in accordance with local, state, and federal safety and environmental regulations.
- If you need to start the vehicle, you should ensure that the parking brake is firmly applied; if necessary, use wheel chocks to prevent the vehicle from moving when the vehicle is started to verify the completion of these tasks.

CDX Tasksheet Number: A6031

Student/Intern Information

Name _________________________________ Date _____________ Class _________________________

Vehicle, Customer, and Service Information

Vehicle used for this activity:

Year _______________ Make _____________________________ Model _________________________

Odometer _________________________________ VIN _________________________________

Materials Required

- Blank work order
- Vehicle with available service history records
- Depending on the type of concern, special diagnostic/hand tools may be required. See your supervisor/instructor for instructions to identify what tools may be required.
- Service information database
- Personal protective equipment
- General hand tools
- Brake pedal depressor

Task-Specific Safety Considerations

- Comply with personal and environmental safety practices associated with clothing; eye protection; hand tools; power equipment; proper ventilation; and the handling, storage, and disposal of chemicals/materials in accordance with local, state, and federal safety and environmental regulations.

Time off_______________

Time on_______________

Total time_______________

▶ **TASK** Inspect interior and exterior lamps and sockets, including headlights and auxiliary lights (fog lights/driving lights), and replace as needed.

MLR
A6E1

Student Instructions: Read through the entire procedure prior to starting. Prepare your workspace and any tools or parts that may be needed to complete the task. When directed by your instructor, begin the procedure to complete the task and check the box as each step is finished. Track your time on this procedure for later comparison to the standard completion time (i.e., "flat rate" or customer pay time).

Procedure:	Step Completed
1. Inspect the exterior lights.	
a. Turn the vehicle's lights on. Inspect the low-beam headlights.	☐
b. Turn the high-beam headlights on and verify operation.	☐
c. If applicable, inspect the vehicle's fog lights.	☐

d. Inspect the parking/running lights.	☐
e. Turn on the left turn signal and inspect the front and rear lights. Then, test the right turn signals.	☐
f. Use the hazard button to ensure that all four-way flashers turn on.	☐
g. Inspect the tail lights.	☐
h. Install a brake pedal depressor. Inspect the brake lights, including the high-mounted center brake light.	☐
i. Inspect the license plate lights.	☐
2. Inspect the interior lights.	
a. Turn the interior light switch to the ON position.	☐
b. Inspect the overhead lights.	☐
c. Turn the map lights on and inspect their operation.	☐
d. Inspect the glovebox light.	☐
e. Open the trunk/hatch and inspect the lighting.	☐
3. Replace light bulbs as needed.	
a. When replacing bulbs, ensure that the replacement bulbs are the same as the original equipment. Verify correct operation of the bulbs once they have been replaced.	☐

Non-Task-Specific Evaluations:	Step Completed
1. Tools and equipment were used as directed and returned in good working order.	☐
2. Complied with all general and task-specific safety standards, including proper use of any personal protective equipment.	☐
3. Completed the task in an appropriate time frame (Recommendation: 1.5 or 2 times flat rate).	☐
4. Left the work space clean and orderly.	☐
5. Cared for customer property and returned it undamaged.	☐

Student signature _________________________________ Date _________________________

Comments:

Have your supervisor/instructor verify satisfactory completion of this procedure, any observations found, and any necessary action(s) recommended.

Evaluation Instructions: The scoring box below is intended to act as a guide for both student and instructor. Each criterion listed will help students understand what is expected from them and evaluators articulate success at a particular task. The scoring is set up to allow a second attempt at each task (see the "Test" and "Retest" columns). Scoring is designed to reward students for correct completion of the task. Points are lost for failure to complete the employability requirements (see "Non-Task-Specific" criteria). When all criteria are evaluated, tally the points for a total at the bottom of each column.

Tasksheet Scoring

	Test		Retest	
Evaluation Items	**Pass**	**Fail**	**Pass**	**Fail**
Task-Specific Evaluation	**(1 pt)**	**(0 pts)**	**(1 pt)**	**(0 pts)**
Inspect the exterior lights				
Inspect the interior lights.				
Replace bulbs as needed.				
Verify bulb operation after replacement.				
Non-Task-Specific Evaluation	**(0 pts)**	**(−1 pt)**	**(0 pts)**	**(−1 pt)**
Student successfully completed at least three of the non-task-specific steps.				
Student successfully completed all five of the non-task-specific steps.				
Total Score: <total # of points / 4 = %>				

Supervisor:

Supervisor/instructor signature _________________________ Date _______________

Comments:

Retest supervisor/instructor signature _________________________ Date _______________

Comments:

CDX Tasksheet Number: A6032

Student/Intern Information

Name _________________________________ Date ___________ Class _____________________

Vehicle, Customer, and Service Information

Vehicle used for this activity:

Year _______________ Make ______________________________ Model _____________________

Odometer _________________________________ VIN _____________________________________

Task-Specific Safety Considerations

- Comply with personal and environmental safety practices associated with clothing; eye protection; hand tools; power equipment; proper ventilation; and the handling, storage, and disposal of chemicals/materials in accordance with local, state, and federal safety and environmental regulations.

▶ **TASK** Aim headlights.

MLR
A6E2

Time off_______________

Time on_______________

Total time_______________

Student Instructions: Read through the entire procedure prior to starting. Prepare your workspace and any tools or parts that may be needed to complete the task. When directed by your instructor, begin the procedure to complete the task and check the box as each step is finished. Track your time on this procedure for later comparison to the standard completion time (i.e., "flat rate" or customer pay time).

Procedure:	Step Completed
1. Research the procedure for aiming headlights.	
a. List the procedure for aiming the headlights.	☐

b. List any specifications given in regards to vehicle positioning and measurements.	☐
2. Aim the headlights.	
a. Position the vehicle so that it is the recommended distance from a wall/door.	☐
b. Turn the headlights on.	☐
c. Observe and describe the headlight aim. Do the headlights require adjustment?	☐
d. Using the proper tools, aim the headlights according to specifications.	☐

Non-Task-Specific Evaluations:	**Step Completed**
1. Tools and equipment were used as directed and returned in good working order.	☐
2. Complied with all general and task-specific safety standards, including proper use of any personal protective equipment.	☐
3. Completed the task in an appropriate time frame (Recommendation: 1.5 or 2 times flat rate).	☐
4. Left the work space clean and orderly.	☐
5. Cared for customer property and returned it undamaged.	☐

Student signature _________________________________ Date _____________________

Comments:

Have your supervisor/instructor verify satisfactory completion of this procedure, any observations found, and any necessary action(s) recommended.

Evaluation Instructions: The scoring box below is intended to act as a guide for both student and instructor. Each criterion listed will help students understand what is expected from them and evaluators articulate success at a particular task. The scoring is set up to allow a second attempt at each task (see the "Test" and "Retest" columns). Scoring is designed to reward students for correct completion of the task. Points are lost for failure to complete the employability requirements (see "Non-Task-Specific" criteria). When all criteria are evaluated, tally the points for a total at the bottom of each column.

Tasksheet Scoring

	Test		Retest	
Evaluation Items	**Pass**	**Fail**	**Pass**	**Fail**
Task-Specific Evaluation	**(1 pt)**	**(0 pts)**	**(1 pt)**	**(0 pts)**
Research the procedure for aiming the headlights.				
Determine the vehicle's positioning when checking the headlight aim.				
Position the vehicle and observe the headlight aim.				
Adjust the headlight aim.				
Non-Task-Specific Evaluation	**(0 pts)**	**(−1 pt)**	**(0 pts)**	**(−1 pt)**
Student successfully completed at least three of the non-task-specific steps.				
Student successfully completed all five of the non-task-specific steps.				
Total Score: <total # of points / 4 = %>				

Supervisor:

Supervisor/instructor signature _________________________________ Date _______________________

Comments:

Retest supervisor/instructor signature _________________________________ Date _______________________

Comments:

CDX Tasksheet Number: A6033

Student/Intern Information

Name _________________________________ Date ___________ Class _____________________

Vehicle, Customer, and Service Information

Vehicle used for this activity:

Year _______________ Make _____________________________ Model _____________________

Odometer _________________________________ VIN _____________________________________

Materials Required

- Blank work order
- Vehicle with available service history records
- Depending on the type of concern, special diagnostic/hand tools may be required. See your supervisor/instructor for instructions to identify what tools may be required.
- Service information database
- Personal protective equipment

Task-Specific Safety Considerations

- Comply with personal and environmental safety practices associated with clothing; eye protection; hand tools; power equipment; proper ventilation; and the handling, storage, and disposal of chemicals/materials in accordance with local, state, and federal safety and environmental regulations.

▶ **TASK** Identify system voltage and safety precautions associated with high-intensity discharge (HID) headlights.

MLR
A6E3

Student Instructions: Read through the entire procedure prior to starting. Prepare your workspace and any tools or parts that may be needed to complete the task. When directed by your instructor, begin the procedure to complete the task and check the box as each step is finished. Track your time on this procedure for later comparison to the standard completion time (i.e., "flat rate" or customer pay time).

Time off___________

Time on___________

Total time___________

Procedure:	Step Completed
1. Research and identify HID headlight voltage specifications for your vehicle.	☐

2. Describe what safety precautions must be taken when replacing an HID bulb.	☐
3. Research and describe any handling procedures associated with replacing an HID bulb.	☐
4. Explain how the high voltage needed to illuminate an HID bulb in a vehicle is created and supplied.	☐
5. Research and describe the life expectancy of an HID bulb.	☐

Non-Task-Specific Evaluations:	Step Completed
1. Tools and equipment were used as directed and returned in good working order.	☐
2. Complied with all general and task-specific safety standards, including proper use of any personal protective equipment.	☐
3. Completed the task in an appropriate time frame (Recommendation: 1.5 or 2 times flat rate).	☐
4. Left the work space clean and orderly.	☐
5. Cared for customer property and returned it undamaged.	☐

Student signature _________________________ Date _________________________

Comments:

Have your supervisor/instructor verify satisfactory completion of this procedure, any observations found, and any necessary action(s) recommended.

Evaluation Instructions: The scoring box below is intended to act as a guide for both student and instructor. Each criterion listed will help students understand what is expected from them and evaluators articulate success at a particular task. The scoring is set up to allow a second attempt at each task (see the "Test" and "Retest" columns). Scoring is designed to reward students for correct completion of the task. Points are lost for failure to complete the employability requirements (see "Non-Task-Specific" criteria). When all criteria are evaluated, tally the points for a total at the bottom of each column.

Taksheet Scoring

	Test		Retest	
Evaluation Items	Pass	Fail	Pass	Fail
Task-Specific Evaluation	**(1 pt)**	**(0 pts)**	**(1 pt)**	**(0 pts)**
Research voltage specifications.				
Research safety precautions regarding HID bulb service and handling procedures.				
Describe the high voltage supply to HID bulbs.				
Describe bulb lifespans.				
Non-Task-Specific Evaluation	**(0 pts)**	**(−1 pt)**	**(0 pts)**	**(−1 pt)**
Student successfully completed at least three of the non-task-specific steps.				
Student successfully completed all five of the non-task-specific steps.				
Total Score: <total # of points / 4 = %>				

Supervisor:

Supervisor/instructor signature ______________________________ Date ________________

Comments:

Retest supervisor/instructor signature ______________________________ Date ________________

Comments:

CDX Tasksheet Number: A6034

Student/Intern Information

Name _________________________________ Date ____________ Class _________________________

Vehicle, Customer, and Service Information

Vehicle used for this activity:

Year _______________ Make _____________________________ Model _____________________________

Odometer _____________________________ VIN _____________________________

Task-Specific Safety Considerations
- Comply with personal and environmental safety practices associated with clothing; eye protection; hand tools; power equipment; proper ventilation; and the handling, storage, and disposal of chemicals/materials in accordance with local, state, and federal safety and environmental regulations.
- **Caution:** Working with airbags in the supplemental restraint systems (SRSs) can cause serious injury or even death if not handled with extreme care. Preventive measures must be taken at all time.

▶ TASK Disable and enable SRS, and verify indicator lamp operation.

MLR
A6E4

Time off_____________

Time on_____________

Total time_____________

Student Instructions: Read through the entire procedure prior to starting. Prepare your workspace and any tools or parts that may be needed to complete the task. When directed by your instructor, begin the procedure to complete the task and check the box as each step is finished. Track your time on this procedure for later comparison to the standard completion time (i.e., "flat rate" or customer pay time).

Procedure:	Step Completed
1. Research the SRS system in your vehicle.	
a. List the main components of the SRS system in your vehicle.	☐

b. What color is the wiring that leads up to airbag components?	☐
c. According to your service information, how is the SRS system disabled in order to service components in your vehicle?	☐
d. How long must you wait after disabling the SRS system to service the vehicle?	☐
e. As the driver of the vehicle, are you able to disable the passenger side SRS system manually? If so, how is that done?	☐
2. Verify correct operation of the airbag indicator light on the instrument cluster.	
a. Turn the ignition to the ON position. Verify that the airbag light comes on. Did the light come on and then turn off after 3–5 seconds? Is this a normal condition?	☐
b. Verify that the seatbelt light goes out as you buckle the seat belt.	☐
c. If the airbag light remains on, what does this mean?	☐
3. Disable the SRS system in your vehicle.	
a. Following proper procedure, disable the SRS system. Explain the steps you took to disable the system.	☐

<table>
<tr><td>b. If you were to remove the driver airbag, explain how you would carry it and how you would place it on the bench for service. DO NOT remove the airbag.</td><td>☐</td></tr>
<tr><td>4. Enable the SRS system in your vehicle.</td><td></td></tr>
<tr><td>a. Following proper procedure, enable the SRS system. Explain the steps you took to enable the system.</td><td>☐</td></tr>
<tr><td>b. After enabling the SRS, verify correct operation by cycling the ignition key once more. Verify that the airbag light comes on momentarily and then goes out, indicating that the self-check is complete.</td><td>☐</td></tr>
</table>

Non-Task-Specific Evaluations:	Step Completed
1. Tools and equipment were used as directed and returned in good working order.	☐
2. Complied with all general and task-specific safety standards, including proper use of any personal protective equipment.	☐
3. Completed the task in an appropriate time frame (Recommendation: 1.5 or 2 times flat rate).	☐
4. Left the work space clean and orderly.	☐
5. Cared for customer property and returned it undamaged.	☐

Student signature _________________________ Date _____________________

Comments:

Have your supervisor/instructor verify satisfactory completion of this procedure, any observations found, and any necessary action(s) recommended.

Evaluation Instructions: The scoring box below is intended to act as a guide for both student and instructor. Each criterion listed will help students understand what is expected from them and evaluators articulate success at a particular task. The scoring is set up to allow a second attempt at each task (see the "Test" and "Retest" columns). Scoring is designed to reward students for correct completion of the task. Points are lost for failure to complete the employability requirements (see "Non-Task-Specific" criteria). When all criteria are evaluated, tally the points for a total at the bottom of each column.

Tasksheet Scoring

Evaluation Items	Test		Retest	
	Pass	Fail	Pass	Fail
Task-Specific Evaluation	**(1 pt)**	**(0 pts)**	**(1 pt)**	**(0 pts)**
Research the SRS system in your vehicle. List the SRS components, wiring color, and safety precautions.				
Verify correct operation of the airbag light.				
Disable the SRS system.				
Enable the SRS system.				
Non-Task-Specific Evaluation	**(0 pts)**	**(−1 pt)**	**(0 pts)**	**(−1 pt)**
Student successfully completed at least three of the non-task-specific steps.				
Student successfully completed all five of the non-task-specific steps.				
Total Score: <total # of points / 4 = %>				

Supervisor:

Supervisor/instructor signature _________________________________ Date _____________________

Comments:

Retest supervisor/instructor signature _________________________________ Date _____________________

Comments:

CDX Tasksheet Number: A6035

Student/Intern Information

Name _________________________________ Date ___________ Class _____________________

Vehicle, Customer, and Service Information

Vehicle used for this activity:

Year _______________ Make _____________________________ Model _____________________

Odometer _____________________________ VIN _____________________________________

Materials Required
- Blank work order
- Vehicle with available service history records
- Depending on the type of concern, special diagnostic/hand tools may be required. See your supervisor/instructor for instructions to identify what tools may be required.
- Service information database
- Personal protective equipment
- Panel clip pliers
- General hand tools

Task-Specific Safety Considerations
- Comply with personal and environmental safety practices associated with clothing; eye protection; hand tools; power equipment; proper ventilation; and the handling, storage, and disposal of chemicals/materials in accordance with local, state, and federal safety and environmental regulations.

▶ TASK Remove and reinstall door panel. **MLR A6E5**

Time off_______________

Time on_______________

Student Instructions: Read through the entire procedure prior to starting. Prepare your workspace and any tools or parts that may be needed to complete the task. When directed by your instructor, begin the procedure to complete the task and check the box as each step is finished. Track your time on this procedure for later comparison to the standard completion time (i.e., "flat rate" or customer pay time).

Total time_______________

Procedure:	Step Completed
1. Research the procedure for removing the door panel.	
a. Describe how the door panel is held on. Is the panel held in by clips? Does the door panel need to be lifted out of position?	☐

b. How many clips are used to hold the door panel in place?	☐
c. Are any special tools required for this task?	☐
2. Remove the door panel.	
a. Remove necessary covers and switches to access all retaining hardware.	☐
b. Remove the retaining hardware.	☐
c. Lift the panel out of position. Use panel clip pliers to aid in the removal of stubborn clips. **Tip:** It may be helpful to roll the window down prior to panel removal.	☐
d. Remove any electrical connections, handles, locks, or linkages that may be attached to the door panel.	☐
3. Install the door panel.	
a. Replace any broken panel clips.	☐
b. Plug in any electrical connections and attach linkages that were removed.	☐
c. Install the door panel. Ensure that all panel clips are attached securely.	☐
d. Install any retaining hardware that was removed. Connect any switches that may have been removed.	☐
e. Verify that the window operates normally and there is no interference with the door panel.	☐

Non-Task-Specific Evaluations:	Step Completed
1. Tools and equipment were used as directed and returned in good working order.	☐
2. Complied with all general and task-specific safety standards, including proper use of any personal protective equipment.	☐
3. Completed the task in an appropriate time frame. (Recommendation: 1.5 or 2 times flat rate).	☐
4. Left the work space clean and orderly.	☐
5. Cared for customer property and returned it undamaged.	☐

Student signature _______________________________ Date _____________________________

Comments:

Have your supervisor/instructor verify satisfactory completion of this procedure, any observations found,

and any necessary action(s) recommended.

Tasksheet Scoring

		Test		Retest	
Evaluation Items	**Pass**	**Fail**	**Pass**	**Fail**
Task-Specific Evaluation	**(1 pt)**	**(0 pts)**	**(1 pt)**	**(0 pts)**
Research the procedure for door panel removal.				
Remove door panel accessories to expose the retaining hardware. Remove the door panel.				
Replace any damaged clips. Install the door panel.				
Verify window operation after panel replacement.				
Non-Task-Specific Evaluation	**(0 pts)**	**(−1 pt)**	**(0 pts)**	**(−1 pt)**
Student successfully completed at least three of the non-task-specific steps.				
Student successfully completed all five of the non-task-specific steps.				
Total Score: <total # of points / 4 = %>				

Supervisor:

Supervisor/instructor signature ________________________________ Date ____________________

Comments:

Retest supervisor/instructor signature ________________________________ Date ____________

Comments:

CDX Tasksheet Number: A6036

Student/Intern Information

Name _________________________________ Date ____________ Class _________________________

Vehicle, Customer, and Service Information

Vehicle used for this activity:

Year _______________ Make _______________________ Model _____________________

Odometer _______________________ VIN _________________________

Materials Required

- Blank work order
- Vehicle with available service history records
- Depending on the type of concern, special diagnostic/hand tools may be required. See your supervisor/instructor for instructions to identify what tools may be required.
- Service information database
- Personal protective equipment

Task-Specific Safety Considerations

- Comply with personal and environmental safety practices associated with clothing; eye protection; hand tools; power equipment; proper ventilation; and the handling, storage, and disposal of chemicals/materials in accordance with local, state, and federal safety and environmental regulations.

▶ TASK Describe the operation of keyless entry/remote-start systems.

MLR
A6E6

Time off________________

Student Instructions: Read through the entire procedure prior to starting. Prepare your workspace and any tools or parts that may be needed to complete the task. When directed by your instructor, begin the procedure to complete the task and check the box as each step is finished. Track your time on this procedure for later comparison to the standard completion time (i.e., "flat rate" or customer pay time).

Time on________________

Total time________________

Procedure:	Step Completed
1. Research the operation of keyless entry systems.	
a. Using service information, determine the estimated range of the remote key transmitter.	☐

b. Determine where the receivers are located on the vehicle.	☐
c. Research and describe the procedure for programming a new key fob.	☐
d. How many keys/transmitters are able to be programmed to your specific vehicle?	☐
e. Determine the operating frequency of the remote keyless entry system.	☐
2. Research the operation of remote-start systems.	
a. Determine the location of the remote-start control module.	☐
b. Describe what sequence of buttons you must press to activate the remote engine starter.	☐
c. Research and describe the procedure for programming the remote-start system.	☐
d. Explain what happens when the vehicle is remotely started. How do you enter the vehicle? Does the vehicle shut off if the doors are opened?	☐

Non-Task-Specific Evaluations:	Step Completed
1. Tools and equipment were used as directed and returned in good working order.	☐
2. Complied with all general and task-specific safety standards, including proper use of any personal protective equipment.	☐
3. Completed the task in an appropriate time frame (Recommendation: 1.5 or 2 times flat rate).	☐
4. Left the work space clean and orderly.	☐
5. Cared for customer property and returned it undamaged.	☐

Student signature ______________________________ Date ______________________________

Comments:

Have your supervisor/instructor verify satisfactory completion of this procedure, any observations found, and any necessary action(s) recommended.

Taksheet Scoring

Evaluation Items	Test		Retest	
	Pass	**Fail**	**Pass**	**Fail**
Task-Specific Evaluation	**(1 pt)**	**(O pts)**	**(1 pt)**	**(O pts)**
Determine the range of the remote transmitter.				
Determine the location of the receiver on the vehicle.				
Describe the programming procedure for remote keyless entry/remote-start.				
Describe remote-start operation.				
Non-Task-Specific Evaluation	**(O pts)**	**(−1 pt)**	**(O pts)**	**(−1 pt)**
Student successfully completed at least three of the non-task-specific steps.				
Student successfully completed all five of the non-task-specific steps.				
Total Score: <total # of points / 4 = %>				

Supervisor:

Supervisor/instructor signature _________________________ Date _______________

Comments:

Retest supervisor/instructor signature _________________________ Date _______________

Comments:

CDX Tasksheet Number: A6037

Student/Intern Information

Name _________________________________ Date ____________ Class _________________________

Vehicle, Customer, and Service Information

Vehicle used for this activity:

Year _______________ Make _______________________________ Model _____________________

Odometer _________________________________ VIN _______________________________

> **Materials Required**
> - Blank work order
> - Vehicle with available service history records
> - Depending on the type of concern, special hand tool/diagnostic tools may be required. See your supervisor/instructor for instructions to identify what tools may be required.
> - Vehicle service information database

Task-Specific Safety Considerations
- Shop rules and procedures are critical to your safety. Please give these your utmost attention.
- When running any vehicles in the shop, make sure you use the shop's exhaust ventilation system to discharge all exhaust gas safely outside.

▶ TASK Verify operation of the instrument panel engine warning indicators, and reset maintenance indicators.

MLR A6E7

Time off_____________

Time on_____________

Total time_____________

Student Instructions: Read through the entire procedure prior to starting. Prepare your workspace and any tools or parts that may be needed to complete the task. When directed by your instructor, begin the procedure to complete the task and check the box as each step is finished. Track your time on this procedure for later comparison to the standard completion time (i.e., "flat rate" or customer pay time).

Procedure:	Step Completed
1. Cycle the ignition to the ON position but do not start the vehicle.	
a. List all of the warning indicators that illuminate when the key is cycled on. Verify that the check engine light warning indicator is illuminated.	☐

b. Start the vehicle. Verify that the warning indicators turn off after the vehicle is started. List any indicators that remain on once the vehicle has been started.	☐
2. List the instrument panel gauges that are equipped with your specific vehicle.	
a. Start the vehicle. Verify that the gauges are in proper working condition.	☐
b. List any gauges that are not working correctly.	☐
c. Operate the turn signals in the vehicle. Verify that the turn indicators are working correctly.	☐
d. Operate the high-beam headlights. Verify that the high-beam indicator is working correctly.	☐
3. Research the procedure for resetting the maintenance indicator.	
a. List or print off the procedure for resetting the maintenance indicator.	☐
b. Perform the maintenance indicator reset procedure.	☐

Non-Task-Specific Evaluations:	Step Completed
1. Tools and equipment were used as directed and returned in good working order.	☐
2. Complied with all general and task-specific safety standards, including proper use of any personal protective equipment.	☐
3. Completed the task in an appropriate time frame (Recommendation: 1.5 or 2 times flat rate).	☐
4. Left the work space clean and orderly.	☐
5. Cared for customer property and returned it undamaged.	☐

Student signature ________________________ Date ________________________

Comments:

Have your supervisor/instructor verify satisfactory completion of this procedure, any observations found, and any necessary action(s) recommended.

Tasksheet Scoring

	Test		Retest	
Evaluation Items	**Pass**	**Fail**	**Pass**	**Fail**
Task-Specific Evaluation	**(1 pt)**	**(0 pts)**	**(1 pt)**	**(0 pts)**
List the warning indicators illuminated on the dash and verify the operation of the check engine light.				
Verify the operation of the gauges.				
Research the procedure for resetting the maintenance light.				
Perform the maintenance light reset procedure.				
Non-Task-Specific Evaluation	**(0 pts)**	**(−1 pt)**	**(0 pts)**	**(−1 pt)**
Student successfully completed at least three of the non-task-specific steps.				
Student successfully completed all five of the non-task-specific steps.				
Total Score: <total # of points / 4 = %>				

Supervisor:

Supervisor/instructor signature ________________________ Date ________________

Comments:

Retest supervisor/instructor signature ________________________ Date ________________

Comments:

CDX Tasksheet Number: A6038

Student/Intern Information

Name _______________________________ Date ____________ Class _______________________

Vehicle, Customer, and Service Information

Vehicle used for this activity:

Year ______________ Make ______________________ Model _______________

Odometer _________________________ VIN _______________________

Materials Required
- Blank work order
- Vehicle with available service history records
- Depending on the type of concern, special diagnostic/hand tools may be required. See your supervisor/instructor for instructions to identify what tools may be required.
- Service information database
- Personal protective equipment

Task-Specific Safety Considerations
- Comply with personal and environmental safety practices associated with clothing; eye protection; hand tools; power equipment; proper ventilation; and the handling, storage, and disposal of chemicals/materials in accordance with local, state, and federal safety and environmental regulations.

▶ TASK Verify windshield wiper and washer operation, and replace wiper blades.

MLR
A6E8

Time off____________

Time on____________

Total time____________

Student Instructions: Read through the entire procedure prior to starting. Prepare your workspace and any tools or parts that may be needed to complete the task. When directed by your instructor, begin the procedure to complete the task and check the box as each step is finished. Track your time on this procedure for later comparison to the standard completion time (i.e., "flat rate" or customer pay time).

Procedure:	Step Completed
1. Verify correct wiper and washer operation.	
a. Activate the washer motor. Verify that both washer nozzles spray an even pattern directed at the correct location on the windshield. **Tip:** Do not use the wiper blades on a dry windshield.	☐
b. Verify that the wipers turn on when the washer motor is activated.	☐
c. Inspect all speeds of wiper motor operation. Periodically, use the washer pump to spray the windshield.	☐

d. If the vehicle is equipped with a rear wiper/washer, verify that it is operating correctly.	☐
2. Replace the wiper blades.	
a. Explain how the wiper blade is held into place on the wiper arm.	☐
b. Following procedure, remove the wiper blade. **Tip:** Do not leave the wiper arm in the upright position, as it may accidentally contact and break the windshield.	☐
c. Install the new wiper blades. Ensure that they are held securely in place.	☐
d. Activate the wiper and washer functions. Observe the wiper pattern and ensure the wipers are not streaking or missing any portion of the windshield.	☐

Non-Task-Specific Evaluations:	Step Completed
1. Tools and equipment were used as directed and returned in good working order.	☐
2. Complied with all general and task-specific safety standards, including proper use of any personal protective equipment.	☐
3. Completed the task in an appropriate time frame (Recommendation: 1.5 or 2 times flat rate).	☐
4. Left the work space clean and orderly.	☐
5. Cared for customer property and returned it undamaged.	☐

Student signature _________________________ Date _____________________

Comments:

Have your supervisor/instructor verify satisfactory completion of this procedure, any observations found, and any necessary action(s) recommended.

Evaluation Instructions: The scoring box below is intended to act as a guide for both student and instructor. Each criterion listed will help students understand what is expected from them and evaluators articulate success at a particular task. The scoring is set up to allow a second attempt at each task (see the "Test" and "Retest" columns). Scoring is designed to reward students for correct completion of the task. Points are lost for failure to complete the employability requirements (see "Non-Task-Specific" criteria). When all criteria are evaluated, tally the points for a total at the bottom of each column.

Tasksheet Scoring

	Test		Retest	
Evaluation Items	**Pass**	**Fail**	**Pass**	**Fail**
Task-Specific Evaluation	**(1 pt)**	**(0 pts)**	**(1 pt)**	**(0 pts)**
Verify correct washer spray pattern and wiper function on all speeds.				
Describe how the wiper blades are held into place.				
Remove the wiper blades.				
Replace the wiper blades and verify proper operation.				
Non-Task-Specific Evaluation	**(0 pts)**	**(−1 pt)**	**(0 pts)**	**(−1 pt)**
Student successfully completed at least three of the non-task-specific steps.				
Student successfully completed all five of the non-task-specific steps.				
Total Score: <total # of points / 4 = %>				

Supervisor:

Supervisor/instructor signature _______________________________ Date _______________________

Comments:

Retest supervisor/instructor signature _______________________________ Date _______________________

Comments:

7A. Heating, Ventilation, and Air Conditioning

Learning Objective/Task	CDX Tasksheet Number	ASE Education Foundation Reference Number; Priority Level
• Research vehicle service information, including refrigerant/oil type, vehicle service history, service precautions, and technical service bulletins (TSBs).	A7001	A7A1, P-1
• Identify heating, ventilation, and air conditioning (HVAC) components and configuration.	A7002	A7A2, P-1

Materials Required

- Blank work order
- Vehicle with available service history records
- Depending on the type of concern, special diagnostic/hand tools may be required. See your supervisor/instructor for instructions to identify what tools may be required.
- Service information database
- Personal protective equipment

Safety Considerations

- Diagnosis of this fault may require test-driving the vehicle on the school grounds. Attempt this task only with full permission from your instructor and follow all the guidelines exactly.
- When running any vehicles in the shop, make sure you use the shop's exhaust ventilation system to discharge all exhaust gas safely outside.
- Extreme caution must be exercised when working around rotating components.
- Refrigerant can cause serious damage if it comes in contact with a person's unprotected skin and eyes.
- When operating, the AC system is normally subject to very high pressure in the system. Extreme caution must be exercised when working on an operating system.
- Comply with personal and environmental safety practices associated with clothing; eye protection; hand tools; power equipment; proper ventilation; and the handling, storage, and disposal of chemicals/materials in accordance with local, state, and federal safety and environmental regulations.

CDX Tasksheet Number: A7OO1

Student/Intern Information

Name ___________________________ Date __________ Class ___________________

Vehicle, Customer, and Service Information

Vehicle used for this activity:

Year ______________ Make ___________________ Model ___________________

Odometer ___________________ VIN ___________________

Materials Required

- Blank work order
- Vehicle with available service history records
- Depending on the type of concern, special diagnostic/hand tools may be required. See your supervisor/instructor for instructions to identify what tools may be required.
- Service information database
- Personal protective equipment

Task-Specific Safety Considerations

- Comply with personal and environmental safety practices associated with clothing; eye protection; hand tools; power equipment; proper ventilation; and the handling, storage, and disposal of chemicals/materials in accordance with local, state, and federal safety and environmental regulations.

► TASK Research vehicle service information, including refrigerant/oil type, vehicle service history, service precautions, and TSBs.

MLR
A7A1

Time off____________

Time on____________

Total time____________

Student Instructions: Read through the entire procedure prior to starting. Prepare your work space and any tools or parts that may be needed to complete the task. When directed by your instructor, begin the procedure to complete the task and check the box as each step is finished. Track your time on this procedure for later comparison to the standard completion time (i.e., "flat rate" or customer pay time).

Procedure:	Step Completed
1. Research your vehicle's service information.	
a. List the refrigerant type and capacity used in your vehicle.	☐

b. List the oil type and capacity used in the AC system.	☐
c. List the coolant type and capacity used in your vehicle.	☐
2. Research your vehicle's service history.	
a. Determine any routine maintenance that has been completed.	☐
b. Determine any outstanding/completed recall work.	☐
c. Determine any major repair in the service history.	☐
d. Identify service precautions related to any maintenance- or recall-related task.	☐
3. Determine the TSBs that have been issued for your specific vehicle.	
a. Identify one HVAC-related TSB and list the TSB number, vehicle concern, and repair procedure.	☐

Non-Task-Specific Evaluations:	Step Completed
1. Tools and equipment were used as directed and returned in good working order.	☐
2. Complied with all general and task-specific safety standards, including proper use of any personal protective equipment.	☐
3. Completed the task in an appropriate time frame (Recommendation: 1.5 or 2 times flat rate).	☐
4. Left the work space clean and orderly.	☐
5. Cared for customer property and returned it undamaged.	☐

Student signature _______________________________ Date _______________________________

Comments:

Have your supervisor/instructor verify satisfactory completion of this procedure, any observations found, and any necessary action(s) recommended.

Tasksheet Scoring

		Test		Retest	
Evaluation Items	**Pass**	**Fail**	**Pass**	**Fail**
Task-Specific Evaluation	**(1 pt)**	**(0 pts)**	**(1 pt)**	**(0 pts)**
Research refrigerant, oil, and coolant types and capacities. Identify HVAC-related TSBs.				
Research the vehicle's service history.				
Identify any service precautions.				
Determine any completed/outstanding recalls.				
Non-Task-Specific Evaluation	**(0 pts)**	**(−1 pt)**	**(0 pts)**	**(−1 pt)**
Student successfully completed at least three of the non-task-specific steps.				
Student successfully completed all five of the non-task-specific steps.				
Total Score: <total # of points / 4 = %>				

Supervisor:

Supervisor/instructor signature _______________________________ Date _______________________

Comments:

Retest supervisor/instructor signature _______________________________ Date _______________________

Comments:

CDX Tasksheet Number: A7002

Student/Intern Information

Name _______________________________ Date ____________ Class _______________________

Vehicle, Customer, and Service Information

Vehicle used for this activity:

Year _____________ Make _____________________ Model _____________________

Odometer _____________________ VIN _____________________________

Task-Specific Safety Considerations

- Comply with personal and environmental safety practices associated with clothing; eye protection; hand tools; power equipment; proper ventilation; and the handling, storage, and disposal of chemicals/materials in accordance with local, state, and federal safety and environmental regulations.

Time off_____________

Time on_____________

Total time_____________

▶ **TASK** Identify heating, ventilation, and air conditioning (HVAC) components and configuration.

MLR
A7A2

Student Instructions: Read through the entire procedure prior to starting. Prepare your work space and any tools or parts that may be needed to complete the task. When directed by your instructor, begin the procedure to complete the task and check the box as each step is finished. Track your time on this procedure for later comparison to the standard completion time (i.e., "flat rate" or customer pay time).

Procedure:	Step Completed
1. Research HVAC components and their configurations. Use service information to determine the configuration of the AC components in your vehicle. Determine the following:	
a. Is the AC compressor a cycling clutch or variable displacement type?	☐

b. Is the AC compressor electric or belt driven?	☐
c. Is your vehicle equipped with an orifice tube or a thermal expansion valve? Describe where these components are located.	☐
d. Is your vehicle equipped with a receiver drier or an accumulator? Describe where these components are located.	☐
e. Describe the configuration of the HVAC controls. (Auto climate control, dual-zone climate control, and rear passenger climate control)	☐
f. Describe how the blend and mode doors are controlled inside the vehicle.	☐
2. Describe the location of the following components:	
a. Heater core	☐
b. Evaporator	☐
c. Condenser	☐
d. Heater control valve	☐

<table>
<tr><td>e. Cabin air filter</td><td>☐</td></tr>
<tr><td>f. AC compressor</td><td>☐</td></tr>
<tr><td>g. Blower motor</td><td>☐</td></tr>
</table>

Non-Task-Specific Evaluations:	**Step Completed**
1. Tools and equipment were used as directed and returned in good working order.	☐
2. Complied with all general and task-specific safety standards, including proper use of any personal protective equipment.	☐
3. Completed the task in an appropriate time frame (Recommendation: 1.5 or 2 times flat rate).	☐
4. Left the work space clean and orderly.	☐
5. Cared for customer property and returned it undamaged.	☐

Student signature _______________________________ Date _______________________________

Comments:

Have your supervisor/instructor verify satisfactory completion of this procedure, any observations found, and any necessary action(s) recommended.

Tasksheet Scoring

Evaluation Items	Test		Retest	
	Pass	**Fail**	**Pass**	**Fail**
Task-Specific Evaluation	**(1 pt)**	**(0 pts)**	**(1 pt)**	**(0 pts)**
Determine the AC compressor and expansion device type.				
Determine the HVAC control configuration.				
Determine blend and mode door operation.				
Determine the location of various HVAC components.				
Non-Task-Specific Evaluation	**(0 pts)**	**(−1 pt)**	**(0 pts)**	**(−1 pt)**
Student successfully completed at least three of the non-task-specific steps.				
Student successfully completed all five of the non-task-specific steps.				
Total Score: <total # of points / 4 = %>				

Supervisor:

Supervisor/instructor signature _______________________________ Date _______________

Comments:

Retest supervisor/instructor signature _______________________ Date _______________

Comments:

Learning Objective/Task	CDX Tasksheet Number	ASE Education Foundation Reference Number; Priority Level
• Inspect and replace A/C compressor drive belts, pulleys, and tensioners; visually inspect A/C components for signs of leaks and determine necessary action.	A7003	A7B1, P-1
• Identify hybrid-vehicle A/C system electrical circuits and the service/safety precautions.	A7004	A7B2, P-2
• Inspect A/C condenser for airflow restrictions and determine necessary action.	A7005	A7B3, P-1

Materials Required

- Blank work order
- Vehicle with available service history records
- Depending on the type of concern, special diagnostic/hand tools may be required. See your supervisor/instructor for instructions to identify what tools may be required.
- Service information database
- Personal protective equipment

Safety Considerations

- Diagnosis of this fault may require test-driving the vehicle on the school grounds. Attempt this task only with full permission from your instructor and follow all the guidelines exactly.
- When running any vehicles in the shop, make sure you use the shop's exhaust ventilation system to discharge all exhaust gas safely outside.
- Extreme caution must be exercised when working around rotating components.
- Refrigerant can cause serious damage if it comes in contact with a person's unprotected skin and eyes.
- When operating, the air conditioning system is normally subject to very high pressure in the system. Extreme caution must be exercised when working on an operating system.
- Comply with personal and environmental safety practices associated with clothing; eye protection; hand tools; power equipment; proper ventilation; and the handling, storage, and disposal of chemicals/materials in accordance with local, state, and federal safety and environmental regulations.

CDX Tasksheet Number: A7003

Student/Intern Information

Name ___________________________ Date ___________ Class ___________________

Vehicle, Customer, and Service Information

Vehicle used for this activity:

Year _______________ Make ___________________________ Model ___________________

Odometer ___________________________ VIN ___________________________

Materials Required
- Blank work order
- Vehicle with available service history records
- Depending on the type of concern, special diagnostic/hand tools may be required. See your supervisor/instructor for instructions to identify what tools may be required.
- Service information database
- Personal protective equipment
- Serpentine belt tools
- Refrigerant detector

Task-Specific Safety Considerations
- Comply with personal and environmental safety practices associated with clothing; eye protection; hand tools; power equipment; proper ventilation; and the handling, storage, and disposal of chemicals/materials in accordance with local, state, and federal safety and environmental regulations.

▶ TASK Inspect and replace A/C compressor drive belts, pulleys, and tensioners; visually inspect A/C components for signs of leaks and determine necessary action.

MLR
A7B1

Time off____________

Time on____________

Total time____________

Student Instructions: Read through the entire procedure prior to starting. Prepare your work space and any tools or parts that may be needed to complete the task. When directed by your instructor, begin the procedure to complete the task and check the box as each step is finished. Track your time on this procedure for later comparison to the standard completion time (i.e., "flat rate" or customer pay time).

Procedure:	Step Completed
1. Inspect and replace the A/C compressor drive belt.	
a. Obtain a belt routing diagram.	☐
b. Inspect the drive belt for cracks and fraying. Inspect the belt tension. Is there one or multiple belts used to drive the engine accessories? What type of tensioner is being used?	☐
c. Use the serpentine belt tools to remove the compressor drive belt.	☐
d. Thoroughly inspect the drive belt and pulleys at this time. Describe any action required.	☐
e. Install the belt using the serpentine belt tools. Ensure that the belt is tensioned properly.	☐
2. Inspect the A/C components for signs of leaks.	
a. Describe the different methods you could use to pinpoint refrigerant leaks.	☐
b. Visually inspect all A/C components for signs of leaks. **Tip:** Typically, when there is a refrigerant leak, the source of the leak will appear to accumulate dirt/dust, as the oil in the refrigerant collects dust and debris.	☐
c. If available, use a refrigerant detector to check all lines, connections, and components for leaking refrigerant. **Tip:** Do not forget to inspect vehicles rear A/C lines and related components if applicable. Check the A/C drain tubes for signs of leaking refrigerant.	☐
3. On the basis of your inspection, what is your recommendation for service?	☐

Non-Task-Specific Evaluations:	Step Completed
1. Tools and equipment were used as directed and returned in good working order.	☐
2. Complied with all general and task-specific safety standards, including proper use of any personal protective equipment.	☐
3. Completed the task in an appropriate time frame (Recommendation: 1.5 or 2 times flat rate).	☐
4. Left the work space clean and orderly.	☐
5. Cared for customer property and returned it undamaged.	☐

Student signature _______________________________ Date _______________________________

Comments:

Have your supervisor/instructor verify satisfactory completion of this procedure, any observations found, and any necessary action(s) recommended.

Tasksheet Scoring

	Test		Retest	
Evaluation Items	**Pass**	**Fail**	**Pass**	**Fail**
Task-Specific Evaluation	**(1 pt)**	**(0 pts)**	**(1 pt)**	**(0 pts)**
Obtain a belt routing diagram. Inspect, remove, and replace A/C drive belt. Inspect pulleys for wear.				
Describe the different methods you could use to pinpoint refrigerant leaks.				
Perform a visual inspection of all A/C components.				
Make recommendations for service based on your inspection.				
Non-Task-Specific Evaluation	**(0 pts)**	**(−1 pt)**	**(0 pts)**	**(−1 pt)**
Student successfully completed at least three of the non-task-specific steps.				
Student successfully completed all five of the non-task-specific steps.				
Total Score: <total # of points / 4 = %>				

Supervisor:

Supervisor/instructor signature ______________________________ Date ________________

Comments:

Retest supervisor/instructor signature ______________________________ Date ________________

Comments:

CDX Tasksheet Number: A7004

Student/Intern Information

Name _________________________________ Date ____________ Class _________________________

Vehicle, Customer, and Service Information

Vehicle used for this activity:

Year _______________ Make _____________________________ Model _____________________________

Odometer _________________________________ VIN ____

Materials Required

- Blank work order
- Vehicle with available service history records
- Depending on the type of concern, special diagnostic/hand tools may be required. See your supervisor/instructor for instructions to identify what tools may be required.
- Service information database
- Personal protective equipment

Task-Specific Safety Considerations

- Comply with personal and environmental safety practices associated with clothing; eye protection; hand tools; power equipment; proper ventilation; and the handling, storage, and disposal of chemicals/materials in accordance with local, state, and federal safety and environmental regulations.

▶ **TASK** Identify hybrid-vehicle A/C system electrical circuits and the service/safety precautions.

MLR
A7B2

Time off_____________

Time on_____________

Total time_____________

Student Instructions: Read through the entire procedure prior to starting. Prepare your work space and any tools or parts that may be needed to complete the task. When directed by your instructor, begin the procedure to complete the task and check the box as each step is finished. Track your time on this procedure for later comparison to the standard completion time (i.e., "flat rate" or customer pay time).

Procedure:	Step Completed
1. Research hybrid-vehicle A/C system service and safety precautions.	
a. Explain how the A/C compressor on the hybrid vehicle is different from a traditional belt-driven A/C compressor.	☐

b. How much voltage is supplied to the A/C compressor? Where is this voltage supplied from?	☐
c. What color wiring indicates high voltage to the A/C compressor?	☐
d. Is the A/C compressor able to operate with the engine off?	☐
e. Where is the A/C compressor located?	☐
f. Use service information to determine any special service precautions needed when working on an electric compressor. List any precautions.	☐

Non-Task-Specific Evaluations:	Step Completed
1. Tools and equipment were used as directed and returned in good working order.	☐
2. Complied with all general and task-specific safety standards, including proper use of any personal protective equipment.	☐
3. Completed the task in an appropriate time frame (Recommendation: 1.5 or 2 times flat rate).	☐
4. Left the work space clean and orderly.	☐
5. Cared for customer property and returned it undamaged.	☐

Student signature _______________________________ Date _______________________________

Comments:

Have your supervisor/instructor verify satisfactory completion of this procedure, any observations found, and any necessary action(s) recommended.

Evaluation Instructions: The scoring box below is intended to act as a guide for both student and instructor. Each criterion listed will help students understand what is expected from them and evaluators articulate success at a particular task. The scoring is set up to allow a second attempt at each task (see the "Test" and "Retest" columns). Scoring is designed to reward students for correct completion of the task. Points are lost for failure to complete the employability requirements (see "Non-Task-Specific" criteria). When all criteria are evaluated, tally the points for a total at the bottom of each column.

Tasksheet Scoring

	Test		Retest	
Evaluation Items	**Pass**	**Fail**	**Pass**	**Fail**
Task-Specific Evaluation	**(1 pt)**	**(0 pts)**	**(1 pt)**	**(0 pts)**
Explain the differences between electric- and belt-driven compressors. Determine the high-voltage wiring color.				
Determine the voltage requirements for electric compressor operation.				
Determine the compressor's location.				
Identify any service/safety precautions.				
Non-Task-Specific Evaluation	**(0 pts)**	**(−1 pt)**	**(0 pts)**	**(−1 pt)**
Student successfully completed at least three of the non-task-specific steps.				
Student successfully completed all five of the non-task-specific steps.				
Total Score: <total # of points / 4 = %>				

Supervisor:

Supervisor/instructor signature _______________________________ Date _______________________

Comments:

Retest supervisor/instructor signature _______________________ Date _______________________

Comments:

Student/Intern Information

Name _________________________________ Date _____________ Class _________________________

Vehicle, Customer, and Service Information

Vehicle used for this activity:

Year _______________ Make _____________________________ Model _____________________________

Odometer _________________________________ VIN ____

Materials Required

- Blank work order
- Vehicle with available service history records
- Depending on the type of concern, special diagnostic/hand tools may be required. See your supervisor/instructor for instructions to identify what tools may be required.
- Service information database
- Personal protective equipment

Task-Specific Safety Considerations

- Comply with personal and environmental safety practices associated with clothing; eye protection; hand tools; power equipment; proper ventilation; and the handling, storage, and disposal of chemicals/materials in accordance with local, state, and federal safety and environmental regulations.

▶ **TASK** Inspect A/C condenser for airflow restrictions and determine necessary action.

MLR
A7B3

Time off______________

Time on______________

Total time______________

Student Instructions: Read through the entire procedure prior to starting. Prepare your work space and any tools or parts that may be needed to complete the task. When directed by your instructor, begin the procedure to complete the task and check the box as each step is finished. Track your time on this procedure for later comparison to the standard completion time (i.e., "flat rate" or customer pay time).

Procedure:	Step Completed
1. Identify problems related to A/C condenser airflow restrictions.	

a. Explain how the air conditioner may operate if the condenser airflow is restricted.	☐
b. Where is the A/C condenser located on your vehicle?	☐
2. Inspect the A/C condenser.	
a. Remove any shielding to access the A/C condenser.	☐
b. Inspect both the front and back of the condenser for debris, such as dirt, leaves, or foreign matter, that may have been picked up from the roadways.	☐
c. Inspect the condenser fins. Ensure that the fins are not deformed and blocking airflow.	☐
3. On the basis of your inspection, what is your recommendation for service?	☐

Non-Task-Specific Evaluations:	Step Completed
1. Tools and equipment were used as directed and returned in good working order.	☐
2. Complied with all general and task-specific safety standards, including proper use of any personal protective equipment.	☐
3. Completed the task in an appropriate time frame (Recommendation: 1.5 or 2 times flat rate).	☐
4. Left the work space clean and orderly.	☐
5. Cared for customer property and returned it undamaged.	☐

Student signature _________________________ Date _________________________

Comments:

Have your supervisor/instructor verify satisfactory completion of this procedure, any observations found, and any necessary action(s) recommended.

Tasksheet Scoring

	Test		Retest	
Evaluation Items	**Pass**	**Fail**	**Pass**	**Fail**
Task-Specific Evaluation	**(1 pt)**	**(O pts)**	**(1 pt)**	**(O pts)**
Explain the effects of airflow restrictions on A/C performance.				
Determine the A/C condenser's location.				
Remove the necessary components to access the A/C condenser. Inspect the condenser for signs of airflow restrictions.				
Make recommendations for service based on your inspection.				
Non-Task-Specific Evaluation	**(O pts)**	**(−1 pt)**	**(O pts)**	**(−1 pt)**
Student successfully completed at least three of the non-task-specific steps.				
Student successfully completed all five of the non-task-specific steps.				
Total Score: <total # of points / 4 = %>				

Supervisor:

Supervisor/instructor signature ______________________________ Date ______________________

Comments:

Retest supervisor/instructor signature ______________________________ Date ______________________

Comments:

Learning Objective/Task	CDX Tasksheet Number	ASE Education Foundation Reference Number; Priority Level
• Inspect engine cooling and heater systems hoses and pipes, and determine necessary action.	A7006	A7C1, P-1

Materials Required

- Blank work order
- Vehicle with available service history records
- Depending on the type of concern, special diagnostic/hand tools may be required. See your supervisor/instructor for instructions to identify what tools may be required.
- Service information database
- Personal protective equipment

Safety Considerations

- Diagnosis of this fault may require test-driving the vehicle on the school grounds. Attempt this task only with full permission from your instructor and follow all the guidelines exactly.
- When running any vehicles in the shop, make sure you use the shop's exhaust ventilation system to discharge all exhaust gas safely outside.
- Extreme caution must be exercised when working around rotating components.
- Refrigerant can cause serious damage if it comes in contact with a person's unprotected skin and eyes.
- When operating, the air conditioning system is normally subject to very high pressure in the system. Extreme caution must be exercised when working on an operating system.
- Comply with personal and environmental safety practices associated with clothing; eye protection; hand tools; power equipment; proper ventilation; and the handling, storage, and disposal of chemicals/materials in accordance with local, state, and federal safety and environmental regulations.

CDX Tasksheet Number: A7006

Student/Intern Information

Name ________________________________ Date ____________ Class ____________________

Vehicle, Customer, and Service Information

Vehicle used for this activity:

Year ______________ Make ____________________________ Model ____________________

Odometer ____________________________ VIN ________________________________

Materials Required

- Blank work order
- Vehicle with available service history records
- Depending on the type of concern, special diagnostic/hand tools may be required. See your supervisor/instructor for instructions to identify what tools may be required.
- Service information database
- Personal protective equipment

Task-Specific Safety Considerations

- Comply with personal and environmental safety practices associated with clothing; eye protection; hand tools; power equipment; proper ventilation; and the handling, storage, and disposal of chemicals/materials in accordance with local, state, and federal safety and environmental regulations.
- **Caution:** Never open a radiator/cooling system when the vehicle is warm or at operating temperature. The cooling system is under pressure. Opening the system when hot may cause immediate boiling and cause serious injury.

▶ TASK Inspect engine cooling and heater systems hoses and pipes, and determine necessary action.

MLR
A7C1

Time off____________

Time on____________

Total time____________

Student Instructions: Read through the entire procedure prior to starting. Prepare your work space and any tools or parts that may be needed to complete the task. When directed by your instructor, begin the procedure to complete the task and check the box as each step is finished. Track your time on this procedure for later comparison to the standard completion time (i.e., "flat rate" or customer pay time).

Procedure:	Step Completed
1. Inspect the hoses and pipes of the engine cooling and heater systems.	
a. Begin by checking the coolant level in the overflow tank and radiator. Describe the coolant level.	☐
b. Inspect the following components:	
i. Upper and lower radiator hose	☐
ii. Radiator	☐
iii. Engine core plugs	☐
iv. Heater hoses and connections	☐
v. Hose clamps	☐
vi. Heater core	☐
vii. Water pump	☐
viii. Intake gasket area	☐
ix. Cylinder head gasket area	☐
c. If the vehicle is equipped with rear heat, inspect the rear heater core and hoses.	☐
2. On the basis of on your inspection, what is your recommendation for service?	☐

Non-Task-Specific Evaluations:	Step Completed
1. Tools and equipment were used as directed and returned in good working order.	☐
2. Complied with all general and task-specific safety standards, including proper use of any personal protective equipment.	☐
3. Completed the task in an appropriate time frame (Recommendation: 1.5 or 2 times flat rate).	☐
4. Left the work space clean and orderly.	☐
5. Cared for customer property and returned it undamaged.	☐

Student signature ____________________________ Date ____________________________

Comments:

Have your supervisor/instructor verify satisfactory completion of this procedure, any observations found, and any necessary action(s) recommended.

Evaluation Instructions: The scoring box below is intended to act as a guide for both student and instructor. Each criterion listed will help students understand what is expected from them and evaluators articulate success at a particular task. The scoring is set up to allow a second attempt at each task (see the "Test" and "Retest" columns). Scoring is designed to reward students for correct completion of the task. Points are lost for failure to complete the employability requirements (see "Non-Task-Specific" criteria). When all criteria are evaluated, tally the points for a total at the bottom of each column.

Tasksheet Scoring

	Test		Retest	
Evaluation Items	**Pass**	**Fail**	**Pass**	**Fail**
Task-Specific Evaluation	**(1 pt)**	**(0 pts)**	**(1 pt)**	**(0 pts)**
Inspect the coolant level.				
Locate the cooling system's components.				
Inspect the cooling system components for signs of leaks.				
Make recommendations for service based on your inspection.				
Non-Task-Specific Evaluation	**(0 pts)**	**(−1 pt)**	**(0 pts)**	**(−1 pt)**
Student successfully completed at least three of the non-task-specific steps.				
Student successfully completed all five of the non-task-specific steps.				
Total Score: <total # of points / 4 = %>				

Supervisor:

Supervisor/instructor signature _________________________________ Date _______________________

Comments:

Retest supervisor/instructor signature _________________________________ Date _______________________

Comments:

7D. Operating Systems and Related Controls

Learning Objective/Task	CDX Tasksheet Number	ASE Education Foundation Reference Number; Priority Level
• Inspect A/C-heater ducts, doors, hoses, cabin filters, and outlets, and determine necessary action.	A7007	A7D1, P-1
• Identify the source of A/C system odors.	A7008	A7D2, P-2

Materials Required

- Blank work order
- Vehicle with available service history records
- Depending on the type of concern, special diagnostic/hand tools may be required. See your supervisor/instructor for instructions to identify what tools may be required.
- Service information database
- Personal protective equipment

Safety Considerations

- Diagnosis of this fault may require test-driving the vehicle on the school grounds. Attempt this task only with full permission from your instructor and follow all the guidelines exactly.
- When running any vehicles in the shop, make sure you use the shop's exhaust ventilation system to discharge all exhaust gas safely outside.
- Extreme caution must be exercised when working around rotating components.
- Refrigerant can cause serious damage if it comes in contact with a person's unprotected skin and eyes.
- When operating, the air conditioning system is normally subject to very high pressure in the system. Extreme caution must be exercised when working on an operating system.
- Comply with personal and environmental safety practices associated with clothing; eye protection; hand tools; power equipment; proper ventilation; and the handling, storage, and disposal of chemicals/materials in accordance with local, state, and federal safety and environmental regulations.

CDX Tasksheet Number: A7007

Student/Intern Information

Name ______________________________ Date ___________ Class ___________________

Vehicle, Customer, and Service Information

Vehicle used for this activity:

Year ______________ Make ___________________________ Model __________________

Odometer ________________________ VIN ____________________________

Materials Required

- Blank work order
- Vehicle with available service history records
- Depending on the type of concern, special diagnostic/hand tools may be required. See your supervisor/instructor for instructions to identify what tools may be required.
- Service information database
- Personal protective equipment
- General hand tools

Task-Specific Safety Considerations

- Comply with personal and environmental safety practices associated with clothing; eye protection; hand tools; power equipment; proper ventilation; and the handling, storage, and disposal of chemicals/materials in accordance with local, state, and federal safety and environmental regulations.

▶ TASK Inspect A/C-heater ducts, doors, hoses, cabin filters, and outlets, and determine necessary action.

MLR
A7D1

Time off______________

Time on______________

Student Instructions: Read through the entire procedure prior to starting. Prepare your work space and any tools or parts that may be needed to complete the task. When directed by your instructor, begin the procedure to complete the task and check the box as each step is finished. Track your time on this procedure for later comparison to the standard completion time (i.e., "flat rate" or customer pay time).

Total time______________

Procedure:	Step Completed
1. Inspect the cabin air filter.	
a. Describe the location of the cabin air filter.	☐

b. Research the procedure for replacing the cabin air filter.	☐
c. Using service information as a guide, remove the components necessary to gain access to the cabin air filter.	☐
d. Remove and inspect the filter. Does the filter require replacement?	☐
e. Install the filter and replace any components that were removed earlier.	☐
2. Inspect the heater ducts, doors, and outlets.	
a. Cycle the ignition key to the ON position.	☐
b. Test the blower motor for operation on all fan speeds.	☐
c. Cycle the temperature slowly from cold to warm. Ensure that the temperature changes according to switch position. If the temperature does not change, explain what components inside the HVAC box could be at fault.	☐
d. Use the mode switch to select each possible mode for airflow. Inspect to make sure that air is directed to the position indicated by the switch and air blows from all positions.	☐
3. On the basis of your inspection, what is your recommendation for service?	☐

Non-Task-Specific Evaluations:	**Step Completed**
1. Tools and equipment were used as directed and returned in good working order.	☐
2. Complied with all general and task-specific safety standards, including proper use of any personal protective equipment.	☐
3. Completed the task in an appropriate time frame (Recommendation: 1.5 or 2 times flat rate).	☐
4. Left the work space clean and orderly.	☐
5. Cared for customer property and returned it undamaged.	☐

Student signature _______________________________ Date _______________________________

Comments:

Have your supervisor/instructor verify satisfactory completion of this procedure, any observations found, and any necessary action(s) recommended.

Tasksheet Scoring

	Test		Retest	
Evaluation Items	**Pass**	**Fail**	**Pass**	**Fail**
Task-Specific Evaluation	**(1 pt)**	**(0 pts)**	**(1 pt)**	**(0 pts)**
Research the procedure for replacing the cabin air filter, and describe its location. Remove, inspect, and replace the cabin air filter.				
Inspect the blower fan operation.				
Inspect the temperature controls.				
Inspect the airflow mode positions.				
Non-Task-Specific Evaluation	**(0 pts)**	**(−1 pt)**	**(0 pts)**	**(−1 pt)**
Student successfully completed at least three of the non-task-specific steps.				
Student successfully completed all five of the non-task-specific steps.				
Total Score: <total # of points / 4 = %>				

Supervisor:

Supervisor/instructor signature _________________________ Date _________________

Comments:

Retest supervisor/instructor signature _________________________ Date _________________

Comments:

CDX Tasksheet Number: A7008

Student/Intern Information

Name _________________________________ Date ____________ Class _________________________

Vehicle, Customer, and Service Information

Vehicle used for this activity:

Year _______________ Make _______________________ Model _____________________

Odometer _________________________ VIN _______________________________

Task-Specific Safety Considerations
- Comply with personal and environmental safety practices associated with clothing; eye protection; hand tools; power equipment; proper ventilation; and the handling, storage, and disposal of chemicals/materials in accordance with local, state, and federal safety and environmental regulations.

▶ TASK Identify the source of A/C system odors.　　　**MLR** **A7D2**

Time off__________

Time on__________

Student Instructions: Read through the entire procedure prior to starting. Prepare your work space and any tools or parts that may be needed to complete the task. When directed by your instructor, begin the procedure to complete the task and check the box as each step is finished. Track your time on this procedure for later comparison to the standard completion time (i.e., "flat rate" or customer pay time).

Total time__________

Procedure:	Step Completed
1. Research the technical service bulletins (TSBs) related to HVAC odors.	
a. List any TSBs that relate to HVAC odors.	☐
b. List the causes and remedies of unwanted odors in the HVAC system.	☐
2. Identify the cause of A/C odors.	
a. Touch the front floor mats to see if they are wet. **Note:** Sometimes the A/C drain can clog and cause condensation to buildup inside the evaporator core housing. This moisture can buildup and spill over into the passenger compartment.	☐
b. If the A/C drain appears to be clogged, apply shop air up through the drain to release the blockage. Did any water come out of the drain?	☐
c. Remove the cabin filter. Describe its condition. Is this the source of the odor?	☐
d. With the cabin filter removed, inspect inside the HVAC housing and blower motor fan. Are there signs of debris or rodent infestation?	☐
3. On the basis of on your inspection, what is your recommendation for service?	☐

Non-Task-Specific Evaluations:	Step Completed
1. Tools and equipment were used as directed and returned in good working order.	☐
2. Complied with all general and task-specific safety standards, including proper use of any personal protective equipment.	☐
3. Completed the task in an appropriate time frame (Recommendation: 1.5 or 2 times flat rate).	☐
4. Left the work space clean and orderly.	☐
5. Cared for customer property and returned it undamaged.	☐

Student signature _______________________________ Date _______________________________

Comments:

Have your supervisor/instructor verify satisfactory completion of this procedure, any observations found,

and any necessary action(s) recommended.

Evaluation Instructions: The scoring box below is intended to act as a guide for both student and instructor. Each criterion listed will help students understand what is expected from them and evaluators articulate success at a particular task. The scoring is set up to allow a second attempt at each task (see the "Test" and "Retest" columns). Scoring is designed to reward students for correct completion of the task. Points are lost for failure to complete the employability requirements (see "Non-Task-Specific" criteria). When all criteria are evaluated, tally the points for a total at the bottom of each column.

Taksheet Scoring

	Test		Retest	
Evaluation Items	**Pass**	**Fail**	**Pass**	**Fail**
Task-Specific Evaluation	**(1 pt)**	**(0 pts)**	**(1 pt)**	**(0 pts)**
Research the TSBs related to HVAC odors.				
Inspect the A/C drain.				
Inspect the cabin filter.				
Make recommendations for service based on your inspection.				
Non-Task-Specific Evaluation	**(0 pts)**	**(−1 pt)**	**(0 pts)**	**(−1 pt)**
Student successfully completed at least three of the non-task-specific steps.				
Student successfully completed all five of the non-task-specific steps.				
Total Score: <total # of points / 4 = %>				

Supervisor:

Supervisor/instructor signature ______________________________ Date ______________________

Comments:

Retest supervisor/instructor signature ______________________________ Date ______________________

Comments:

8A. Engine Performance

Learning Objective/Task	CDX Tasksheet Number	ASE Education Foundation Reference Number; Priority Level
• Research vehicle service information, including fluid type, vehicle service history, service precautions, and technical service bulletins (TSBs).	A8001	A8A1, P-1
• Perform engine absolute manifold pressure tests (vacuum/boost), and document results.	A8002	A8A2, P-2
• Perform cylinder power balance test, and document results.	A8003	A8A3, P-2
• Perform cylinder cranking and running compression tests, and document results.	A8004	A8A4, P-2
• Perform cylinder leakage test, and document results.	A8005	A8A5, P-2
• Verify engine operating temperature.	A8006	A8A6, P-1
• Remove and replace spark plugs, and inspect secondary ignition components for wear and damage.	A8007	A8A7, P-1

Materials Required

- Blank work order
- Vehicle with available service history records
- Depending on the type of concern, special hand tool/diagnostic tools may be required. See your supervisor/instructor for instructions to identify what tools may be required.
- Vehicle service information database

Safety Considerations

- When running any vehicles in the shop, make sure you use the shop's exhaust ventilation system to discharge all exhaust gas safely outside.
- Extreme caution must be exercised when working around rotating components.
- Lifting equipment, such as vehicle jacks and stands, vehicle hoists, and engine hoists, are important tools that increase productivity and make the job easier. However, they can also cause severe injury or death if used improperly. Make sure you follow the manufacturer's operation procedures. Also, make sure you have your supervisor's/instructor's permission to use any particular type of lifting equipment.
- Comply with personal and environmental safety practices associated with clothing; eye protection; hand tools; power equipment; proper ventilation; and the handling, storage, and disposal of chemicals/materials in accordance with local, state, and federal safety and environmental regulations.

CDX Tasksheet Number: A8001

Student/Intern Information

Name ________________________________ Date ____________ Class ________________________

Vehicle, Customer, and Service Information

Vehicle used for this activity:

Year ______________ Make ____________________________ Model ______________________________

Odometer ________________________________ VIN ________________________________

Materials Required

- Blank work order
- Vehicle with available service history records
- Depending on the type of concern, special hand tool/diagnostic tools may be required. See your supervisor/instructor for instructions to identify what tools may be required.
- Vehicle service information database

Task-Specific Safety Considerations

- Comply with personal and environmental safety practices associated with clothing; eye protection; hand tools; power equipment; proper ventilation; and the handling, storage, and disposal of chemicals/materials in accordance with local, state, and federal safety and environmental regulations.

Time off________________

Time on________________

Total time________________

▶ **TASK** Research vehicle service information, including fluid type, vehicle service history, service precautions, and TSBs.

MLR
A8A1

Student Instructions: Read through the entire procedure prior to starting. Prepare your work space and any tools or parts that may be needed to complete the task. When directed by your instructor, begin the procedure to complete the task and check the box as each step is finished. Track your time on this procedure for later comparison to the standard completion time (i.e., "flat rate" or customer pay time).

Procedure:	Step Completed
1. Research vehicle fluid types and list the following criteria:	
a. Determine engine oil type and capacity with oil filter change.	☐
b. Determine coolant/antifreeze type and full capacity.	☐

c. Determine the fuel type required for your vehicle. What is the fuel tank capacity?	☐
2. Locate and list vehicle's service history, including:	
a. Determine the routine maintenance completed.	☐
b. Determine whether there is any outstanding/completed recall work.	☐
c. Determine whether there are any major repairs in the service history.	☐
d. Identify any service precautions related to the recall work.	☐
3. Determine any TSBs issued for the specific vehicle.	
a. Identify one engine performance-related TSB and list the TSB number, vehicle concern, and repair procedure.	☐

Non-Task-Specific Evaluations:	Step Completed
1. Tools and equipment were used as directed and returned in good working order.	☐
2. Complied with all general and task-specific safety standards, including proper use of any personal protective equipment.	☐
3. Completed the task in an appropriate time frame (Recommendation: 1.5 or 2 times flat rate).	☐
4. Left the work space clean and orderly.	☐
5. Cared for customer property and returned it undamaged.	☐

Student signature _________________________________ Date _________________________________

Comments:

Have your supervisor/instructor verify satisfactory completion of this procedure, any observations found, and any necessary action(s) recommended.

Evaluation Instructions: The scoring box below is intended to act as a guide for both student and instructor. Each criterion listed will help students understand what is expected from them and evaluators articulate success at a particular task. The scoring is set up to allow a second attempt at each task (see the "Test" and "Retest" columns). Scoring is designed to reward students for correct completion of the task. Points are lost for failure to complete the employability requirements (see "Non-Task-Specific" criteria). When all criteria are evaluated, tally the points for a total at the bottom of each column.

Tasksheet Scoring

	Test		Retest	
Evaluation Items	**Pass**	**Fail**	**Pass**	**Fail**
Task-Specific Evaluation	**(1 pt)**	**(0 pts)**	**(1 pt)**	**(0 pts)**
Determine the vehicle fluid types and capacities.				
Determine the vehicle's service history.				
Determine the vehicle's service precautions.				
Determine the vehicle's TSBs.				
Non-Task-Specific Evaluation	**(0 pts)**	**(−1 pt)**	**(0 pts)**	**(−1 pt)**
Student successfully completed at least three of the non-task-specific steps.				
Student successfully completed all five of the non-task-specific steps.				
Total Score: <total # of points / 4 = %>				

Supervisor:

Supervisor/instructor signature ___________________________ Date ___________________

Comments:

Retest supervisor/instructor signature ___________________________ Date ___________________

Comments:

CDX Tasksheet Number: A8002

Student/Intern Information

Name _________________________________ Date ____________ Class _________________________

Vehicle, Customer, and Service Information

Vehicle used for this activity:

Year _______________ Make _____________________________ Model _______________________

Odometer _________________________________ VIN _________________________________

Materials Required

- Blank work order
- Vehicle with available service history records
- Depending on the type of concern, special hand tool/diagnostic tools may be required. See your supervisor/instructor for instructions to identify what tools may be required.
- Vehicle service information database
- Vacuum gauge
- Scan tool

Task-Specific Safety Considerations

- When running any vehicles in the shop, make sure you use the shop's exhaust ventilation system to discharge all exhaust gas safely outside.
- Extreme caution must be exercised when working around rotating components.
- Lifting equipment, such as vehicle jacks and stands, vehicle hoists, and engine hoists, are important tools that increase productivity and make the job easier. However, they can also cause severe injury or death if used improperly. Make sure you follow the manufacturer's operation procedures. Also, make sure you have your supervisor's/instructor's permission to use any particular type of lifting equipment.
- Comply with personal and environmental safety practices associated with clothing; eye protection; hand tools; power equipment; proper ventilation; and the handling, storage, and disposal of chemicals/materials in accordance with local, state, and federal safety and environmental regulations.

Time off____________

Time on____________

Total time____________

▶ TASK Perform engine absolute manifold pressure tests (vacuum/boost), and document results.

MLR
A8A2

Student Instructions: Read through the entire procedure prior to starting. Prepare your work space and any tools or parts that may be needed to complete the task. When directed by your instructor, begin the procedure to complete the task and check the box as each step is finished. Track your time on this procedure for later comparison to the standard completion time (i.e., "flat rate" or customer pay time).

Procedure:	Step Completed
1. Research specifications for manifold pressure.	
a. How much vacuum should be present at the manifold when the vehicle is idling?	☐
b. How much vacuum should be present at the manifold when the engine is running at 2500 revolutions per minute (RPM)?	☐
2. Perform absolute manifold pressure tests.	
a. Install a scan tool and follow prompts to select the correct vehicle that you are working on. Check for history, pending, and present trouble codes. Are there any DTCs stored at this time? If so, list them.	☐
b. Select the engine data and view the following parameters: MAP sensor voltage, short-term fuel trim (STFT), and long-term fuel trim (LTFT).	☐
c. Start the vehicle. Record the MAP sensor voltage, STFT, and LTFT at idle. Are these readings normal?	☐
d. With the engine running, create a small vacuum leak. Observe the scan tool readings. Did any readings change? **Tip:** Creating a vacuum leak that is too large may cause the engine to stall.	☐
e. Turn the engine off. Connect a vacuum gauge to the intake manifold.	☐
f. Start the engine. Observe the vacuum gauge. Record the vacuum. Is this reading normal? Is the gauge fluctuating?	☐

g. With the engine running, disconnect another small vacuum line. Observe the vacuum gauge. Did the reading change?	☐
h. Reconnect the vacuum line. Raise the engine to 2500 RPM. Record the gauge reading. Quickly, apply full throttle and let off the accelerator pedal completely. What does the gauge do during this test?	☐
i. Turn the engine off. Disable the fuel and ignition system in preparation for a cranking vacuum test.	☐
j. With the vacuum gauge still installed at the manifold, crank the engine for 5 seconds and observe the gauge reading. List your results.	☐
k. Are the results of the cranking vacuum test normal? Describe what type of gauge readings you might see if the engine mechanical timing was incorrect.	☐
l. Disconnect the vacuum gauge and scan tool. Enable the fuel and ignition system.	☐

<table>
<tr><td>Non-Task-Specific Evaluations:</td><td>Step Completed</td></tr>
<tr><td>1. Tools and equipment were used as directed and returned in good working order.</td><td>☐</td></tr>
<tr><td>2. Complied with all general and task-specific safety standards, including proper use of any personal protective equipment.</td><td>☐</td></tr>
<tr><td>3. Completed the task in an appropriate time frame (Recommendation: 1.5 or 2 times flat rate).</td><td>☐</td></tr>
<tr><td>4. Left the work space clean and orderly.</td><td>☐</td></tr>
<tr><td>5. Cared for customer property and returned it undamaged.</td><td>☐</td></tr>
</table>

Student signature _______________________________ Date _______________________________

Comments:

Have your supervisor/instructor verify satisfactory completion of this procedure, any observations found, and any necessary action(s) recommended.

Evaluation Instructions: The scoring box below is intended to act as a guide for both student and instructor. Each criterion listed will help students understand what is expected from them and evaluators articulate success at a particular task. The scoring is set up to allow a second attempt at each task (see the "Test" and "Retest" columns). Scoring is designed to reward students for correct completion of the task. Points are lost for failure to complete the employability requirements (see "Non-Task-Specific" criteria). When all criteria are evaluated, tally the points for a total at the bottom of each column.

Tasksheet Scoring

	Test		Retest	
Evaluation Items	**Pass**	**Fail**	**Pass**	**Fail**
Task-Specific Evaluation	**(1 pt)**	**(0 pts)**	**(1 pt)**	**(0 pts)**
Research specifications for manifold pressure.				
Use a scan tool to check for history, present, and pending codes. Use a scan tool to view MAP sensor, STFT, and LTFT data.				
Create vacuum leaks and observe scan tool data.				
Create vacuum leaks and observe vacuum gauge readings.				
Non-Task-Specific Evaluation	**(0 pts)**	**(−1 pt)**	**(0 pts)**	**(−1 pt)**
Student successfully completed at least three of the non-task-specific steps.				
Student successfully completed all five of the non-task-specific steps.				
Total Score: <total # of points / 4 = %>				

Supervisor:

Supervisor/instructor signature ________________________________ Date ________________

Comments:

Retest supervisor/instructor signature ________________________________ Date ________________

Comments:

CDX Tasksheet Number: A8003

Student/Intern Information

Name _________________________________ Date ___________ Class _________________________

Vehicle, Customer, and Service Information

Vehicle used for this activity:

Year _______________ Make _____________________________ Model _____________________

Odometer _________________________________ VIN _______________________________________

Materials Required

- Blank work order
- Vehicle with available service history records
- Depending on the type of concern, special hand tool/diagnostic tools may be required. See your supervisor/instructor for instructions to identify what tools may be required.
- Vehicle service information database
- Scan tool

Task-Specific Safety Considerations

- Comply with personal and environmental safety practices associated with clothing; eye protection; hand tools; power equipment; proper ventilation; and the handling, storage, and disposal of chemicals/materials in accordance with local, state, and federal safety and environmental regulations.

▶ **TASK** Perform cylinder power balance test, and document results. **MLR A8A3**

Time off____________

Time on____________

Student Instructions: Read through the entire procedure prior to starting. Prepare your work space and any tools or parts that may be needed to complete the task. When directed by your instructor, begin the procedure to complete the task and check the box as each step is finished. Track your time on this procedure for later comparison to the standard completion time (i.e., "flat rate" or customer pay time).

Total time____________

Procedure:	Step Completed
1. Perform a cylinder balance test.	
a. Research the procedure for performing a cylinder balance test. There are many ways to accomplish this task. Describe the testing procedure you will use.	☐

b. Install a scan tool. Inspect for history, present, and pending trouble codes. List any trouble codes. Follow the prompts and select engine data. Select the engine revolutions per minute (RPM) parameter.	☐
c. Start the vehicle and allow it to reach operating temperature.	☐
d. Follow the procedure you outlined in step 1a and perform a cylinder balance test. **Tip:** If disabling fuel, allow the engine to run for at least 30 seconds in between tests so that the catalytic converter does not overheat.	☐
e. List the RPM drop on each cylinder.	☐
2. On the basis of your testing, what can you conclude about the condition of your engine's mechanical, fuel, and ignition systems?	☐

Non-Task-Specific Evaluations:	Step Completed
1. Tools and equipment were used as directed and returned in good working order.	☐
2. Complied with all general and task-specific safety standards, including proper use of any personal protective equipment.	☐
3. Completed the task in an appropriate time frame (Recommendation: 1.5 or 2 times flat rate).	☐
4. Left the work space clean and orderly.	☐
5. Cared for customer property and returned it undamaged.	☐

Student signature _________________________________ Date _________________________________

Comments:

Have your supervisor/instructor verify satisfactory completion of this procedure, any observations found, and any necessary action(s) recommended.

Evaluation Instructions: The scoring box below is intended to act as a guide for both student and instructor. Each criterion listed will help students understand what is expected from them and evaluators articulate success at a particular task. The scoring is set up to allow a second attempt at each task (see the "Test" and "Retest" columns). Scoring is designed to reward students for correct completion of the task. Points are lost for failure to complete the employability requirements (see "Non-Task-Specific" criteria). When all criteria are evaluated, tally the points for a total at the bottom of each column.

Tasksheet Scoring

	Test		Retest	
Evaluation Items	**Pass**	**Fail**	**Pass**	**Fail**
Task-Specific Evaluation	**(1 pt)**	**(O pts)**	**(1 pt)**	**(O pts)**
Research the procedure for performing a cylinder power balance test.				
Install a scan tool and check for DTCs.				
Perform a cylinder power balance test and record all data from the testing procedure.				
Make conclusions on engine's mechanical, fuel, and ignition systems.				
Non-Task-Specific Evaluation	**(O pts)**	**(−1 pt)**	**(O pts)**	**(−1 pt)**
Student successfully completed at least three of the non-task-specific steps.				
Student successfully completed all five of the non-task-specific steps.				
Total Score: <total # of points / 4 = %>				

Supervisor:

Supervisor/instructor signature ________________________ Date ____________

Comments:

Retest supervisor/instructor signature ________________________ Date ____________

Comments:

CDX Tasksheet Number: A8004

Student/Intern Information

Name _____________________________ Date __________ Class _______________________

Vehicle, Customer, and Service Information

Vehicle used for this activity:

Year ______________ Make _____________________ Model _____________________

Odometer _____________________ VIN _________________________________

Task-Specific Safety Considerations

- When running any vehicles in the shop, make sure you use the shop's exhaust ventilation system to discharge all exhaust gas safely outside.
- Extreme caution must be exercised when working around rotating components.
- Comply with personal and environmental safety practices associated with clothing; eye protection; hand tools; power equipment; proper ventilation; and the handling, storage, and disposal of chemicals/materials in accordance with local, state, and federal safety and environmental regulations.

Time off__________

Time on__________

Total time__________

▶ **TASK** Perform cylinder cranking and running compression tests, and document results.

MLR
A8A4

Student Instructions: Read through the entire procedure prior to starting. Prepare your work space and any tools or parts that may be needed to complete the task. When directed by your instructor, begin the procedure to complete the task and check the box as each step is finished. Track your time on this procedure for later comparison to the standard completion time (i.e., "flat rate" or customer pay time).

Procedure:	Step Completed
1. Perform a cylinder cranking compression test.	
a. Start the vehicle and allow it to reach operating temperature.	☐
b. Turn the vehicle off. Disable the fuel and ignition systems.	☐
c. Remove one spark plug at a time. Connect the compression tester. Hold the throttle wide open and crank the engine for 5 seconds. Record the compression reading.	☐
d. Remove the compression tester, install the spark plug, and move to the next cylinder. Repeat the process on all cylinders. Record the compression reading on all cylinders.	☐
e. Check the service information for compression specifications. Are the cylinders within specification? Describe what two adjacent cylinders with low compression readings may indicate.	☐
2. Perform a wet cranking compression test.	
a. If a cylinder's compression reading was low, you can use a wet test to determine if the piston rings may be at fault.	☐
b. Remove the spark plug from the cylinder in question. Squirt a few ounces of clean engine oil into the cylinder.	☐
c. Install the compression gauge. Crank the engine for 5 seconds at wide open throttle. Compare the reading from the dry and wet tests. What do the readings indicate: Is the engine at fault for the low compression reading?	☐
3. Perform a running compression test.	
a. Enable the fuel and ignition systems.	☐
b. Remove one spark plug at a time. Install the compression gauge.	☐
c. Start the engine and record the compression reading. Repeat this process and test all cylinders. List all compression readings.	☐

d. Explain why the compression readings are lower when the engine is running compared to when it's cranking.	☐
e. On the basis of the results of the cranking and running compression tests, what can you infer about the condition of the engine?	☐

Non-Task-Specific Evaluations:	Step Completed
1. Tools and equipment were used as directed and returned in good working order.	☐
2. Complied with all general and task-specific safety standards, including proper use of any personal protective equipment.	☐
3. Completed the task in an appropriate time frame (Recommendation: 1.5 or 2 times flat rate).	☐
4. Left the work space clean and orderly.	☐
5. Cared for customer property and returned it undamaged.	☐

Student signature _________________________________ Date _________________________________

Comments:

Have your supervisor/instructor verify satisfactory completion of this procedure, any observations found, and any necessary action(s) recommended.

Evaluation Instructions: The scoring box below is intended to act as a guide for both student and instructor. Each criterion listed will help students understand what is expected from them and evaluators articulate success at a particular task. The scoring is set up to allow a second attempt at each task (see the "Test" and "Retest" columns). Scoring is designed to reward students for correct completion of the task. Points are lost for failure to complete the employability requirements (see "Non-Task-Specific" criteria). When all criteria are evaluated, tally the points for a total at the bottom of each column.

Tasksheet Scoring

Evaluation Items	Test		Retest	
	Pass	**Fail**	**Pass**	**Fail**
Task-Specific Evaluation	**(1 pt)**	**(0 pts)**	**(1 pt)**	**(0 pts)**
Disable the fuel and ignition systems.				
Perform a dry cranking compression test and record results. Determine engine condition based on test results.				
Perform a wet compression test and record results. Determine engine condition based on test results.				
Perform a running compression test and record results. Determine engine condition based on test results.				
Non-Task-Specific Evaluation	**(0 pts)**	**(−1 pt)**	**(0 pts)**	**(−1 pt)**
Student successfully completed at least three of the non-task–specific steps.				
Student successfully completed all five of the non-task-specific steps.				
Total Score: <total # of points / 4 = %>				

Supervisor:

Supervisor/instructor signature ___________________________ Date ___________________

Comments:

Retest supervisor/instructor signature ___________________________ Date ___________________

Comments:

CDX Tasksheet Number: A8005

Student/Intern Information

Name _________________________________ Date ___________ Class _____________________

Vehicle, Customer, and Service Information

Vehicle used for this activity:

Year _______________ Make ___________________________ Model ________________________

Odometer _________________________________ VIN _____________________________________

Materials Required

- Blank work order
- Vehicle with available service history records
- Depending on the type of concern, special hand tool/diagnostic tools may be required. See your supervisor/instructor for instructions to identify what tools may be required.
- Vehicle service information database
- Cylinder leakage tester
- Spark plug socket
- General hand tools
- TDC indicator

Task-Specific Safety Considerations

- When running any vehicles in the shop, make sure you use the shop's exhaust ventilation system to discharge all exhaust gas safely outside.
- Extreme caution must be exercised when working around rotating components.
- Comply with personal and environmental safety practices associated with clothing; eye protection; hand tools; power equipment; proper ventilation; and the handling, storage, and disposal of chemicals/materials in accordance with local, state, and federal safety and environmental regulations.

▶ TASK Perform cylinder leakage test, and document results.

MLR
A8A5

Time off_____________

Time on_____________

Total time_____________

Student Instructions: Read through the entire procedure prior to starting. Prepare your work space and any tools or parts that may be needed to complete the task. When directed by your instructor, begin the procedure to complete the task and check the box as each step is finished. Track your time on this procedure for later comparison to the standard completion time (i.e., "flat rate" or customer pay time).

Procedure:	Step Completed
1. Perform a cylinder leakage test.	
a. Remove the spark plug from cylinder number 1.	☐
b. Install the TDC indicator. Rotate the engine by hand until the piston reaches TDC on the compression stroke.	☐
c. Describe what is happening with the valves during the compression stroke.	☐
d. Remove the TDC indicator and install the leakage tester.	☐
e. Slowly apply shop air to the cylinder. **Caution:** Keep hands clear of the engine accessories. The engine may spin over during this step.	☐
f. Apply 100 psi to the cylinder and record the percentage of leakage. Is the leakage within specification?	☐
g. If a leak is greater than normal, inspect the following areas for signs of leaking air:	
• Exhaust tailpipe.	☐
• Throttle body.	☐
• Radiator opening.	☐
• Oil cap opening or PCV opening.	☐
h. Continue to test all cylinders following the procedure. Record all test data. Are any cylinders leaking more than others? If so, what may be causing the leakage?	☐
2. On the basis of your testing, what can you infer about the condition of your engine?	☐

Non-Task-Specific Evaluations:	Step Completed
1. Tools and equipment were used as directed and returned in good working order.	☐
2. Complied with all general and task-specific safety standards, including proper use of any personal protective equipment.	☐
3. Completed the task in an appropriate time frame (Recommendation: 1.5 or 2 times flat rate).	☐
4. Left the work space clean and orderly.	☐
5. Cared for customer property and returned it undamaged.	☐

Student signature _________________________ Date _____________________________

Comments:

Have your supervisor/instructor verify satisfactory completion of this procedure, any observations found, and any necessary action(s) recommended.

Evaluation Instructions: The scoring box below is intended to act as a guide for both student and instructor. Each criterion listed will help students understand what is expected from them and evaluators articulate success at a particular task. The scoring is set up to allow a second attempt at each task (see the "Test" and "Retest" columns). Scoring is designed to reward students for correct completion of the task. Points are lost for failure to complete the employability requirements (see "Non-Task-Specific" criteria). When all criteria are evaluated, tally the points for a total at the bottom of each column.

Tasksheet Scoring

	Test		Retest	
Evaluation Items	**Pass**	**Fail**	**Pass**	**Fail**
Task-Specific Evaluation	**(1 pt)**	**(0 pts)**	**(1 pt)**	**(0 pts)**
Remove the spark plug and locate TDC on the compression stroke.				
Install the cylinder leakage tester.				
Observe and record any cylinder leakage. Determine the causes of cylinder leakage.				
Determine the condition of the engine based on your test results.				
Non-Task-Specific Evaluation	**(0 pts)**	**(−1 pt)**	**(0 pts)**	**(−1 pt)**
Student successfully completed at least three of the non-task-specific steps.				
Student successfully completed all five of the non-task-specific steps.				
Total Score: <total # of points / 4 = %>				

Supervisor:

Supervisor/instructor signature ________________________________ Date ___________________

Comments:

Retest supervisor/instructor signature ________________________________ Date ___________________

Comments:

CDX Tasksheet Number: A8006

Student/Intern Information

Name _________________________________ Date _____________ Class _____________________

Vehicle, Customer, and Service Information

Vehicle used for this activity:

Year _______________ Make _________________________ Model _______________________

Odometer _________________________ VIN _________________________________

Task-Specific Safety Considerations

- When running any vehicles in the shop, make sure you use the shop's exhaust ventilation system to discharge all exhaust gas safely outside.
- Extreme caution must be exercised when working around rotating components.
- Comply with personal and environmental safety practices associated with clothing; eye protection; hand tools; power equipment; proper ventilation; and the handling, storage, and disposal of chemicals/materials in accordance with local, state, and federal safety and environmental regulations.

Time off_______________

Time on_______________

Total time_______________

▶ **TASK** Verify engine operating temperature.　　　　　　　**MLR** **A8A6**

Student Instructions: Read through the entire procedure prior to starting. Prepare your work space and any tools or parts that may be needed to complete the task. When directed by your instructor, begin the procedure to complete the task and check the box as each step is finished. Track your time on this procedure for later comparison to the standard completion time (i.e., "flat rate" or customer pay time).

Procedure:	Step Completed
1. Research engine operating temperature specifications.	
a. Use service information to determine engine operating temperature specifications. **Tip:** You may find thermostat temperature specifications; but remember that this rating is when the thermostat begins to open.	☐
2. Verify the engine's operating temperature.	
a. Install a scan tool. Follow the prompts and select engine data. Select the "Coolant Temperature" parameter.	☐
b. Before starting the vehicle, record the temperature reading.	☐
c. Start the vehicle and allow it to run until operating temperature is achieved. Watch the scan tool data. Did the temperature increase to the specified temperature?	☐
d. Use an infrared temperature gun to measure the temperature at four places: the thermostat housing, upper radiator hose, radiator, and lower radiator hose. Record your readings. Is there a difference in temperature where coolant enters the radiator and where it exits? Does this temperature reflect the specifications you researched?	☐
e. Verify that the cooling fans turned on as operating temperature was achieved.	☐
f. What cooling system component controls minimum coolant temperature? How does this device work?	☐
g. What cooling system component controls maximum coolant temperature? How does this device work?	☐
3. On the basis of your observations, what can you conclude about the cooling system?	☐

Non-Task-Specific Evaluations:	Step Completed
1. Tools and equipment were used as directed and returned in good working order.	☐
2. Complied with all general and task-specific safety standards, including proper use of any personal protective equipment.	☐
3. Completed the task in an appropriate time frame (Recommendation: 1.5 or 2 times flat rate).	☐
4. Left the work space clean and orderly.	☐
5. Cared for customer property and returned it undamaged.	☐

Student signature _________________________ Date _________________________

Comments:

Have your supervisor/instructor verify satisfactory completion of this procedure, any observations found, and any necessary action(s) recommended.

Evaluation Instructions: The scoring box below is intended to act as a guide for both student and instructor. Each criterion listed will help students understand what is expected from them and evaluators articulate success at a particular task. The scoring is set up to allow a second attempt at each task (see the "Test" and "Retest" columns). Scoring is designed to reward students for correct completion of the task. Points are lost for failure to complete the employability requirements (see "Non-Task-Specific" criteria). When all criteria are evaluated, tally the points for a total at the bottom of each column.

Tasksheet Scoring

Evaluation Items	Test		Retest	
	Pass	**Fail**	**Pass**	**Fail**
Task-Specific Evaluation	**(1 pt)**	**(0 pts)**	**(1 pt)**	**(0 pts)**
Research engine operating temperature specifications.				
Use a scan tool to select the correct parameters and observe engine operating temperature. Use an infrared temperature gun to measure coolant temperature at various points.				
Verify cooling fan operation. Explain the different cooling system components.				
Make conclusions about the cooling system based on your testing procedures.				
Non-Task-Specific Evaluation	**(0 pts)**	**(−1 pt)**	**(0 pts)**	**(−1 pt)**
Student successfully completed at least three of the non-task-specific steps.				
Student successfully completed all five of the non-task-specific steps.				
Total Score: <total # of points / 4 = %>				

Supervisor:

Supervisor/instructor signature _________________________ Date _______________

Comments:

Retest supervisor/instructor signature _________________________ Date _______________

Comments:

CDX Tasksheet Number: A8007

Student/Intern Information

Name _________________________________ Date ____________ Class _____________________

Vehicle, Customer, and Service Information

Vehicle used for this activity:

Year _______________ Make _______________________ Model ____________________________

Odometer _______________________ VIN _________________________________

Materials Required

- Blank work order
- Vehicle with available service history records
- Depending on the type of concern, special hand tool/diagnostic tools may be required. See your supervisor/instructor for instructions to identify what tools may be required.
- Vehicle service information database
- Spark plug sockets
- Torque wrench
- General hand tools

Task-Specific Safety Considerations

- Comply with personal and environmental safety practices associated with clothing; eye protection; hand tools; power equipment; proper ventilation; and the handling, storage, and disposal of chemicals/materials in accordance with local, state, and federal safety and environmental regulations.

Time off____________

Time on____________

Total time____________

▶ TASK Remove and replace spark plugs, and inspect secondary ignition components for wear and damage.

MLR
A8A7

Student Instructions: Read through the entire procedure prior to starting. Prepare your work space and any tools or parts that may be needed to complete the task. When directed by your instructor, begin the procedure to complete the task and check the box as each step is finished. Track your time on this procedure for later comparison to the standard completion time (i.e., "flat rate" or customer pay time).

Procedure:	Step Completed
1. Research the spark plug replacement procedure.	
a. Determine the maintenance interval for replacing the spark plugs on your vehicle.	☐
b. What types of spark plugs are used in your vehicle?	☐
c. What is the torque specification for the spark plugs?	☐
d. Is there a gap specification for your spark plugs?	☐
e. Describe the type of ignition system used on your vehicle, that is, distributor, coil-on-plug, and coil-near-plug.	☐
2. Replace the spark plugs.	
a. Remove any components necessary to gain access to the spark plugs.	☐
b. Use shop air to remove any debris that has built-up around the coil pack/plug wire.	☐
c. Remove the coil pack or spark plug wire.	☐
d. Use a spark plug socket to remove the spark plug.	☐
e. Inspect the old spark plug. Describe the condition. What can you determine about the way the engine is running on each cylinder that the spark plug was removed from?	☐
f. Check that the new spark plug is the same type, heat range, and length as the original plug.	☐
g. Check the spark plug gap prior to installation. **Tip:** Some types of spark plugs are pregapped. Damage to the plug may occur if you try to change the gap.	☐
h. Using a spark plug socket, thread the new plug into the cylinder by hand. Use a torque wrench to tighten the plug to specification.	☐

i. Inspect the spark plug wire for signs of deterioration. If one wire is damaged, it is good practice to replace all wires.	☐
j. Repeat this procedure for all remaining spark plugs.	☐
k. With all components put back together, start the vehicle. Ensure that the engine is operating correctly with no warning lights illuminated on the gauges.	☐

Non-Task-Specific Evaluations:	Step Completed
1. Tools and equipment were used as directed and returned in good working order.	☐
2. Complied with all general and task-specific safety standards, including proper use of any personal protective equipment.	☐
3. Completed the task in an appropriate time frame (Recommendation: 1.5 or 2 times flat rate).	☐
4. Left the work space clean and orderly.	☐
5. Cared for customer property and returned it undamaged.	☐

Student signature _______________________________ Date _______________________________

Comments:

Have your supervisor/instructor verify satisfactory completion of this procedure, any observations found, and any necessary action(s) recommended.

Evaluation Instructions: The scoring box below is intended to act as a guide for both student and instructor. Each criterion listed will help students understand what is expected from them and evaluators articulate success at a particular task. The scoring is set up to allow a second attempt at each task (see the "Test" and "Retest" columns). Scoring is designed to reward students for correct completion of the task. Points are lost for failure to complete the employability requirements (see "Non-Task-Specific" criteria). When all criteria are evaluated, tally the points for a total at the bottom of each column.

Tasksheet Scoring

Evaluation Items	Test		Retest	
	Pass	**Fail**	**Pass**	**Fail**
Task-Specific Evaluation	**(1 pt)**	**(0 pts)**	**(1 pt)**	**(0 pts)**
Determine the maintenance interval for replacing the spark plugs on your vehicle.				
Determine spark plug specifications and the type of ignition system used in your vehicle.				
Replace spark plugs. Determine engine condition based on spark plug condition.				
Torque spark plugs to specification. Verify that the engine is running correctly after replacement of spark plugs.				
Non-Task-Specific Evaluation	**(0 pts)**	**(−1 pt)**	**(0 pts)**	**(−1 pt)**
Student successfully completed at least three of the non-task-specific steps.				
Student successfully completed all five of the non-task-specific steps.				
Total Score: <total # of points / 4 = %>				

Supervisor:

Supervisor/instructor signature _______________________________ Date _________________

Comments:

Retest supervisor/instructor signature _______________________ Date _________________

Comments:

8B. Computerized Controls

Learning Objective/Task	CDX Tasksheet Number	ASE Education Foundation Reference Number; Priority Level
• Retrieve and record diagnostic trouble codes (DTC), OBD monitor status, and freeze frame data, and clear codes when applicable.	A8008	A8B1, P-1
• Describe the use of the OBD monitors for repair verification.	A8009	A8B2, P-1

Materials Required

- Blank work order
- Vehicle with available service history records
- Depending on the type of concern, special hand tool/diagnostic tools may be required. See your supervisor/instructor for instructions to identify what tools may be required.
- Vehicle service information database

Safety Considerations

- When running any vehicles in the shop, make sure you use the shop's exhaust ventilation system to discharge all exhaust gas safely outside.
- Extreme caution must be exercised when working around rotating components.
- Lifting equipment, such as vehicle jacks and stands, vehicle hoists, and engine hoists, are important tools that increase productivity and make the job easier. However, they can also cause severe injury or death if used improperly. Make sure you follow the manufacturer's operation procedures. Also, make sure you have your supervisor's/instructor's permission to use any particular type of lifting equipment.
- Comply with personal and environmental safety practices associated with clothing; eye protection; hand tools; power equipment; proper ventilation; and the handling, storage, and disposal of chemicals/materials in accordance with local, state, and federal safety and environmental regulations.

CDX Tasksheet Number: A8008

Student/Intern Information

Name _________________________________ Date ____________ Class _____________________

Vehicle, Customer, and Service Information

Vehicle used for this activity:

Year ________________ Make ________________________ Model ___________________

Odometer ____________________________ VIN ___________________________

Materials Required

- Blank work order
- Vehicle with available service history records
- Depending on the type of concern, special hand tool/diagnostic tools may be required. See your supervisor/instructor for instructions to identify what tools may be required.
- Vehicle service information database
- Scan tool

Task-Specific Safety Considerations

- Comply with personal and environmental safety practices associated with clothing; eye protection; hand tools; power equipment; proper ventilation; and the handling, storage, and disposal of chemicals/materials in accordance with local, state, and federal safety and environmental regulations.

▶ TASK Retrieve and record DTC, OBD monitor status, and freeze frame data, and clear codes when applicable.

MLR A8B1

Time off____________

Time on____________

Student Instructions: Read through the entire procedure prior to starting. Prepare your work space and any tools or parts that may be needed to complete the task. When directed by your instructor, begin the procedure to complete the task and check the box as each step is finished. Track your time on this procedure for later comparison to the standard completion time (i.e., "flat rate" or customer pay time).

Total time____________

Procedure:	Step Completed
1. Use a scan tool to retrieve and clear DTCs, inspect freeze frame and live data, and observe OBD monitors.	
a. Describe the location of your vehicles OBD II data link connector (DLC).	☐

b. Plug the scan tool into the DLC. Confirm that the scan tool is receiving power from the DLC. What is one possible reason that a DLC would not supply power?	☐
c. Cycle the ignition key to the ON position and turn the scan tool on.	☐
d. Follow the prompts on the scan tool and select "Diagnostic Trouble Codes". Check for pending, history, and present codes. Record any codes found in each category. **Tip:** Some scan tools have the ability to do a full system "health check," which will retrieve fault codes from each onboard computer.	
e. Explain the differences between pending, history, and present trouble codes.	☐
f. If a trouble codes is stored, follow the scan tool prompts to retrieve any stored freeze frame data. Describe the conditions when the fault occurred. Explain how this information can be helpful when diagnosing a vehicle. **Note:** Do not clear any trouble codes until freeze frame and monitor status has been recorded.	☐
g. Follow the prompts and select "monitor status". List the monitors on your vehicle and record the status of each. Are there any monitors that have run? Any that run all the time? Any that cannot run because of a trouble code being present?	☐
h. Once all information has been recorded, attempt to use the scan tool to clear the trouble code. Cycle the ignition key and start the vehicle.	☐
i. Did the check engine light turn off? Did all codes clear from the computer? Explain what causes a code to remain stored as a "Present" code.	☐

<table>
<tr><td>j. Describe what happened to the freeze frame and monitor status data when you cleared the trouble codes.</td><td>☐</td></tr>
</table>

Non-Task-Specific Evaluations:	Step Completed
1. Tools and equipment were used as directed and returned in good working order.	☐
2. Complied with all general and task-specific safety standards, including proper use of any personal protective equipment.	☐
3. Completed the task in an appropriate time frame (Recommendation: 1.5 or 2 times flat rate).	☐
4. Left the work space clean and orderly.	☐
5. Cared for customer property and returned it undamaged.	☐

Student signature _________________________ Date _________________________

Comments:

Have your supervisor/instructor verify satisfactory completion of this procedure, any observations found, and any necessary action(s) recommended.

Tasksheet Scoring

Evaluation Items	Test		Retest	
	Pass	**Fail**	**Pass**	**Fail**
Task-Specific Evaluation	**(1 pt)**	**(0 pts)**	**(1 pt)**	**(0 pts)**
Describe the location of the DLC. Install a scan tool and perform a system "health check" to retrieve trouble codes from all systems.				
List any stored trouble codes. Describe the difference between present, history, and pending codes.				
Retrieve freeze frame data. Explain how freeze frame data is helpful when diagnosing vehicle faults.				
Retrieve the system monitor status.				
Non-Task-Specific Evaluation	**(0 pts)**	**(−1 pt)**	**(0 pts)**	**(−1 pt)**
Student successfully completed at least three of the non-task-specific steps.				
Student successfully completed all five of the non-task-specific steps.				
Total Score: <total # of points / 4 = %>				

Supervisor:

Supervisor/instructor signature _______________________________ Date _______________

Comments:

Retest supervisor/instructor signature _______________________________ Date _______________

Comments:

CDX Tasksheet Number: A8009

Student/Intern Information

Name _________________________________ Date ___________ Class _________________________

Vehicle, Customer, and Service Information

Vehicle used for this activity:

Year _____________ Make _____________________ Model _________________

Odometer _____________________________ VIN _________________________

Materials Required

- Blank work order
- Vehicle with available service history records
- Depending on the type of concern, special hand tool/diagnostic tools may be required. See your supervisor/instructor for instructions to identify what tools may be required.
- Vehicle service information database
- Scan tool

Task-Specific Safety Considerations

- Comply with personal and environmental safety practices associated with clothing; eye protection; hand tools; power equipment; proper ventilation; and the handling, storage, and disposal of chemicals/materials in accordance with local, state, and federal safety and environmental regulations.

▶ TASK Describe the use of the OBD monitors for repair verification.

MLR
A8B2

Time off_____________

Time on_____________

Student Instructions: Read through the entire procedure prior to starting. Prepare your work space and any tools or parts that may be needed to complete the task. When directed by your instructor, begin the procedure to complete the task and check the box as each step is finished. Track your time on this procedure for later comparison to the standard completion time (i.e., "flat rate" or customer pay time).

Total time_____________

Procedure:	Step Completed
1. Retrieve the monitor status of onboard computers.	
a. Install a scan tool. Follow the prompts and select "vehicle system monitors".	☐
b. How many monitors are present on your vehicle?	☐

c. List the systems that use a continuous monitor. Explain what conditions need to be met in order for this monitor to run.	☐
d. List the systems that use a non-continuous monitor. Explain what conditions need to be met in order for this monitor to run.	☐
e. Say, for instance, that you just replaced a purge solenoid in the evaporative emissions system. Describe how you can use a monitor to verify that your repair was successful.	☐
f. Using service information, research how to complete a drive cycle to run a monitor for the EVAP system. Verify correct operation of all components in that system. List or print off and attach to this sheet the procedure for completing a drive cycle.	☐
g. Explain how to clear a trouble code using a drive cycle rather than using a scan tool to clear the code.	☐

<table>
<tr><td>Non-Task-Specific Evaluations:</td><td>Step Completed</td></tr>
<tr><td>1. Tools and equipment were used as directed and returned in good working order.</td><td>☐</td></tr>
<tr><td>2. Complied with all general and task-specific safety standards, including proper use of any personal protective equipment.</td><td>☐</td></tr>
<tr><td>3. Completed the task in an appropriate time frame (Recommendation: 1.5 or 2 times flat rate).</td><td>☐</td></tr>
<tr><td>4. Left the work space clean and orderly.</td><td>☐</td></tr>
<tr><td>5. Cared for customer property and returned it undamaged.</td><td>☐</td></tr>
</table>

Student signature _________________________________ Date _________________________________

Comments:

Have your supervisor/instructor verify satisfactory completion of this procedure, any observations found, and any necessary action(s) recommended.

Tasksheet Scoring

Evaluation Items	Test		Retest	
	Pass	**Fail**	**Pass**	**Fail**
Task-Specific Evaluation	**(1 pt)**	**(0 pts)**	**(1 pt)**	**(0 pts)**
Use a scan tool to access vehicle monitors.				
List all available monitors on your vehicle.				
Distinguish the differences between continuous and non-continuous monitors, and describe how monitors are used to verify vehicle repairs.				
Use service information to research drive cycle procedures.				
Non-Task-Specific Evaluation	**(0 pts)**	**(−1 pt)**	**(0 pts)**	**(−1 pt)**
Student successfully completed at least three of the non-task-specific steps.				
Student successfully completed all five of the non-task-specific steps.				
Total Score: <total # of points / 4 = %>				

Supervisor:

Supervisor/instructor signature _______________________________ Date _____________________

Comments:

Retest supervisor/instructor signature _______________________________ Date _____________________

Comments:

8C. Fuel, Air Induction, and Exhaust Systems

Learning Objective/Task	CDX Tasksheet Number	ASE Education Foundation Reference Number; Priority Level
• Replace fuel filter(s) where applicable.	A8010	A8C1, P-2
• Inspect, service, or replace air filters, filter housings, and intake duct work.	A8011	A8C2, P-1
• Inspect integrity of the exhaust manifold, exhaust pipes, muffler(s), catalytic converter(s), resonator(s), tail pipe(s), and heat shields, and determine necessary action.	A8012	A8C3, P-1
• Inspect condition of exhaust system hangers, brackets, clamps, and heat shields, and determine necessary action.	A8013	A8C4, P-1
• Check and refill diesel exhaust fluid.	A8014	A8C5, P-2

Materials Required

- Blank work order
- Vehicle with available service history records
- Depending on the type of concern, special hand tool/diagnostic tools may be required. See your supervisor/instructor for instructions to identify what tools may be required.
- Vehicle service information database

Safety Considerations

- When running any vehicles in the shop, make sure you use the shop's exhaust ventilation system to discharge all exhaust gas safely outside.
- Extreme caution must be exercised when working around rotating components.
- Lifting equipment, such as vehicle jacks and stands, vehicle hoists, and engine hoists, are important tools that increase productivity and make the job easier. However, they can also cause severe injury or death if used improperly. Make sure you follow the manufacturer's operation procedures. Also, make sure you have your supervisor's/instructor's permission to use any particular type of lifting equipment.
- Comply with personal and environmental safety practices associated with clothing; eye protection; hand tools; power equipment; proper ventilation; and the handling, storage, and disposal of chemicals/materials in accordance with local, state, and federal safety and environmental regulations.

CDX Tasksheet Number: A8O10

Student/Intern Information

Name _________________________________ Date __________ Class _________________________

Vehicle, Customer, and Service Information

Vehicle used for this activity:

Year _____________ Make _____________________ Model ___________________________

Odometer ______________________ VIN _______________________________

Task-Specific Safety Considerations

- Lifting equipment, such as vehicle jacks and stands, vehicle hoists, and engine hoists, are important tools that increase productivity and make the job easier. However, they can also cause severe injury or death if used improperly. Make sure you follow the manufacturer's operation procedures. Also, make sure you have your supervisor's/instructor's permission to use any particular type of lifting equipment.
- Comply with personal and environmental safety practices associated with clothing; eye protection; hand tools; power equipment; proper ventilation; and the handling, storage, and disposal of chemicals/materials in accordance with local, state, and federal safety and environmental regulations.
- Use extreme caution when working with fuel, as it is extremely flammable.

Time off________________

Time on________________

Total time________________

▶ TASK Replace fuel filter(s) where applicable.

MLR
A8C1

Student Instructions: Read through the entire procedure prior to starting. Prepare your work space and any tools or parts that may be needed to complete the task. When directed by your instructor, begin the procedure to complete the task and check the box as each step is finished. Track your time on this procedure for later comparison to the standard completion time (i.e., "flat rate" or customer pay time).

<table>
<tr><th>Procedure:</th><th>Step Completed</th></tr>
<tr><td>1. Research the fuel filter replacement procedure.</td><td></td></tr>
<tr><td>   a. Determine and list the maintenance service interval for replacing the fuel filter on your vehicle.</td><td>☐</td></tr>
<tr><td>   b. Describe the location of the fuel filter on your vehicle.</td><td>☐</td></tr>
<tr><td>   c. Are there any special service precautions when performing this task?</td><td>☐</td></tr>
<tr><td>   d. Describe the procedure for depressurizing the fuel system.</td><td>☐</td></tr>
<tr><td>2. Replace the fuel filter.</td><td></td></tr>
<tr><td>   a. Following proper procedure, depressurize the fuel system.</td><td>☐</td></tr>
<tr><td>   b. Locate the fuel filter and describe its condition. Do the lines appear to have excessive rust?</td><td>☐</td></tr>
<tr><td>   c. Remove the fuel line from both ends of the fuel filter. Use the fuel line disconnect tools if needed.</td><td>☐</td></tr>
<tr><td>   d. Remove the filter bracket and filter. Be cautious, as fuel may still be present in the filter.</td><td>☐</td></tr>
<tr><td>   e. Compare the new filter with the old. Install the new fuel filter and bracket. Tip: Fuel filters are directional. Be sure to install in the correct orientation.</td><td>☐</td></tr>
<tr><td>   f. Install the fuel lines. If an o-ring is used to seal the line, inspect the o-ring before installation. Replace as necessary.</td><td>☐</td></tr>
<tr><td>   g. Pressurize the fuel system. Cycle the ignition key to the ON position several times and inspect the filter for leaks prior to starting the vehicle.</td><td>☐</td></tr>
<tr><td>   h. Start the vehicle and verify the engine is running correctly and there are no leaks at the filter.</td><td>☐</td></tr>
</table>

Non-Task-Specific Evaluations:	Step Completed
1. Tools and equipment were used as directed and returned in good working order.	☐
2. Complied with all general and task-specific safety standards, including proper use of any personal protective equipment.	☐
3. Completed the task in an appropriate time frame (Recommendation: 1.5 or 2 times flat rate).	☐
4. Left the work space clean and orderly.	☐
5. Cared for customer property and returned it undamaged.	☐

Student signature _________________________ Date _________________________

Comments:

Have your supervisor/instructor verify satisfactory completion of this procedure, any observations found, and any necessary action(s) recommended.

Tasksheet Scoring

	Test		Retest	
Evaluation Items	**Pass**	**Fail**	**Pass**	**Fail**
Task-Specific Evaluation	**(1 pt)**	**(0 pts)**	**(1 pt)**	**(0 pts)**
Determine the maintenance interval for replacing the fuel filter.				
Describe and perform the procedure for depressurizing the fuel system.				
Replace the fuel filter in the correct orientation.				
Inspect for fuel leaks after the filter is replaced.				
Non-Task-Specific Evaluation	**(0 pts)**	**(−1 pt)**	**(0 pts)**	**(−1 pt)**
Student successfully completed at least three of the non-task-specific steps.				
Student successfully completed all five of the non-task-specific steps.				
Total Score: <total # of points / 4 = %>				

Supervisor:

Supervisor/instructor signature _________________________________ Date _________________

Comments:

Retest supervisor/instructor signature _________________________ Date _________________

Comments:

CDX Tasksheet Number: A8011

Student/Intern Information

Name _________________________________ Date ___________ Class _________________________

Vehicle, Customer, and Service Information

Vehicle used for this activity:

Year _______________ Make _________________________ Model _________________________

Odometer _____________________________ VIN _________________________________

Materials Required

- Blank work order
- Vehicle with available service history records
- Depending on the type of concern, special hand tool/diagnostic tools may be required. See your supervisor/instructor for instructions to identify what tools may be required.
- Vehicle service information database
- General hand tools

Task-Specific Safety Considerations

- Comply with personal and environmental safety practices associated with clothing; eye protection; hand tools; power equipment; proper ventilation; and the handling, storage, and disposal of chemicals/materials in accordance with local, state, and federal safety and environmental regulations.

Time off_________________

Time on_________________

Total time_________________

▶ TASK Inspect, service, or replace air filters, filter housings, and intake duct work.

MLR
A8C2

Student Instructions: Read through the entire procedure prior to starting. Prepare your work space and any tools or parts that may be needed to complete the task. When directed by your instructor, begin the procedure to complete the task and check the box as each step is finished. Track your time on this procedure for later comparison to the standard completion time (i.e., "flat rate" or customer pay time).

Procedure:	Step Completed
1. Research the maintenance interval and replacement procedure for the air filter.	
a. List or print off and attach to this sheet the procedure for replacing the air filter.	☐

b. What is the maintenance interval for inspection and replacement of the air filter?	☐
c. Describe the location of the air filter.	☐
2. Inspect and replace the air filter and all related filter housing and duct work.	
a. Following procedure, remove the air filter housing.	☐
b. Remove and inspect the air filter. Does it require replacement?	☐
c. Inspect the related duct work and filter housing for cracks, tears, or loose-fitting connections.	☐
d. Clean the air filter housing. Remove any debris prior to replacing the air filter.	☐
e. Install the new air filter. Replace the filter housing. Be sure to tighten all components/clamps that were removed.	☐
f. Start the vehicle and ensure that the engine runs correctly.	☐

<table>
<tr><td>Non-Task-Specific Evaluations:</td><td>Step Completed</td></tr>
<tr><td>1. Tools and equipment were used as directed and returned in good working order.</td><td>☐</td></tr>
<tr><td>2. Complied with all general and task-specific safety standards, including proper use of any personal protective equipment.</td><td>☐</td></tr>
<tr><td>3. Completed the task in an appropriate time frame (Recommendation: 1.5 or 2 times flat rate).</td><td>☐</td></tr>
<tr><td>4. Left the work space clean and orderly.</td><td>☐</td></tr>
<tr><td>5. Cared for customer property and returned it undamaged.</td><td>☐</td></tr>
</table>

Student signature _________________________ Date _________________________

Comments:

Have your supervisor/instructor verify satisfactory completion of this procedure, any observations found, and any necessary action(s) recommended.

Tasksheet Scoring

Evaluation Items	Test		Retest	
	Pass	Fail	Pass	Fail
Task-Specific Evaluation	**(1 pt)**	**(0 pts)**	**(1 pt)**	**(0 pts)**
Research the procedure for replacing the air filter, and determine the air filter maintenance interval.				
Remove and inspect the air filter.				
Inspect the air filter housing and duct work/hoses.				
Replace the air filter and confirm that the vehicle runs as designed.				
Non-Task-Specific Evaluation	**(0 pts)**	**(−1 pt)**	**(0 pts)**	**(−1 pt)**
Student successfully completed at least three of the non-task-specific steps.				
Student successfully completed all five of the non-task-specific steps.				
Total Score: <total # of points / 4 = %>				

Supervisor:

Supervisor/instructor signature _________________________________ Date _________________

Comments:

Retest supervisor/instructor signature _________________________________ Date _________________

Comments:

CDX Tasksheet Number: A8012

Student/Intern Information

Name _______________________________ Date ___________ Class _____________________

Vehicle, Customer, and Service Information

Vehicle used for this activity:

Year _______________ Make _______________________ Model _____________________

Odometer _______________________ VIN _________________________________

> **Materials Required**
> - Blank work order
> - Vehicle with available service history records
> - Depending on the type of concern, special hand tool/diagnostic tools may be required. See your supervisor/instructor for instructions to identify what tools may be required.
> - Vehicle service information database

Task-Specific Safety Considerations
- Lifting equipment, such as vehicle jacks and stands, vehicle hoists, and engine hoists, are important tools that increase productivity and make the job easier. However, they can also cause severe injury or death if used improperly. Make sure you follow the manufacturer's operation procedures. Also, make sure you have your supervisor's/instructor's permission to use any particular type of lifting equipment.
- Comply with personal and environmental safety practices associated with clothing; eye protection; hand tools; power equipment; proper ventilation; and the handling, storage, and disposal of chemicals/materials in accordance with local, state, and federal safety and environmental regulations.
- **Caution:** Exhaust systems reach very high temperatures. Use with caution when inspecting and servicing these systems to avoid personal injury.

Time off_____________

Time on_____________

Total time_____________

▶ TASK Inspect integrity of the exhaust manifold, exhaust pipes, muffler(s), catalytic converter(s), resonator(s), tail pipe(s), and heat shields, and determine necessary action.

MLR
A8C3

Student Instructions: Read through the entire procedure prior to starting. Prepare your work space and any tools or parts that may be needed to complete the task. When directed by your instructor, begin the procedure to complete the task and check the box as each step is finished. Track your time on this procedure for later comparison to the standard completion time (i.e., "flat rate" or customer pay time).

Procedure:	Step Completed
1. Inspect the exhaust system components.	
a. Begin inspection with a cold vehicle. Start the vehicle and listen for any exhaust leaks. Are there any leaks that can be heard at this time? **Tip:** A cold vehicle tends to reveal exhaust leaks better than a vehicle at operating temperature.	☐
b. Lift the vehicle.	☐
c. Inspect the exhaust manifolds. Are all fasteners in place? Are the manifolds free of excessive rust and free of cracks?	☐
d. Inspect the gaskets at the flange where the manifolds meet the downpipe. Are the gaskets in good condition and all fasteners in place?	☐
e. Continue to inspect the exhaust system components. Is the catalytic converter held securely in place and free of holes or damage?	☐
f. Inspect the muffler for rust, leaks, or damage. List any action required.	☐
g. Inspect all heat shields. Ensure that they are installed properly.	☐
2. On the basis of your inspection, what is your recommendation for service?	☐

Non-Task-Specific Evaluations:	Step Completed
1. Tools and equipment were used as directed and returned in good working order.	☐
2. Complied with all general and task-specific safety standards, including proper use of any personal protective equipment.	☐
3. Completed the task in an appropriate time frame (Recommendation: 1.5 or 2 times flat rate).	☐
4. Left the work space clean and orderly.	☐
5. Cared for customer property and returned it undamaged.	☐

Student signature _________________________ Date _________________________

Comments:

Have your supervisor/instructor verify satisfactory completion of this procedure, any observations found, and any necessary action(s) recommended.

Evaluation Instructions: The scoring box below is intended to act as a guide for both student and instructor. Each criterion listed will help students understand what is expected from them and evaluators articulate success at a particular task. The scoring is set up to allow a second attempt at each task (see the "Test" and "Retest" columns). Scoring is designed to reward students for correct completion of the task. Points are lost for failure to complete the employability requirements (see "Non-Task-Specific" criteria). When all criteria are evaluated, tally the points for a total at the bottom of each column.

Tasksheet Scoring

	Test		Retest	
Evaluation Items	**Pass**	**Fail**	**Pass**	**Fail**
Task-Specific Evaluation	**(1 pt)**	**(0 pts)**	**(1 pt)**	**(0 pts)**
Start the vehicle and listen for exhaust leaks.				
Inspect the exhaust manifolds.				
Inspect the catalytic converter, resonator, muffler, and heat shields.				
Make recommendations for service based on your inspection.				
Non-Task-Specific Evaluation	**(0 pts)**	**(−1 pt)**	**(0 pts)**	**(−1 pt)**
Student successfully completed at least three of the non-task-specific steps.				
Student successfully completed all five of the non-task-specific steps.				
Total Score: <total # of points / 4 = %>				

Supervisor:

Supervisor/instructor signature ___________________________ Date _______________

Comments:

Retest supervisor/instructor signature ___________________________ Date _______________

Comments:

CDX Tasksheet Number: A8013

Student/Intern Information

Name _________________________________ Date ___________ Class _______________________

Vehicle, Customer, and Service Information

Vehicle used for this activity:

Year _______________ Make _______________________ Model _______________________

Odometer _______________________ VIN _______________________

Task-Specific Safety Considerations

- Lifting equipment, such as vehicle jacks and stands, vehicle hoists, and engine hoists, are important tools that increase productivity and make the job easier. However, they can also cause severe injury or death if used improperly. Make sure you follow the manufacturer's operation procedures. Also, make sure you have your supervisor's/instructor's permission to use any particular type of lifting equipment.
- Comply with personal and environmental safety practices associated with clothing; eye protection; hand tools; power equipment; proper ventilation; and the handling, storage, and disposal of chemicals/materials in accordance with local, state, and federal safety and environmental regulations.

Time off____________

Time on____________

Total time____________

▶ **TASK** Inspect condition of exhaust system hangers, brackets, clamps, and heat shields, and determine necessary action.

MLR
A8C4

Student Instructions: Read through the entire procedure prior to starting. Prepare your work space and any tools or parts that may be needed to complete the task. When directed by your instructor, begin the procedure to complete the task and check the box as each step is finished. Track your time on this procedure for later comparison to the standard completion time (i.e., "flat rate" or customer pay time).

Procedure:	Step Completed
1. Inspect the exhaust system hangers, brackets, clamps, and heat shields.	
a. Lift the vehicle.	☐
b. Inspect all exhaust system hangers. Ensure that the rubber hangers are not cracking or broken.	☐
c. Inspect exhaust system brackets. Ensure that all brackets are free of rust or damage.	☐
d. Inspect all exhaust system clamps or flanges. Inspect for rust or damage.	☐
e. Inspect all heat shields. Ensure that no heat shields are missing and that shields are held securely in place.	☐
2. On the basis of your inspection, what is your recommendation for service?	☐

Non-Task-Specific Evaluations:	Step Completed
1. Tools and equipment were used as directed and returned in good working order.	☐
2. Complied with all general and task-specific safety standards, including proper use of any personal protective equipment.	☐
3. Completed the task in an appropriate time frame (Recommendation: 1.5 or 2 times flat rate).	☐
4. Left the work space clean and orderly.	☐
5. Cared for customer property and returned it undamaged.	☐

Student signature _______________________________ Date _____________________________

Comments:

Have your supervisor/instructor verify satisfactory completion of this procedure, any observations found, and any necessary action(s) recommended.

Tasksheet Scoring

Evaluation Items	Test		Retest	
	Pass	**Fail**	**Pass**	**Fail**
Task-Specific Evaluation	**(1 pt)**	**(0 pts)**	**(1 pt)**	**(0 pts)**
Lift the vehicle correctly.				
Inspect the exhaust hangers, brackets, clamps, and flanges.				
Inspect the exhaust system heat shields.				
Make recommendations for service based on your inspection.				
Non-Task-Specific Evaluation	**(0 pts)**	**(–1 pt)**	**(0 pts)**	**(–1 pt)**
Student successfully completed at least three of the non-task-specific steps.				
Student successfully completed all five of the non-task-specific steps.				
Total Score: <total # of points / 4 = %>				

Supervisor:

Supervisor/instructor signature _________________________________ Date _________________

Comments:

Retest supervisor/instructor signature _________________________ Date _________________

Comments:

CDX Tasksheet Number: A8014

Student/Intern Information

Name _________________________________ Date _____________ Class _______________________

Vehicle, Customer, and Service Information

Vehicle used for this activity:

Year _______________ Make _______________________ Model _______________________

Odometer _______________________ VIN _______________________

> **Materials Required**
> - Blank work order
> - Vehicle with available service history records
> - Depending on the type of concern, special hand tool/diagnostic tools may be required. See your supervisor/instructor for instructions to identify what tools may be required.
> - Vehicle service information database

Task-Specific Safety Considerations
- Comply with personal and environmental safety practices associated with clothing; eye protection; hand tools; power equipment; proper ventilation; and the handling, storage, and disposal of chemicals/materials in accordance with local, state, and federal safety and environmental regulations.

▶ **TASK** Check and refill diesel exhaust fluid (DEF).

MLR
A8C5

Time off_______________

Time on_______________

Total time_______________

Student Instructions: Read through the entire procedure prior to starting. Prepare your work space and any tools or parts that may be needed to complete the task. When directed by your instructor, begin the procedure to complete the task and check the box as each step is finished. Track your time on this procedure for later comparison to the standard completion time (i.e., "flat rate" or customer pay time).

Procedure:	Step Completed
1. Check and refill the DEF.	
a. Use service information to determine the procedure for checking and filling the DEF. Where is the DEF container located on your vehicle?	☐

b. How would a driver know when the DEF is getting low?	☐
c. Explain the purpose of DEF.	☐
d. Describe how the vehicle may operate if the DEF level is too low.	☐
e. How do you check the DEF level on your vehicle?	☐
f. Inspect the instrument cluster for a DEF warning light. Are any lights illuminated?	☐
g. Identify and describe the location of the DEF reservoir.	☐
h. Inspect the DEF fluid level. Top off the fluid.	☐
i. Did the warning light turn off?	☐

Non-Task-Specific Evaluations:	Step Completed
1. Tools and equipment were used as directed and returned in good working order.	☐
2. Complied with all general and task-specific safety standards, including proper use of any personal protective equipment.	☐
3. Completed the task in an appropriate time frame (Recommendation: 1.5 or 2 times flat rate).	☐
4. Left the work space clean and orderly.	☐
5. Cared for customer property and returned it undamaged.	☐

Student signature _________________________ Date _________________________

Comments:

Have your supervisor/instructor verify satisfactory completion of this procedure, any observations found, and any necessary action(s) recommended.

Evaluation Instructions: The scoring box below is intended to act as a guide for both student and instructor. Each criterion listed will help students understand what is expected from them and evaluators articulate success at a particular task. The scoring is set up to allow a second attempt at each task (see the "Test" and "Retest" columns). Scoring is designed to reward students for correct completion of the task. Points are lost for failure to complete the employability requirements (see "Non-Task-Specific" criteria). When all criteria are evaluated, tally the points for a total at the bottom of each column.

Tasksheet Scoring

Evaluation Items	Test		Retest	
	Pass	**Fail**	**Pass**	**Fail**
Task-Specific Evaluation	**(1 pt)**	**(0 pts)**	**(1 pt)**	**(0 pts)**
Explain the purpose of DEF.				
Explain the driving limitations that occur when the DEF level is too low.				
Determine the location of the DEF reservoir and inspect the instrument cluster for DEF warning indicators.				
Check the DEF level and top off as needed.				
Non-Task-Specific Evaluation	**(0 pts)**	**(−1 pt)**	**(0 pts)**	**(−1 pt)**
Student successfully completed at least three of the non-task-specific steps.				
Student successfully completed all five of the non-task-specific steps.				
Total Score: <total # of points / 4 = %>				

Supervisor:

Supervisor/instructor signature _________________________________ Date _________________________

Comments:

Retest supervisor/instructor signature _________________________________ Date _________________________

Comments:

Learning Objective/Task	CDX Tasksheet Number	ASE Education Foundation Reference Number; Priority Level
• Inspect, test, and service positive crankcase ventilation (PCV) filter/breather, valve, tubes, orifices, and hoses, and perform necessary action.	A8015	A8D1, P-2

Materials Required

- Blank work order
- Vehicle with available service history records
- Depending on the type of concern, special hand tool/diagnostic tools may be required. See your supervisor/instructor for instructions to identify what tools may be required.
- Vehicle service information database

Safety Considerations

- When running any vehicles in the shop, make sure you use the shop's exhaust ventilation system to discharge all exhaust gas safely outside.
- Extreme caution must be exercised when working around rotating components.
- Lifting equipment, such as vehicle jacks and stands, vehicle hoists, and engine hoists, are important tools that increase productivity and make the job easier. However, they can also cause severe injury or death if used improperly. Make sure you follow the manufacturer's operation procedures. Also, make sure you have your supervisor's/instructor's permission to use any particular type of lifting equipment.
- Comply with personal and environmental safety practices associated with clothing; eye protection; hand tools; power equipment; proper ventilation; and the handling, storage, and disposal of chemicals/materials in accordance with local, state, and federal safety and environmental regulations.

CDX Tasksheet Number: A8015

Student/Intern Information

Name ____________________ Date ________ Class ____________________

Vehicle, Customer, and Service Information

Vehicle used for this activity:

Year ____________ Make ____________________ Model ____________________

Odometer ____________________ VIN ____________________

<table>
<tr><td>

Materials Required

- Blank work order
- Vehicle with available service history records
- Depending on the type of concern, special hand tool/diagnostic tools may be required. See your supervisor/instructor for instructions to identify what tools may be required.
- Vehicle service information database
- General hand tools

</td></tr>
</table>

Task-Specific Safety Considerations

- Comply with personal and environmental safety practices associated with clothing; eye protection; hand tools; power equipment; proper ventilation; and the handling, storage, and disposal of chemicals/materials in accordance with local, state, and federal safety and environmental regulations.

Time off___________

Time on___________

Total time___________

▶ **TASK** Inspect, test, and service PCV filter/breather, valve, tubes, orifices, and hoses, and perform necessary action.

MLR
A8D1

Student Instructions: Read through the entire procedure prior to starting. Prepare your work space and any tools or parts that may be needed to complete the task. When directed by your instructor, begin the procedure to complete the task and check the box as each step is finished. Track your time on this procedure for later comparison to the standard completion time (i.e., "flat rate" or customer pay time).

Procedure:	Step Completed
1. Research the PCV system.	
a. Use service information and determine the location of the PCV valve.	☐

b. Describe the purpose of the PCV valve.	☐
c. What is the maintenance interval for replacing the PCV valve?	☐
d. Describe symptoms of a faulty PCV valve/system.	☐
2. Inspect and test the PCV valve.	
a. Start the vehicle. Does the engine seem to be running smoothly? Is there any smoke being emitted from the tailpipe?	☐
b. Shut the engine off. Open the hood and identify and describe the location of the PCV valve.	☐
c. Describe the appearance of the PCV valve, grommet, and tubing.	☐
d. Remove the PCV valve. Shake the valve. Does it rattle or appear to be clogged?	☐
e. Start the vehicle and check for the presence of vacuum at the PCV hose/tubing.	☐
f. Reinstall the PCV valve. Inspect the engine valve covers, timing covers, engine main seals, and oil pan for the presence of leaks. List any leaks found.	☐

<table>
<tr><td>g. Describe how a faulty PCV system may cause oil leaks.</td><td>☐</td></tr>
</table>

Non-Task-Specific Evaluations:	**Step Completed**
1. Tools and equipment were used as directed and returned in good working order.	☐
2. Complied with all general and task-specific safety standards, including proper use of any personal protective equipment.	☐
3. Completed the task in an appropriate time frame (Recommendation: 1.5 or 2 times flat rate).	☐
4. Left the work space clean and orderly.	☐
5. Cared for customer property and returned it undamaged.	☐

Student signature ______________________________ Date ______________________________

Comments:

Have your supervisor/instructor verify satisfactory completion of this procedure, any observations found, and any necessary action(s) recommended.

Tasksheet Scoring

		Test		Retest	
Evaluation Items	**Pass**	**Fail**	**Pass**	**Fail**
Task-Specific Evaluation	**(1 pt)**	**(0 pts)**	**(1 pt)**	**(0 pts)**
Describe the purpose and function of the PCV valve.				
Determine the maintenance interval for replacing the PCV valve.				
Determine the location of the PCV valve, and remove the PCV valve.				
Inspect the PCV valve and related systems.				
Non-Task-Specific Evaluation	**(0 pts)**	**(−1 pt)**	**(0 pts)**	**(−1 pt)**
Student successfully completed at least three of the non-task-specific steps.				
Student successfully completed all five of the non-task-specific steps.				
Total Score: <total # of points / 4 = %>				

Supervisor:

Supervisor/instructor signature _________________________________ Date _____________________

Comments:

Retest supervisor/instructor signature _________________________________ Date _____________________

Comments: